T0192440

Plasma Simulations
by Example

Plasma Simulations by Example

Lubos Brieda

CRC Press
Taylor & Francis Group
Boca Raton London New York

CRC Press is an imprint of the
Taylor & Francis Group, an **informa** business

CRC Press
Taylor & Francis Group
6000 Broken Sound Parkway NW, Suite 300
Boca Raton, FL 33487-2742

First issued in paperback 2021

© 2019 by Taylor & Francis Group, LLC
CRC Press is an imprint of Taylor & Francis Group, an Informa business

No claim to original U.S. Government works

ISBN-13: 978-1-138-34232-3 (hbk)
ISBN-13: 978-1-03-217614-7 (pbk)
DOI: 10.1201/9780429439780

This book contains information obtained from authentic and highly regarded sources. Reasonable efforts have been made to publish reliable data and information, but the author and publisher cannot assume responsibility for the validity of all materials or the consequences of their use. The authors and publishers have attempted to trace the copyright holders of all material reproduced in this publication and apologize to copyright holders if permission to publish in this form has not been obtained. If any copyright material has not been acknowledged please write and let us know so we may rectify in any future reprint.

Except as permitted under U.S. Copyright Law, no part of this book may be reprinted, reproduced, transmitted, or utilized in any form by any electronic, mechanical, or other means, now known or hereafter invented, including photocopying, microfilming, and recording, or in any information storage or retrieval system, without written permission from the publishers.

For permission to photocopy or use material electronically from this work, please access www.copyright.com (http://www.copyright.com/) or contact the Copyright Clearance Center, Inc. (CCC), 222 Rosewood Drive, Danvers, MA 01923, 978-750-8400. CCC is a not-for-profit organization that provides licenses and registration for a variety of users. For organizations that have been granted a photocopy license by the CCC, a separate system of payment has been arranged.

Trademark Notice: Product or corporate names may be trademarks or registered trademarks, and are used only for identification and explanation without intent to infringe.

Publisher's Note

The publisher has gone to great lengths to ensure the quality of this reprint but points out that some imperfections in the original copies may be apparent.

Visit the Taylor & Francis Web site at
http://www.taylorandfrancis.com

and the CRC Press Web site at
http://www.crcpress.com

To Sandra and April.

Contents

Preface

THE OBJECTIVE of this book is to teach you how to develop plasma simulation codes through easy to follow examples. While many excellent books already exist on the mathematical formulation of plasma simulation methods, they contain few, if any, practical code samples. As the saying goes, the devil is in the details. I have been developing plasma and rarefied gas codes for the past fifteen years. This book is my attempt to capture some of the intricacies that are encountered along the way. But, in order to keep the page count reasonable, it is not a detailed survey. Entire books have been written on topics covered in individual chapters. References included in the "Background" section include texts that I have found invaluable in my past work. These books contain discussions of alternate schemes or higher order methods that should be straightforward to implement once you grasp the basics covered in this text.

We focus on the kinetic Particle in Cell (PIC) method, although fluid methods are also discussed. The first chapter starts by introducing fundamental concepts, including the governing equations for the electrostatic Particle in Cell (ES-PIC) method. This is also where we discuss kinetic versus fluid modeling approaches. We then develop a simple particle integrator based on the Finite Difference Method to simulate an electron trapped in a potential well. The reminder of the book focuses on various forms of "a sphere in plasma" simulations. My background is in electric (plasma) propulsion, and the examples come from this research area. We mainly focus on low density discharges interacting with objects. The sphere is a placeholder that could represent a miniature Langmuir probe or a massive planet interacting with solar wind. In Chapter 2, we put together the basic code framework and use it to perform a fully-kinetic simulation of ions and electrons interacting in a grounded box. We add the sphere in Chapter 3. Here we also learn about hybrid approaches utilizing fluid electrons, and implement a faster potential solver based on the Preconditioned Conjugate Gradient (PCG) and Newton-Raphson (NR) methods. In Chapter 4, we incorporate surface interactions, and also implement inter-particle collisions through the Monte Carlo Collisions (MCC) and Direct Simulation Monte Carlo (DSMC) methods. Chapter 5 discusses methods for reducing simulation run times by taking advantage of problem symmetry. We develop two-dimensional planar and axisymmetric codes based on the Finite Volume Method. In chapter 6, we modify the code to operate on an unstructured grid and learn about the Finite Element Method.

Electromagnetic effects are discussed in Chapter 7. We cover the Boris method for integrating particle velocities, and compute the magnetic field for a magnetostatic problem. We also develop an electromagnetic PIC (EM-PIC) code. Chapter 8 introduces Eulerian methods that can be used stand-alone, or can be coupled with the Lagrangian PIC method in a hybrid simulation. We also develop a 1D-1V solver for the Vlasov equation. Finally, we close the book in Chapter 9 with a review of parallelization strategies. We discuss how to take advantage of multi-core CPU architectures using threads, and how to distribute simulations to multiple computers using MPI. We also cover basics of GPU programming using CUDA.

BACKGROUND

This book is written with the assumption that you are already familiar with basic plasma physics, linear algebra, and numerical analysis. If not, some useful references that I have encountered along the way include the works of Chen[24], Spitzer[52], Lieberman [42], Jackson[36], Sutton [54], Burden[23], Aris[10], and Boyce[17]. The mathematical formulation for kinetic and fluid plasma simulations is discussed in Birdsall [15], Tajima [55], Hockney [34], and Jardin [39]. Numerical methods for neutral gases are discussed by Anderson [8], Ferziger [28], Tu [56], and Bird [12]. The work of Hughes [35] forms the basis of the finite element solver in Chapter 6. Finally, the books of Zhang[59], Jahn[37], and Hastings[33] provide an insight into the microscopic energy transfer, plasma propulsion, and the space environment.

All examples are developed in C++11, which is the 2011 revision of the C++ programming language. These days, Python and MATLAB®tend to be the primary languages taught in university settings. However, even with the help of the optimized Numpy libraries, Python remains many times slower than C++. Furthermore, the lack of variable types can lead to strange bugs, especially when a distinction between integer and floating point values is needed. Various scientific computing libraries are also implemented in C++, or interface with it without the need for external third party wrappers. These include the Message Passing Interface (MPI)[32] used for cluster computing and CUDA used for running code on NVIDIA graphics cards [58]. The downside of C++ is that it can be a bit overwhelming, especially if your background in objected oriented programming is limited. Furthermore, modern C++ introduces numerous new additions, such as standard library storage containers, loop iterators, and multi-threading, that you may not be familiar with if you have been mainly developing in the classic C. While the code is explained as we go along, I found the recent concise work of Stroustrop[53] to be an invaluable review of modern C++ language features.

CODE SAMPLES

Including in print the source code for all examples would be very much impractical and perhaps not very useful. The shorter codes in the first two chapters are replicated mainly in full, however, the rest of the book focuses only on the important pieces. The complete source code can be downloaded from the companion website

https://www.particleincell.com/plasma-by-example

There you will also find color versions of the plots in the book. The example codes were tested with the GNU g++ compiler version 7.4.0, running on Ubuntu 18 Linux. This version natively enables the C++11 language extensions. If you are using an older compiler, you may need to enable the -std=c++11 flag, i.e.

g++ -std=c++11 *.cpp -o sphere

While C++ is a powerful language, it does not include any built-in graphical support. We output simulation results in the Visualization Toolkit (VTK) formats [51]. These files can then be visualized using programs such as Paraview or VisIt. These programs can be downloaded from https://www.paraview.org/ and https://wci.llnl.gov/simulation/computer-codes/visit/

ACKNOWLEDGEMENTS

This book would not be possible if it were not for the many professors, research partners, and colleagues I have had the privilege of working with. I entered the world of numerical plasma physics by a chance when, as an undergraduate student at Virginia Tech, I joined the research group of a new professor Joseph Wang who was looking for students to help rewrite a legacy Fortran plasma plume code. Joining his group helped satisfy a technical elective requirement. Shortly afterwards, we were tasked with developing a new general electric propulsion modeling software based on a Cartesian mesh with cut cells. A team at MIT, led by Prof. Martinez-Sanchez and consisting of Mark Santi and Shannon Cheng, worked on a complimentary solver based on the unstructured mesh. I became the primary developer, but other students also made significant contributions. Raed Kafafy developed an immersed finite element solver, while Julien Pierru worked on a parallel version, and Alex Barrie added surface charging. Later, Randy Spicer performed extensive code validation against experimental data taken by another student from our group, Mike Nakles. After graduating I continued the code development at the Air Force Research Lab. There I collaborated with many brilliant scientists and engineers, including Mike Fife, Doug VanGilder, Justin Koo, Rob Martin, Jun Araki, David Bilyeu, Michelle Scharfe, John Loverich, and Michael Gorrilla. After three years, an opportunity arose to join professor Michael Keidar at the George Washington University for a doctoral program. Some of the most memorable courses from my entire schooling career were taken here. Working with Prof. Keidar, I mainly focused on electron transport in Hall thrusters, but also briefly collaborated with his students Taisen Zhuang on vacuum arcs, and Olga Volotskova on atmospheric plasma jets for medical applications. Many fun moments were shared with my office mates Madhu Kundrapu and Brent Duffy.

While working on the doctorate, I also started a scientific computing blog at `particleincell.com`. Shortly after graduating, an idea arose to try offering an on-line plasma simulation course. I posted a note on the website, and to my surprise, almost twenty students signed up for the inaugural *Fundamentals of the PIC Method*. In the following years, I added a course on *Advanced PIC*, *Distributed Computing*, and *Fluid Modeling of Plasmas*. The feedback from the students was invaluable in shaping the material that eventually morphed into this book. Finally, much thanks goes to the book reviewers, to my parents and sister, and to my wife Sandra for putting up with the many late nights spent working first on the dissertation, and then later, this text.

Lubos Brieda
lubos.brieda@particleincell.com

CHAPTER 1

Fundamentals

1.1 INTRODUCTION

T HIS CHAPTER starts with a review of kinetic and fluid approaches for simulating plasma flows. The governing equations of the Electrostatic Particle in Cell (ES-PIC) method are then introduced. We next discuss the Finite Difference Method for discretizing differential equations. We use the method to simulate an electron trapped in a potential well.

1.2 GAS SIMULATION APPROACHES

Plasma, on the microscopic scale, is simply a gas containing charged particles: ions and electrons. That is at least the simplistic view we take in this book. More complex flows can be found in nature. Plasma may contain negative ions or large dust particulates that collect charge through tribolectric charging or photoemission. There may be multiphase flows involving gaseous components and liquid droplets. We do not consider such cases here, and instead focus solely on the basic combination of neutral atoms, positive ions, and enough electrons to neutralize the space charge. The objective of flow simulations is to predict the evolution of particle velocities and positions given some initial conditions and governing laws. Within the realms of Newtonian physics, we assume that all neutrals, ions, and electrons are rigid bodies governed by the equations of motions,

$$\frac{d\vec{x}}{dt} = \vec{v} \qquad \frac{d\vec{v}}{dt} = \vec{a} \tag{1.1}$$

where $\vec{x}$, $\vec{v}$, and $\vec{a}$ are the position, velocity, and acceleration of a single particle. If the particle mass m remains constant, as is assumed throughout this book, the acceleration can be written in terms of total force, $\vec{a} = \sum \vec{F}/m$ per Newton's Second Law. Some forces, such as gravity, originate from the external environment. Other forces may be intrinsic to the system. Charged particles interact with each other through the *Coulomb force*,

$$\vec{F}_{c,ij} = \frac{1}{4\pi\epsilon_0} \frac{q_i q_j}{|\vec{x}_i - \vec{x}_j|^3} (\vec{x}_i - \vec{x}_j) \tag{1.2}$$

where the i and j are indexes of two unique particles with charges q_i and q_j. $\epsilon_0 \approx 8.8542 \times 10^{-12}$ C/(Vm) is a physical constant known as *permittivity of free space*. This formulation ignores magnetic field effects found with moving charges. The total force acting on a single particle i is

$$\vec{F}_{c,i} = \sum_{j}^{N} \vec{F}_{c,ij} \quad i \neq j \tag{1.3}$$

where the sum is over all particles in the system. Particle velocity can also change through collisions. We thus have

$$\frac{d\vec{v}}{dt} = \frac{1}{m}\left(\vec{F}_g + \vec{F}_c + \ldots\right) + \left(\frac{d\vec{v}}{dt}\right)_{col} \tag{1.4}$$

Although gravity is included in the above formulation, it is customary to ignore it. A quick "back of the envelope" calculation demonstrates why. Large vacuum chambers used in plasma processing operate at pressures p around 10^{-6} Torr (1 Torr is 1/760th of 1 atmosphere). Using the ideal gas law, $p = nk_BT$, where $k_B \approx 1.381 \times 10^{-23}$ J/K is the Boltzmann constant and T is the room temperature, the average number density n is around 3.3×10^{16} particles per cubic meter. The mean inter-particle spacing can be estimated from the volume needed to be assigned to each ion to fill a unit cube. Modeling each particle as a cube with sides a, the particles are $a = (1/n)^{1/3}$ or approximately 3 μm apart. At this distance, the Coulomb force is around 10^{-17} N. Even for the heavy xenon used in plasma propulsion, the force of gravity, $F = mg$, is only 10^{-24} N. The gravitational force is thus seven orders of magnitude smaller than the electrostatic Coulomb force. In the Low Earth Orbit (LEO), the ambient plasma density drops to 10^{12} m^{-3}, but the Coulomb force still dominates by four orders of magnitude. Therefore, unless we are dealing with problems involving massive gravitational fields, it is safe to ignore the force of gravity. Collisions can also be ignored at sufficiently low pressures.

1.2.1 Direct Method

Conceptually, we could model a system containing N particles by looping through the entire list, and for each, evaluating the Coulomb force from

$$\vec{F}_{c,i} = q_i\vec{E}_i \tag{1.5}$$

where

$$\vec{E}_i = \sum_{j} \frac{1}{4\pi\epsilon_0} \frac{q_j}{r_{ij}^3} \left(\vec{r}_i - \vec{r}_j\right) \quad i \neq j \tag{1.6}$$

This approach is inefficient. The Coulomb force needs to be evaluated $N - 1$ times for each particle. Since there are N particles in total, we need to perform $N(N-1) \approx N^2$ Coulomb force calculations. It takes my computer 10^{-8} s to

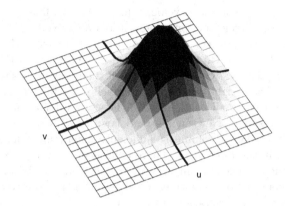

Figure 1.1: Example of a typical gas velocity distribution function.

evaluate Equation 1.2 for a single particle pair. Due to the N^2 dependence, it would take over 3×10^{17} years to compute $\vec{E}$ for all particles in just a single cubing meter at vacuum chamber pressures. This interval is seven orders of magnitude longer than the age of the universe! Clearly, this direct particle-particle approach is intractable. Fortunately, as we see shortly, there is a much faster way to compute the $\vec{E}$ term that scales with N.

Yet, even assuming the linear scaling, the computational run time would be 21 years due to the immense number of particles. Furthermore, even if some sudden breakthrough resulted in a massive increase in computational speeds, the amount of memory required to store the particle data is staggering. Over 700 billion Gb of RAM are needed to store just the six floating point numbers capturing particle position and velocity for the 10^{16} particles in a single cubic meter at vacuum pressures. This is beyond any supercomputer capacity.

1.2.2 Boltzmann Equation

Fortunately, alternatives exist. Imagine you could instantaneously freeze all particles within a small control volume for just long enough to group them by their velocity components. To accomplish the grouping, we define a histogram in the three-dimensional velocity space. Each bin corresponds to a unique combination of x, y, and z velocity components, u, v, and w. The width of each bin is given by Δu, Δv, and Δw. Particles with $u \pm 0.5\Delta u$, $v \pm 0.5\Delta v$, and $v \pm 0.5\Delta v$ all fall into the same bin. In the limit of the bins becoming infinitesimal, the discrete histogram becomes a smooth hyperspace surface. This surface represents the *velocity distribution function*, or VDF for short. Figure 1.1 plots one possible VDF if only the u and v components are considered. The value of the function, represented by the height of the surface and the shading, indicates the number of particles with the corresponding u and v velocity components. We clearly see that the majority of particles are concentrated around some mean velocity pair. The two lines plot $u = 0$ and $v = 0$ that divide each velocity component into a negative and a positive half-space. The v velocities are centered on $v = 0$ and therefore there is no net motion in the y direction. On the other hand, there is a clear drift in x. Even though there are particles with $u < 0$, most of them have a positive u.

What makes this distribution-based approach powerful is realizing that, in the absence of collisions, all particles with the same velocity move in unison. This is a direct result of Newton's second law: $d\vec{v}/dt = \vec{F}/m$. Here we are assuming that the control volume is sufficiently small such that there is no spatial variation in the force term. There is no need to integrate the velocity for more than one particle per bin since the result is identical for all. It is only through collisions that some fraction of the population is affected in a unique way. We thus write a conservation statement

$$\frac{df}{dt} = \left(\frac{\partial f}{\partial t}\right)_{col} \tag{1.7}$$

This relationship indicates that the total derivative of the distribution function f changes only through some yet to be defined collision operator.

The distribution that we have discussed so far is valid only for the small control volume where it was sampled. Different volumes, corresponding to other spatial regions, have a unique combination of particle velocities. The distribution also evolves with time. The VDF is thus a function of seven variables, $f = f(x, y, z, u, v, w, t)$. Here f is the number of particles with velocity components (u, v, w) at position (x, y, z) and at time t. The distribution can be optionally normalized, in which case the function value $\hat{f}$ is the probability of finding such a particle. I prefer to work with the unnormalized version as it carries along all information needed to completely describe the gas state.

Utilizing chain rule to rewrite the derivative in Equation 1.7 leads to the *Boltzmann equation*,

$$\frac{\partial f}{\partial t} + \nabla \cdot f + \frac{F}{m}\nabla_v \cdot f = \left(\frac{\partial f}{\partial t}\right)_{col} \tag{1.8}$$

The collisionless variant, in which the right hand side is zero, is known as the *Vlasov equation*. Instead of considering inter-particle interactions directly, we can model gas evolution by integrating this partial differential equation (PDE). The vast majority of plasma simulation codes do not use this approach. The reason is simple and familiar: computational requirements. As was alluded previously, numerically storing an arbitrary VDF requires a three-dimensional velocity grid. This grid is basically a multi-dimensional array of floats, with the three indexes mapping to a particular u, v, and w bin. Even utilizing an extremely coarse $20 \times 20 \times 20$ resolution supporting only 20 discrete velocities per dimension, we need to store 8,000 floating point values per control volume. In order to resolve the spatial variation in gas properties, we also need to discretize the physical domain into another three dimensional grid. Just as the velocity gridding limits the smallest discrete velocity difference, the volume grid controls the smallest distance that can be resolved. Utilizing a coarse $100 \times 100 \times 100$ discretization for the spatial domain, total of 800 million values need to be stored to capture a single VDF. This translates to almost 30 Gb of RAM. Integration schemes for the Boltzmann equation generally depend on at least one temporary buffer, doubling the memory requirements. Including additional gas species increases the storage correspondingly. While these requirements offer a huge improvement over the 700 billion Gb with the direct particle-particle scheme, they still greatly exceed the capabilities of standard computational systems. These requirements arise from the need to store six-dimensional data if all three spatial and velocity components are retained. At reduced dimensionality, direct solution of the Boltzmann equation can actually be quite attractive. We consider this class of solvers, confusingly called Vlasov solvers even when collisions are retained, in Chapter 6.

1.2.3   Maxwell-Boltzmann Distribution Function

Clearly, alternatives must exist, since three dimensional plasma simulations are carried out routinely with much finer meshes on standard desktop work-stations. Much of the memory requirement arises from the large number of velocity bins needed in each spatial cell. Most of these bins are effectively empty - this can in fact be seen in Figure 1.1. Therefore, some additional simplifications can be made. The first option is to assume that the VDF follows an analytic profile. Specifically, we can let the velocity in each direction be given by

$$f_M(v) = \frac{1}{\sqrt{\pi}v_{th}} \exp\left(\frac{-(v - v_d)^2}{v_{th}^2}\right) \tag{1.9}$$

This very important equation is known as the *Maxwell-Boltzmann distribution function*. Quantum-theory or statistical mechanics can be used to derive that it is in fact the natural highest entropy state for a molecular population [59]. Imagine that you have a box initially filled with gas containing exactly two types of particles: slow molecules moving with speed A, and fast molecules moving with speed B. If there are no external forces, and the walls are perfectly specular, then velocities can change only through collisions. In the momentum exchange collision, the faster molecule slows down a bit while the slower one speeds up. The collision rate also scales inversely with the relative speed. Eventually, after a sufficient time has passed, the two initially discrete populations merge into a single one in which there is no net momentum transfer. The Maxwell-Boltzmann distribution corresponds to this end state and the process of reaching it is called *thermalization*.

The major benefit of utilizing Equation 1.9 is that it reduces the VDF to just two parameters for each dimension: the mean (or drift) velocity v_d, and the thermal velocity v_{th}. Thermal velocity is in turn defined in terms of temperature,$v_{th} = \sqrt{2k_B T/m}$. Therefore, just six floats fully describe the

VDF in each spatial cell. Often, we further assume isotropic temperature, reducing the memory required to just four values: the three components of drift velocity $\vec{v}_d$, and a single magnitude of thermal velocity. This is an immense improvement over the 8,000 values needed with the initial coarse histogram approach! But nothing in life is free, and this approach is not without a major drawback. In order to use Equation 1.9, we require that a "sufficient time" has passed for the gas to thermalize. This requirement is not always satisfied. The *collision rate* scales with number density, and the average distance a molecule travels between collisions is given by the *mean free path*,

$$\lambda = 1/(\sigma n) \tag{1.10}$$

The term σ is the *collision cross-section* and is discussed in more detail in Chapter 4. n is the gas number density. An important factor, known as the *Knudsen number*, relates the mean free path to a characteristic length,

$$K_n = \frac{\lambda}{L} \tag{1.11}$$

This length is problem specific. In a simulation of a vacuum chamber, we can let L be the chamber diameter. If $K_n \ll 1$, implying $\lambda \ll L$, each molecule can be expected to undergo many collisions as it travels between the walls. Such a flow is said to be in *continuum*, and the Maxwellian VDF can be assumed to be valid. On the other hand, if $K_n \gg 1$, $\lambda \gg L$ and the molecules are much more likely to collide with the wall than each other. This state describes *free molecular flow*. Collisions can be completely ignored, and gas dynamics is driven by wall effects. Finally, if $K_n \approx 1$, the distance between collisions is comparable to the characteristic length. In this case, collisions happen but not frequently enough to assume the Maxwellian VDF with certainty. This state, known as *rarefied gas*, is often encountered in plasmas.

1.2.4 Fluid and Kinetic Methods

We can therefore divide gas simulation approaches into two categories: *fluid* and *kinetic*. Fluid approaches assume continuum with the VDF having an analytical form. Familiar gas properties such as density or mean velocity are obtained by evaluating *moments* of the distribution function. These are integrals of f convoluted with velocity raised to an increasingly higher power and evaluated over the entire velocity space,

$$M^k(\vec{x}, t) = \int_{\vec{v}} \vec{v}^k f(\vec{x}, \vec{v}, t) d\vec{v} \tag{1.12}$$

The zeroth, first, and second moments lead to conservation equations for mass, momentum, and energy. For the Maxwellian VDF, these moments give us the standard Navier-Stokes equations used in computational fluid dynamics (CFD).

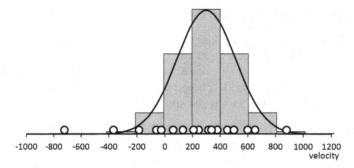

Figure 1.2: Comparison of three approaches for representing a velocity distribution function: analytical function (black line), a histogram of discrete velocities (gray bars), and random samples (white circles).

Kinetic approaches, on the other hand, do not make any assumptions about the shape of the VDF and allow it to evolve self-consistently. The direct and Vlasov solvers discussed previously are two examples of the kinetic approach. Both were found to have excessive computational requirements. Fortunately, there exists another class of methods based on *stochastic* (random) sampling. Many velocity bins contain zero or only a few particles. Instead of attempting to resolve the discretized VDF completely, we can generate a number of random velocity samples with probability following the distribution function. This sampling is illustrated in Figure 1.2. The representation of the one-dimension Maxwell-Boltzmann distribution using the analytical shape is compared to a Vlasov solver velocity grid and the random sampling. We treat these samples as particles and let them move through the computational domain according to the equations of motion. In doing so, they convect their discrete velocity sample into new spatial regions. Each particle is assigned a *macroparticle weight* w_{mp} such that the initial number density is recovered using a much smaller number of samples. The macroparticle (or specific) weight is simply the number of real atoms, ions, or electrons represented by the simulation particle. For the case of constant weight, we have $w_{mp} = N_{real}/N_{sim}$. Two popular schemes based on this approach are the Particle in Cell (PIC) method [15] and Direct Simulation Monte Carlo (DSMC) [12]. PIC considers acceleration only due to electromagnetic forces. DSMC considers neutral gas and only models collisions. The two methods share many similarities and are often combined to model partially ionized gases. We develop a PIC-DSMC simulation in Chapter 4. The stochastic nature introduces noise, which can be reduced by utilizing more samples (particles). The noise can also be reduced by averaging the results over a long interval once the solution stops evolving. Stochastic methods are thus best suited to problems in which the steady-state is achieved.

1.2.5 Lagrangian and Eulerian Descriptions

PIC and Vlasov solvers are both kinetic algorithms since they resolve the VDF self-consistently. They however treat the gas in a different way. While PIC follows individual particles as they move about the domain, Vlasov solvers evolve the gas using a stationary grid. These two models are known as *Lagrangian* and *Eulerian*. We can thus group gas simulation into four categories:

- Fluid/Eulerian: Computational Fluid Dynamics (CFD), MagnetoHydro-Dynamics (MHD)

- Fluid/Lagrangian: Smoothed Particle Hydrodynamics (SPH)

- Kinetic/Eulerian: Vlasov (direct kinetic) solvers

- Kinetic/Lagrangian: Particle in Cell (PIC), Direct Simulation Monte Carlo (DSMC)

The SPH scheme, which is not discussed in this text, is similar to PIC in that it represents the gas by Lagrangian particles. But unlike PIC, in which each particle carries a discrete velocity, SPH particles represent fluid parcels with a Maxwellian drift velocity and temperature. Finally, the PIC and DSMC schemes utilize a Eulerian spatial grid. This grid is used to compute inter-particle interactions. Recently, much progress has been made on grid-free particle methods that use dynamic octrees [26]. Such mesh-free approaches make the methods truly Lagrangian since the dependence on a spatial grid is eliminated.

The vast majority of this book is devoted to the kinetic PIC method. The reason is two fold. First, fluid methods share many similarities with their neutral gas counterparts. Instead of solving the Navier-Stokes equations as is done in computational fluid dynamics (CFD), the magnetohydrodynamics (MHD) equations are solved. The solution schemes are similar as they both involve numerically integrating systems of partial differential equations governing the conservation of mass, momentum, and energy. The main difference arises from the need to consider the evolution of the electromagnetic fields. Many excellent books already offer a comprehensive survey of relevant integration methods. But for completeness, fluid solvers are demonstrated in Chapter 6. Fluid methods are also commonly coupled with PIC to form a hybrid scheme. In my field of electric propulsion, it is not common to simulate electrons directly. Instead, they are represented by fluid models of problem-specific complexity. Secondly, the PIC method is more versatile. The fluid methods are limited to the continuum regime. PIC, on the other hand, can be used across the entire range of Knudsen numbers, although they are best suited to rarefied and free molecular flows. The limitations of PIC arise from efficiency and not some inherent assumption about the VDF. Applying PIC to continuum flows requires much greater computational resources than the fluid approach, but in the end, the

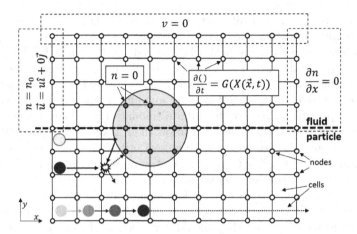

Figure 1.3: Comparison of a fluid (top) and particle (bottom) approach.

same result is obtained. The reverse is not true, as fluid approaches cannot model rarefied gases in which the velocity distribution is non-Maxwellian.

1.2.6 Comparison of Fluid and Kinetic Approaches

To illustrate the difference between fluid and particle methods, let's consider a simulation of neutral gas flow past a sphere. A schematic of this setup is shown in Figure 1.3. Starting with the fluid approach, we can use our understanding of gas dynamics (or look in appropriate reference books) to arrive at partial differential equations for the conservation of mass, momentum, and possibly energy. The energy equation is needed to compute temperature, but often we assume that temperature remains constant. For instance, we may have the following expressions for density and momentum [24]

$$\frac{\partial n}{\partial t} + \nabla \cdot (n\vec{u}) = 0 \tag{1.13}$$

$$mn \left[\frac{\partial \vec{v}}{\partial t} + (\vec{v} \cdot \nabla)\vec{v} \right] = \vec{F} - \nabla p - mn\nu\vec{v} \tag{1.14}$$

Details of these equations are not important right now. Instead we should notice that both contain a time derivative term $\partial()/\partial t$ and one or more terms that depend only on spatial properties (such as the $\vec{F}$ force term or the collisional drag term $mn\nu\vec{v}$) or their derivatives (the pressure gradient term ∇p). The equations can thus be recast into a form

$$\frac{\partial()}{\partial t} = G(X(\vec{x}, t)) \tag{1.15}$$

where G is some function of the spatially-varying gas state X (density, velocity, pressure, etc...) at time t. From this form we see that fluid simulations involve time integration of differential equations for density, velocity, and so on. These governing equations do not have an analytical solution except in trivial cases, and some numerical scheme needs to be utilized. The scheme generally involves dividing the temporal domain into states separated by a small Δt time step. The physical domain is discretized into a spatial grid. This approach is needed, since the finite amount of computer memory does not make it possible to resolve the infinite amount of information available in the continuous real world. The integration then involves *time-marching* the solution from an *initial condition* until an exit criterion is met. This could involve simulating a prescribed real time, or the simulation reaching the *steady state* in which properties of interest no longer change, $\partial()/\partial t = 0$.

The blocks making up the spatial domain are called *cells*. The cell corners are known as *nodes* and are shared among neighbor cells. Cells can generally have an arbitrary shape but they should not overlap. The collection of cells is called a *simulation mesh* or a *simulation grid* if the cell ordering follows a logical structure. Figure 1.3 shows the uniform Cartesian grid with cube-shaped, constant-dimension cells and cell edges aligned with the three coordinate axes. This type of a mesh is ubiquitous in numerical simulations due to its simplicity. The simulation mesh makes it possible to define discrete points at which properties of interest, such as density or velocity, are known. The two obvious choices are the nodes and the cell centers. Unless noted otherwise, all simulations developed in this book utilize the *node-centered* approach. The governing equations are then cast to a form that makes it possible to evaluate them using the mesh-based data. We also specify *boundary conditions* that control how gas enters and leaves the computational domain. For instance, we may assign constant density $n = n_0$ and non-zero flow velocity in the x direction, $\vec{u} = u_0\hat{i}$ on the x_{min} inlet plane. We also specify conditions on embedded boundaries. In Figure 1.3, these nodes are shown by the gray shading. As can be deduced from the plot, smooth geometries turn into degenerate *sugarcube* (or *stair case*) representations. This inability to resolve complex geometries is the primary limitation of Cartesian meshes. Some workarounds involve the use of adaptive mesh refinement [38] or cut-cells [40]. In Chapter 7 we see how to address this issue with the help of non-uniform meshes.

Loss of ions to surface neutralization can be modeled by setting ion density to zero, $n = 0$, on the nodes belonging to the sphere. This is an example of the *Dirichlet* boundary condition. This condition specifies the actual value for the unknown. It is not applicable everywhere. We do not know ahead of time the density on the downstream, $x = x_{max}$ face. Specifying the Dirichlet boundary condition there would be incorrect. Instead, we tell the code that the flow should be fully developed by requiring that there is no variation in the perpendicular (or normal) direction to the boundary, $\partial n/\partial \hat{n} \equiv \partial n/\partial x = 0$. This is an example of the *Neumann* boundary condition. Other types also exist, including Robin and Cauchy, but are encountered much less frequently.

The particle method, visualized in the bottom half, does not utilize any macroscopic governing equations at all. It instead simply injects particles into the computational domain according to some source model. Here we inject particles on the x_{min} face. The number of injected particles is selected such that the desired density is achieved. The initial velocity is sampled from the inlet velocity distribution function. We advance particle velocities and positions by integrating the equations of motion, 1.1 through discrete Δt time steps. We first compute the new velocity of each particle by taking into account forces acting on it. We may also consider collisions. We then push particles to new positions by integrating velocity. Boundary conditions now control the microscopic behavior of individual particles. A neutral particle impacting the sphere may bounce off diffusely, while an ion may recombine into an atom. Particles leaving the computational domain may be removed or reflected back if the simulation models a problem exhibiting spatial symmetry. What makes particle simulations different from their fluid counterpart is that at each time step, the simulation only consists of a large collection of discrete position and velocities pairs. We are generally interested in computing the bulk densities, velocities, and temperatures. Obtaining these macroscopic properties from the microscopic particle data requires some additional averaging. This is generally accomplished with the help of a computational mesh. The local number density could be obtained by computing the number of particles in each cell since number density is just the particle count divided by the cell volume. Gas velocities and temperatures can be obtained in a similar manner.

1.3   ELECTROSTATIC PARTICLE IN CELL METHOD

With this in mind, we now introduce the Particle in Cell (PIC) method. This plasma simulation method was popularized by Birdsall and Langdon [15], and to this day, their book serves as an invaluable reference for anyone working in this field. Let's assume that some small region of space contains N real ions described by a local velocity distribution function f. We can approximate this population by sampling M simulation particles from the VDF. This much smaller sample represents the entire population, and hence each particle corresponds to $w_{mp} = N/M$ real ions or electrons. Each simulation particle is born with a discrete velocity and carries momentum $w_{mp}m\vec{v}$. Similarly, the particle contributes w_{mp} and $w_{mp}q$ count and charge to local number and charge density calculation. Velocity and position is updated according to the mass and charge of the individual physical ion or electron of mass m and charge q. In order to avoid confusion between the charge used to move the particle and the contribution to charge density, it is helpful to envision each macroparticle as a stack of w_{mp} real ions or electrons moving in unison.

1.3.1   Lorentz Force

Particles move according to the equations of motion discussed previously, with force computed per Equation 1.5. However, instead of using the Coulomb law, we obtain the $\vec{E}$ term, called the *electric field*, in a more efficient way. The relationship given by Equation 1.5 is actually incomplete as it neglects the effect of moving charges. The complete form is known as the *Lorentz force*,

$$\vec{F} = q\left(\vec{E} + \vec{v} \times \vec{B}\right) \tag{1.16}$$

where $\vec{v}$ is the particle velocity and $\vec{B}$ is the magnetic field.

1.3.2 Maxwell's Equations

The evolution of the electric and magnetic fields is given by the four fundamental equations of electromagnetics known as *Maxwell's equations*. They are:

$$\text{Gauss' law:} \quad \nabla \cdot \vec{E} = \frac{\rho}{\epsilon_0} \tag{1.17}$$

$$\text{Gauss' law for magnetism:} \quad \nabla \cdot \vec{B} = 0 \tag{1.18}$$

$$\text{Faraday's law:} \quad \nabla \times \vec{E} = -\frac{\partial \vec{B}}{\partial t} \tag{1.19}$$

$$\text{Ampere's law:} \quad \nabla \times \vec{B} = \mu_0\left(\vec{j} + \epsilon_0 \frac{\partial \vec{E}}{\partial t}\right) \tag{1.20}$$

Here $\rho \equiv \sum_s q_s n_s$ is the *charge density*, and $\vec{j} \equiv \sum_s q_s n_s \vec{v}_s$ is the *current density*. Faraday's law tells us that if $\partial \vec{B}/\partial t = 0$, $\nabla \times \vec{E} = 0$. A vector field with a zero curl is said to be irrotational. Helmholtz decomposition, also known as the fundamental law of vector calculus, further tells us that any sufficiently smooth vector field $\vec{F}$ can be decomposed into an irrotational (curl-free) and a solenoidal (divergence-free) part. These two parts can be defined in terms of a scalar (for irrotational) and a vector (for solenoidal) potential,

$$\vec{F} = -\nabla\phi + \nabla \times \vec{A} \tag{1.21}$$

The scalar potential ϕ is included with a negative sign per convention. Therefore, if the magnetic field is time invariant, the solenoidal contribution must vanish, and the electric field can be defined in terms of scalar potential,

$$\vec{E} = -\nabla\phi \tag{1.22}$$

This $\partial \vec{B}/\partial t = 0$ condition is known as the *electrostatic assumption*. Since Ampere's law indicates that magnetic field arises from currents, this condition implies that current density is low enough for the self-induced magnetic field to be negligible. Furthermore, if external magnets are present, the imparted field cannot be varying with time.

1.3.3 Poisson's Equation

The electrostatic assumption vastly simplifies the remaining analysis. We can substitute Equation 1.22 into Gauss' law to obtain

$$\nabla \cdot (-\nabla\phi) = \frac{\rho}{\epsilon_0} \tag{1.23}$$

or

$$\nabla^2\phi = -\frac{\rho}{\epsilon_0} \tag{1.24}$$

This is the Poisson's equation and it forms the basis of the Electrostatic Particle in Cell (ES-PIC) method. It provides a relationship between plasma potential and charge density. Charge density is simply the aggregate of number densities for individual species scaled by their charge. In a system containing only singly and doubly-charged ions and electrons, we have $\rho = e(n_i + 2n_{i2} - n_e)$, where e is the elementary charge. The density of the doubly charged ions can alternatively be written in terms of the singly-charged population, $n_{i2} = kn_i$, where k is some fraction. We then have $\rho = e[(1 + 2k)n_i - n_e] = e(Z_i n_i - n_e)$. The Z_i term is the average ionization state.

What makes this approach attractive is realizing that charge density is a macroscopic property. Practically this implies that we define density in terms of a simulation mesh. The number of mesh nodes G is many times smaller than the number of computational particles, M. Instead of the M^2 dependence required with the direct method, the operation count is reduced roughly to:

- M operations to compute charge density (loop over particles)

- $G \log(G)$ operations to solve Poisson's equation (loop over mesh nodes)

- G operations to compute electric field (loop over mesh nodes)

- M operations to integrate particle positions (loop over particles)

The $G \log(G)$ terms captures the typical convergence rate of solvers for the elliptic Poisson equation. Assuming we have a mesh with 100,000 cells and 20 particles per cell, the total operation count for the PIC scheme is reduced by six orders of magnitude compared to the direct Coulomb force summation. This is the great strength of the PIC method. It allows us to perform kinetic simulations in a manner that is compatible with standard desktop workstations.

1.3.4 Simulation Main Loop

In the PIC method, the computational domain is first discretized into a simulation mesh. At each time step, we use particle positions to compute the number density. We then use the corresponding charge density to compute plasma potential using Equation 1.24. Once the potential is known, the electric field is obtained from Equation 1.22. This electric field is used to update particle velocities according to the Lorentz force. The new velocity is used to push particles to new positions and the whole process repeats. The general ES-PIC algorithm follows the following steps:

```
// initialization
solve initial potential
compute electric field
load particles

// simulation loop
main_loop:
    // fields
    compute charge density
    compute potential
    compute electric field

    // particles
    update particle velocity
    update particle position

    // optional steps
    perform collisions
    inject additional particles
    output diagnostics

// finalization
output results
```

The simulation main loop continues until some exit criterion is met. This could involve completing a preset real time, or reaching the steady state. After the loop finishes, we typically save the final results and perform additional finalization. An electromagnetic PIC (EM-PIC), discussed in Chapter 7, follows a similar outline, but a different method is used to advance the electric and magnetic fields.

1.4 SINGLE PARTICLE MOTION

We now show how the core concepts of the PIC method are actually implemented by developing a simple simulation program. Let's imagine that a region of uniform charge density ρ_0 exists between two grounded electrodes. This setup is shown in Figure 1.4. We also assume that the electrodes are infinitely tall and deep. Since the charge density is uniform, plasma properties vary only in the x direction. This is an example of a one dimensional (1D) problem.

We assume that the region is filled only with ions, $\rho_0 = en_i$. For $\rho_0 > 0$, positive potential develops between the two walls. A single stationary electron placed near one of the two electrodes should start moving towards the top of the potential hill, where it reaches the maximum speed. It then continues down the hill, slowing down along the way until its velocity reaches zero again. The motion then reverses, and the electron continues to oscillate with this back and forth motion. Of course, this model ignores any frictional (or collisional)

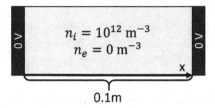

$$n_i = 10^{12} \text{ m}^{-3}$$
$$n_e = 0 \text{ m}^{-3}$$

0.1m

Figure 1.4: Simulation domain for the single particle motion example.

losses. In order to develop the simulation, we assume that the background ion density remains constant and that it is sufficiently high so the charge carried by the single electron has negligible effect on the potential profile. This then allows us to compute the potential just once. The rest of the code simply involves updating the electron position. We have the following pseudo-code:

```
solve potential
compute electric field
load particle

main_loop:
    update particle velocity
    update particle position
    write diagnostics
```

1.4.1 Domain Discretization

As was alluded to before, we use a computational mesh to describe the spatial variation in potential, charge density, and the electric field. Computationally this involves allocating a memory array to store n_i floating point values for each variable, where n_i is the number of mesh nodes. In classic C++, arrays can be dynamically allocated using the **new** operator,

```cpp
// constants
namespace Const {
    const double QE = 1.602176565e-19;   // C, elementary charge
};

int main() {
    const int ni = 21;                   // number of mesh nodes
    double *phi = new double[ni];        // potential
    double *rho = new double[ni];        // charge density
    double *ef = new double[ni];         // electric field

    // initialize the data
    for (int i=0; i<ni; i++) {
        phi[i] = 0;
        rho[i] = Const::QE*1e12;          // charge density
        ef[i] = 0;
    }

    /* rest of the code goes here */
```

```
delete [] phi;                    // memory cleanup
delete [] rho;
delete [] ef;
return 0;                          // normal exit
}
```

This approach has several downsides. First, the allocated data is not initialized by default. Each array component initially contains just some random bytes that existed in the assigned memory location. We need to manually set each entry to zero (or other values that make sense) prior to use. Secondly, we need to remember to use the delete[] operator to release the memory once the array is no longer needed.

A cleaner approach is to use the std::vector container from the C++ standard library,

```
#include <vector>
using namespace std;            // allows skipping std:: in std::vector

int main() {
  const int ni = 21;
  vector<double> phi(ni);      // space for ni doubles, all set to 0
  vector<double> rho(ni,Const::QE*1e12);   // set to QE*1e12
  vector<double> ef(ni);

  /* rest of the code goes here */

  return 0;        // normal exit
}
```

You can notice two main differences: there is no longer the need to initialize the data. It is automatically set to zero, unless a different value is provided as is the case with charge density. We also don't need to free the memory - this happens automatically when the main block exits and the vector variables go out of scope. std::vector is a *templated* container capable of storing arbitrary data types. The <double> template argument tells the compiler that we need a container for double precision floating point values. Sometimes having to repeatedly type these arguments can become cumbersome. To get around this, we can define a new type called dvector with the help of the using directive,

```
#include <vector>
using namespace std;
using dvector = vector<double>;       // define a 'nickname'

int main() {
  const int ni = 21;
  dvector phi(ni);   // ni doubles to hold potential

  /* rest of the code */

  return 0;        // normal exit
}
```

This syntax for the using keyword was introduced with the C++11 language extensions. Alternatively, we can utilize the legacy typedef keyword,

```
typedef dvector vector<double>;   // another pre-C++11 method
```

So far, we only specified the resolution of the mesh. We have yet to include information on the physical region it corresponds to. A uniform Cartesian mesh in the x direction has nodes with positions given by

$$x = x_0 + i\Delta x \qquad i \in [0, c_i] \tag{1.25}$$

where c_i is the number of cells in the i (or x) direction. Alternatively, the mesh spacing required to cover the distance L using c_i cells is

$$\Delta x = L/c_i \tag{1.26}$$

The number of nodes is $n_i = c_i + 1$, and thus the above equation can also be written as

$$\Delta x = \frac{x_m - x_0}{n_i - 1} \tag{1.27}$$

Here $L = x_m - x_0$ is the distance between the point with the highest x value and the origin. Following the schematic from Figure 1.4, we have $x_0 = 0$ and $x_m = 0.1$ m. We add the code to set these parameters. We also include a function for outputting the mesh fields to a comma-separated (.csv) file. This code is found in ch1_01.cpp.

```cpp
#include <vector>
#include <iostream>
#include <fstream>

// constants
namespace Const {
    const double QE = 1.602176565e-19;   // C, electron charge
    const double EPS0 = 8.85418782e-12;  // C/V/m, vac. permittivity
    const double ME = 9.10938215e-31;    // kg, electron mass
};

using namespace std;
using namespace Const;   // to avoid having to write Const::QE
using dvector = vector<double>;

// function prototypes
bool outputCSV(double x0, double dx, const dvector &phi, const
    dvector &rho, const dvector &ef);

// main
int main() {
    const int ni = 21;                  // number of nodes
    const double x0 = 0;                // mesh origin
    const double xm = 0.1;              // opposite end
    double dx = (xm-x0)/(ni-1);         // node spacing

    dvector phi(ni);
    dvector rho(ni,QE*1e12);
```

```
dvector ef(ni);

// ouput to a CSV file for plotting
outputCSV(x0,dx,phi,rho,ef);

return 0;                          // normal exit
}

// outputs the given fields to a CSV file, returns true if ok
bool outputCSV(double x0, double dx, const dvector &phi, const
   dvector &rho, const dvector &ef) {
   ofstream out("results.csv");   // open file for writing
   if (!out) {
      cerr<<"Could not open output file!"<<endl;
      return false;
   }

   out<<"x,phi,rho,ef\n";         // write header
   for (int i=0;i<phi.size();i++) {
      out<<x0+i*dx;   // write i-th position
      out<<","<<phi[i]<<","<< rho[i]<<","<<ef[i];  // write values
      out<<"\n";      // new line, not using endl to avoid buffer flush
   }

   return true;  // file closed automatically when "out" goes out of
      scope
}
```

This function takes as arguments references to the three `vector<double>` fields. The pass-by-reference is specified by the `&` operator. It avoids unnecessary duplication of the vector to create a temporary local data object. The `const` directive tells the compiler that we do not locally modify passed-in data. We then use the `ofstream` class included via the `<fstream>` header to open an output file `results.csv`. The function returns if the file open operation failed. Next the column names are written, followed by positions $x_i = x_0 + i\Delta x$, potential ϕ_i, charge density ρ_i, and electric field E_i. We use the phi vector's `size()` function to get the number of nodes. A more robust version should verify that all three vectors have the same size but here we just trust the caller. Compiling and running with

```
$ g++ ch1_01.cpp
$ ./a.out
```

we obtain a `results.csv` file containing the following

```
x,phi,rho,ef
0,0,1.60218e-07,0
0.005,0,1.60218e-07,0
...
0.095,0,1.60218e-07,0
0.1,0,1.60218e-07,0
```

As expected, the x position ranges from 0 to 0.1. ϕ and E are initialized to zero. $\rho \approx 1.6 \times 10^{-7}$ C/m^{-3}, which is the charge density corresponding to the 10^{12} m^{-3} number density of singly charged ions.

1.4.2 Finite Difference Method

We are now ready to solve Poisson's equation $\epsilon_0 \nabla^2 \phi = -\rho/\epsilon_0$ for plasma potential ϕ. For the case of constant charge density $\rho_0 = en_0$ considered here, the solution can be obtained analytically,

$$\frac{\partial^2 \phi}{\partial x^2} = -\frac{\rho_0}{\epsilon_0} \tag{1.28}$$

$$\frac{\partial \phi}{\partial x} = -\frac{\rho_0}{\epsilon_0} x + A \tag{1.29}$$

$$\phi = -\frac{\rho}{\epsilon_0} \frac{x^2}{2} + Ax + B \tag{1.30}$$

The integration constants are set from the boundary conditions. Substituting $\phi = 0$ for $x = 0$ leads to $B = 0$. Then using $\phi = 0$ for $x = L$ leads to $A = \rho_0 L/(2\epsilon_0)$. The analytical expression for potential is given by

$$\phi = \frac{\rho_0}{2\epsilon_0} x(L - x) \tag{1.31}$$

An expression for electric field, $E = -\partial\phi/\partial x$, can also be written

$$E = \frac{\rho}{\epsilon_0} \left(x - \frac{L}{2} \right) \tag{1.32}$$

Of course, the purpose of numerical simulations is to investigate problems that do not have such an elegant analytical solution. We thus need to perform this integration numerically. This process involves rewriting the differential equation in a form that can be solved by a computer.

There are three main approaches available to us: the Finite Difference Method (FDM), the Finite Volume Method (FVM), and the Finite Element Method (FEM). We learn all three in this book but start with the Finite Difference as it is the easiest of the three. The objective of all three methods is to rewrite the derivatives with a numerical approximation. For FDM, we utilize the *Taylor Series*. This infinite series provides an expression for the value of some function f at a point offset by Δx from another point at which the values of the function and its derivatives are known,

$$f(x + \Delta x) = f(x) + \frac{\Delta x}{1!} \frac{\partial f}{\partial x} + \frac{(\Delta x)^2}{2!} \frac{\partial^2 f}{\partial x^2} + \frac{(\Delta x)^3}{3!} \frac{\partial^3 f}{\partial x^3} + \ldots \tag{1.33}$$

We can write a similar expression for a point offset by $-\Delta x$,

$$f(x - \Delta x) = f(x) - \frac{\Delta x}{1!} \frac{\partial f}{\partial x} + \frac{(\Delta x)^2}{2!} \frac{\partial^2 f}{\partial x^2} - \frac{(\Delta x)^3}{3!} \frac{\partial^3 f}{\partial x^3} + \ldots \tag{1.34}$$

The negative sign appears only on the odd terms since $(-\Delta x)^2$ is positive by definition. These two equations can be added together. For simplicity, we adopt the syntax mentioned previously, and let $f_i \equiv f(x_i)$. Similarly, we let $f_{i+1} \equiv f(x + \Delta x)$ and $f_{i-1} \equiv f(x - \Delta x)$. The resulting form is

$$f_{i+1} + f_{i-1} = 2f_i + \Delta^2 x \frac{\partial^2 f}{\partial x^2} + \text{HOT} \tag{1.35}$$

The HOT acronym is a placeholder for Higher Order Terms and $\Delta^2 x \equiv (\Delta x)^2$. Since generally $\Delta x << 1$ and the factorial in the denominator grows rapidly, these terms quickly become vanishingly small. It is thus customary to discard them, and note that this scheme is second-order accurate. We identify this truncation error using a "big O" notation, $O(2)$. The above equation can next be re-arranged to isolate the derivative term,

$$\frac{\partial^2 f}{\partial x^2} = \frac{f_{i-1} - 2f_i + f_{i+1}}{\Delta^2 x} + O(3) \tag{1.36}$$

This expression is known as the standard central difference for the second derivative.

1.4.3 Potential Solver

Poisson's equation holds everywhere inside the domain. On each internal node we have

$$\frac{\partial^2 \phi}{\partial x^2} = -\frac{\rho_i}{\epsilon_0} \qquad i \in [1, n_i - 2] \tag{1.37}$$

Substituting Equation 1.36 for the left hand side, we obtain

$$\frac{\phi_{i-1} - 2\phi_i + \phi_{i+1}}{\Delta^2 x} = -(\rho_i/\epsilon_0) \qquad i \in [1, n_i - 2] \tag{1.38}$$

There are $n_i - 2$ equations for n_i unknowns. We close the system by specifying boundary conditions for the left-most and the right-most node. The Dirichlet boundary is used, since the potential on the two end nodes is fixed, $\phi_0 = \phi_{ni-1} = 0$ V.

As an example, let's consider a mesh with only seven nodes. The Finite Difference Method formulation for the Poisson's equation results in the following equations:

$$\phi_0 = \phi_{left} \tag{1.39}$$

$$(1/\Delta^2 x)(\phi_0 - 2\phi_1 + \phi_2) = -\rho_1/\epsilon_1 \tag{1.40}$$

$$(1/\Delta^2 x)(\phi_1 - 2\phi_2 + \phi_3) = -\rho_2/\epsilon_2 \tag{1.41}$$

$$(1/\Delta^2 x)(\phi_2 - 2\phi_3 + \phi_4) = -\rho_3/\epsilon_3 \tag{1.42}$$

$$(1/\Delta^2 x)(\phi_3 - 2\phi_4 + \phi_5) = -\rho_4/\epsilon_4 \tag{1.43}$$

$$(1/\Delta^2 x)(\phi_4 - 2\phi_5 + \phi_6) = -\rho_5/\epsilon_5 \tag{1.44}$$

$$\phi_6 = \phi_{right} \tag{1.45}$$

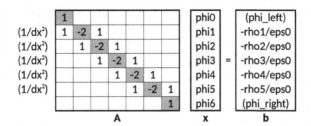

Figure 1.5: One dimensional Poisson equation coefficient matrix for a mesh with 7 nodes and Dirichlet boundary conditions.

We can rewrite this system using matrix notation as depicted in Figure 1.5. The coefficient matrix $\mathbf{A}$ holds the values multiplying the unknowns (the node potentials) from vector $\vec{\phi}$. Obtaining the potentials then "simply" involves solving the matrix system $\mathbf{A}\vec{\phi} = \vec{b}$. Some programming languages, such as MATLAB®, contain built-in matrix solvers. That unfortunately is not the case with C++, although high performance external libraries such as LAPACK [1], PETSc [11], or MUMPS [7] exist that provide this functionality.

Any linear matrix system can be solved by multiplying both sides by the matrix inverse, $\mathbf{A}^{-1}\mathbf{A}\vec{\phi} = \mathbf{A}^{-1}\vec{b}$. This formulation assumes that the inverse exists: the matrix is not singular. Since the product of a matrix with its inverse is the identity matrix, we have $\vec{\phi} = \mathbf{A}^{-1}\vec{b}$. Unfortunately, computing the inverse is computationally expensive. Furthermore, the coefficient matrix for the Poisson's equation consists only of few non-zero bands centered around the main diagonal. This is an example of a *banded* matrix, which in itself is a type of a *sparse* matrix. Sparse matrices can be stored efficiently by recording only the non-zero entries. The inverted matrix loses this banded structure and contains significantly more non-zero entries. Storing a full coefficient matrix for a mesh with just $G = 100,000$ nodes requires over 74 Gb of RAM due to the size scaling with G^2. This can be compared with only 2 Mb required for the banded system.

1.4.3.1 Tridiagonal Algorithm

The approach taken in most codes is to compute the solution of the given system without computing the inverse explicitly. Matrix solvers can be classified into two types: *direct* and *iterative*. Direct solvers take advantage of some inherent structure in the matrix data to produce the exact solution in a fixed number of iterations. You may already be familiar with the tridiagonal, or Thomas, algorithm. Following the implementation from [5], we let $\vec{a}$ and $\vec{c}$ contain coefficients for the band to the left and to the right of the main diagonal $\vec{b}$. We label the right hand side vector $\vec{d}$. The algorithm consists of two steps. First, in a forward sweep, we modify the coefficients according to

$$c_i' = \begin{cases} \frac{c_i}{b_i}, & ; i = 0 \\ \frac{c_i}{b_i - a_i c_{i-1}'} & ; i = 1, \ldots, n-2 \end{cases} \tag{1.46}$$

and

$$d_i' = \begin{cases} \frac{d_i}{b_i}, & ; i = 0 \\ \frac{d_i - a_i d_{i-1}'}{b_i - a_i c_{i-1}'} & ; i = 1, \ldots, n-2 \end{cases} \tag{1.47}$$

This is then followed by backward substitution,

$$x_{n-1} = d_{n-1}' \tag{1.48}$$

$$x_i = d_i' - c_i' x_{i+1} \qquad ; i = n-2, \ldots, 0 \tag{1.49}$$

Implementing this algorithm is quite simple, as the following code illustrates,

```
void solvePotentialDirect(double dx, dvector &phi, const dvector
    &rho) {
  int ni = phi.size();        // number of mesh nodes
  dvector a(ni);              // coefficients for phi[i-1]
  dvector b(ni);              // coefficients for phi[i]
  dvector c(ni);              // coefficients for phi[i+1]
  dvector d(ni);              // right-hand-side (RHS)

  // set coefficients
  for (int i=0;i<ni;i++) {
    if (i==0 || i==ni-1) {    // Dirichlet boundary
      b[i] = 1;               // 1 on the diagonal
      d[i] = 0;               // RHS value: 0 V
    }
    else {  // set the standard stencil
      a[i] = 1/(dx*dx);       // phi[i-1] coefficient
      b[i] = -2/(dx*dx);      // phi[i] coefficient
      c[i] = 1/(dx*dx);       // phi[i+1] coefficient
      d[i] = -rho[i]/EPS_0;   // RHS value
    }
  }

  // initialize
  c[0] = c[0]/b[0];
  d[0] = d[0]/b[0];

  // forward step
  for (int i=1;i<ni;i++) {
    if (i<ni-1)
      c[i] = c[i]/(b[i]-a[i]*c[i-1]);
    d[i] = (d[i] - a[i]*d[i-1])/(b[i] - a[i]*c[i-1]);
  }

  // backward substitution
  phi[ni-1] = d[ni-1];
  for (int i=ni-2;i>=0;i--)
    phi[i] = d[i] - c[i]*phi[i+1];
}
```

The argument list to this function includes references to the phi and rho vectors, &phi and &rho. The rho vector is not modified and is marked as constant. This is not true for the phi vector, as it is updated with the new solution. The function starts by allocating memory space for the three matrix coefficient bands a, b, and c, and the right hand side vector d. The values are initialized to the coefficients from Figure 1.5. Note, normally this part would be done outside the solver in an initialization script. We then loop through the rows and apply Equations 1.46, 1.47, and 1.49. This last step updates values in the phi vector. Since this vector was passed by reference, this step sets data in the object allocated in main. Without the reference, this section of the code would update a temporary local copy that is destroyed on function exit.

1.4.3.2 Jacobi Iteration

For a system with N unknowns, the Thomas algorithm requires exactly N forward passes followed by N back-substitution steps. The algorithm always requires this many steps, regardless of the particular boundary conditions or the values contained in the right hand side vector. Iterative schemes, on the other hand, do not have a predetermined number of operations. As the name suggests, these methods iterate through the matrix, and continuously produce a new of set of estimates for $\vec{\phi}$ that, hopefully, represent an improvement over the prior approximation. This is known as *convergence*. We allow the solver to run until the error between the approximate and the true solution becomes smaller than some tolerance. While the Thomas algorithm offers an efficient direct solver for one-dimensional problems, fast direct solvers for higher dimensional problems are not simple to implement. Therefore, we often end up utilizing iterative solvers for 2D and 3D problems. One of the best known schemes is the Jacobi iteration. To implement it, we simply solve the equation given by the matrix row i for the unknown ϕ_i,

$$\phi_i^{k+1} = \left[b_i - \sum_{j=0}^{nj-1} (1 - \delta_{ij}) a_{ij} \phi_j^k \right] / a_{ii} \qquad (1.50)$$

Here a_{ij} is the value on the i-th row and j-th column of the matrix, and δ_{ij} is the Kronecker delta. It has a value of 1 if the two indexes match, but is zero otherwise. The $(1 - \delta_{ij})$ expression simply skips the diagonal term. The superscript k and $k+1$ corresponds to the current and the new value on node i. Here it also makes sense to take advantage of the sparse structure of the coefficient matrix. Furthermore, for the Poisson's equation all non-boundary coefficients are identical, allowing us to write

$$\phi_i^{k+1} = \left(\phi_{i-1}^k + \phi_{i+1}^k - b\Delta^2 x \right) / 2 \qquad (1.51)$$

A simple code for the Jacobi algorithm is given below:

```
for (int k=0; k<max_it; k++) {
```

```
// compute "k+1" values on internal nodes
for (int i=1; i<ni-1; i++)
  phi_new[i] = (phi[i-1] + phi[i+1] - (dx*dx)b[i])/2;

// apply boundaries
phi_new[0] = phi_left;
phi_new[ni-1] = phi_right;

// copy down the new 'k+1' solution to 'k'
for (int i=0; i<ni; i++)
  phi[i] = phi_new[i];
}
```

1.4.3.3 Gauss-Seidel Scheme

The Jacobi algorithm requires two arrays: one to hold the current data, ϕ^k, and another to store the new approximations, ϕ^{k+1}. At the end of each iteration, we copy down from the new "k+1" array to the "k" array. But what if instead of using this secondary array, we just replace the values in place? This gives us the Gauss-Seidel scheme. It effectively replaces Equation 1.51 with

$$\phi_i^{k+1} = \left(\phi_{i-1}^{k+1} + \phi_{i+1}^k - b\Delta^2 x\right)/2 \qquad (1.52)$$

The difference is the subscript on ϕ_{i-1}. Since we are iterating through the ϕ array in an incremental fashion, nodes with index smaller than i already have the new ϕ^{k+1} value applied. The corresponding pseudocode is

```
// apply boundaries
phi[0] = phi_left;
phi[ni-1] = phi_right;

for (int k=0;k<max_it;k++) {
  // compute 'k+1' values on internal nodes
  for (int i=1;i<ni-1;i++)
    phi[i] = (phi[i-1] + phi[i+1] - (dx*dx)b[i])/2;
}
```

The Gauss-Seidel scheme is simpler and does not require the secondary array. It also typically converges faster than Jacobi, although there are matrix systems for which Gauss-Seidel may diverge while Jacobi does not.

1.4.3.4 Successive Over Relaxation

We can further speed up the convergence by implementing *successive over relaxation*, or SOR. This scheme uses the old and the new approximation to predict the future solution. For any two points y_1 and y_2, we can define another point using a linear fit, $y = y_1 + w(y_2 - y_1)$. Here w is a parametric coordinate. This expression evaluates to y_1 for $w = 0$ and to y_2 for $w = 1$. For $w > 1$, we obtain an interpolated value beyond the y_2 limit. We let y_1 be the previous value of the node potential, and y_2 be the new approximation per

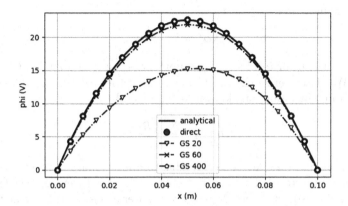

Figure 1.6: Potential between two grounded plates computed by the direct method and after several iterations of Gauss-Seidel.

Equation 1.52. Perhaps an even better approximation can be obtained with $t > 1$. The SOR-accelerated Gauss-Seidel algorithm becomes

```
for (int k=0;k<max_it;k++) {
   // compute "k+1" values on internal nodes
   for (int i=1; i<ni-1; i++) {
      double g = (phi[i-1] + phi[i+1] - (dx*dx)b[i])/2;
      phi[i] = phi[i] + w*(g-phi[i]);   // SOR
   }

   // apply boundaries
   phi[0] = phi_left;
   phi[ni-1] = phi_right;
}
```

The value of the optimal acceleration parameter w needs to be determined by trial-and-error but it is customary to use values around 1.4. Exceedingly large w can lead to convergence issues as initial errors propagate in the wrong direction. Utilizing $w < 1$ is called *under-relaxation* and is used in CFD to stabilize solvers that may otherwise diverge. An adaptive method is sometimes preferred.

1.4.3.5 Convergence Check

Figure 1.6 shows the potential computed by the direct solver and after several iterations of the GS-SOR method. These solutions are also compared to the analytical result from Equation 1.31. An excellent agreement is seen. We can notice that while there is a big difference between the GS solution after 20 and 60 iterations, the solution does not change much afterwards. In other words, the solution is converging to some final state. Each iteration of the Gauss-Seidel scheme produces only an approximate solution. The actual matrix system at iteration k is

$$\mathbf{A}\vec{\phi}^k = \vec{b}^k + \vec{R}^k \qquad (1.53)$$

where $\vec{R}^k$ is the error residue. As ϕ^k approaches the true solution, $\vec{R}$ becomes smaller. We characterize the magnitude of a vector with its norm. There are several norms, including one known as L2, given by,

$$||R|| = \sqrt{\frac{\sum_i^n (R_i)^2}{n}} \qquad (1.54)$$

Since this calculation takes some valuable computational resources, we generally do not compute residue at every iteration. I generally evaluate the residue in Gauss-Seidel solvers every 25 to 50 solver iterations, as this seems to be a good compromise between not performing convergence checks too often, and running extra iterations beyond the desired tolerance. The complete solver code is shown below. It runs for up to `max_it` iterations. Once the norm drops below 10^{-6}, the function terminates with a success code. Otherwise, if the maximum number of solver iterations is exceeded without reaching the desired tolerance, we return `false` to indicate failure.

```
bool solvePotentialGS(double dx, dvector &phi, const dvector &rho,
  int max_it=5000) {
  double L2;
  double dx2 = dx*dx;    // precompute dx*dx
  const double w = 1.4;
  int ni = phi.size();   // number of mesh nodes

  // solve potential
  for (int solver_it=0;solver_it<max_it;solver_it++) {
    phi[0] = 0;          // dirichlet boundary on left
    phi[ni-1] = 0;       // dirichlet boundary on right

    // Gauss Seidel method
    for (int i=1;i<ni-1;i++) {
      double g = 0.5*(phi[i-1] + phi[i+1] + dx2*rho[i]/EPS_0);
      phi[i] = phi[i] + w*(g-phi[i]);   // SOR
    }

    // check for convergence
    if (solver_it%50==0) {
      double sum = 0;

      // only checking internal (non-Dirichlet) nodes
      for (int i=1;i<ni-1;i++) {
        double R = -rho[i]/EPS_0 - (phi[i-1] - 2*phi[i] +
          phi[i+1])/dx2;
        sum+=R*R;
      }

      L2 = sqrt(sum/ni);   // L2 norm
      if (L2<1e-6) {
        cout<<"GS solver converged after "<<solver_it<<"
          iterations"<<endl;
```

```
        return true;
      }
    }
  }
  cout<<"GS solver failed to converge, L2="<<L2<<endl;
  return false;
}
```

1.4.4 Electric Field

Returning back to the pseudo code on Page 15, the next step involves computing the electric field from

$$E = -\frac{\partial \phi}{\partial x} \tag{1.55}$$

In other words, we need to numerically estimate the first derivative. We again utilize the Taylor series defined for a node to the left and right of node i,

$$f_{i+1} = f_i + \frac{\Delta x}{1}\frac{\partial f}{\partial x} + \frac{(\Delta x)^2}{2}\frac{\partial^2 f}{\partial x^2} + \text{HOT} \tag{1.56}$$

$$f_{i-1} = f_i - \frac{\Delta x}{1}\frac{\partial f}{\partial x} + \frac{(\Delta x)^2}{2}\frac{\partial^2 f}{\partial x^2} + \text{HOT} \tag{1.57}$$

Instead of adding the two equations as was done for Equation 1.36, we subtract the second equation from the first. This gives us

$$f_{i+1} - f_{i-1} = 2\Delta x\frac{\partial f}{\partial x} + \text{HOT} \tag{1.58}$$

Re-arranging, we obtain

$$\frac{\partial f}{\partial x} \approx \frac{f_{i+1} - f_{i-1}}{2\Delta x} \tag{1.59}$$

This is the standard central-difference for the first derivative. The $\partial^2 f/\partial x^2$ term naturally drops out, making this scheme second-order accurate. This scheme can be used on the internal nodes. On the boundaries, we can use equations 1.56 or 1.57 directly. If we discard the second derivative term, we obtain

$$\frac{\partial f}{\partial x} \approx \frac{f_{i+1} - f_i}{\Delta x} \tag{1.60}$$

and

$$\frac{\partial f}{\partial x} \approx \frac{f_i - f_{i-1}}{\Delta x} \tag{1.61}$$

These two expressions are called the *forward* and the *backward* finite difference. They are only first-order accurate, since the second derivative term had to be explicitly discarded. Alternatively, the same way we used Taylor series to define the value at node $i + 1$, we can also estimate the value at $i + 2$,

$$f_{i+2} = f_i + 2\Delta x\frac{\partial f}{\partial x} + 2\Delta^2 x\frac{\partial^2 f}{\partial x} \tag{1.62}$$

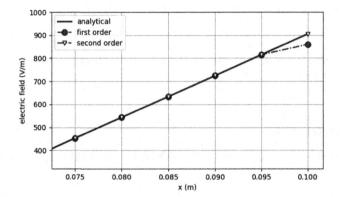

Figure 1.7: Electric field on the right boundary. The first-order backward method results in a discrepancy.

We can next subtract Equation 1.56 multiplied by 4. The second derivative terms then cancel naturally, and we obtained the second-order forward difference scheme,

$$\frac{\partial f}{\partial x} \approx \frac{-3f_i + 4f_{i+1} - f_{i+2}}{2\Delta x} \tag{1.63}$$

We can similarly derive a second order backward difference scheme,

$$\frac{\partial f}{\partial x} \approx \frac{f_{i-2} - 4f_{i-1} + 3f_i}{2\Delta x} \tag{1.64}$$

These functions are implemented in a function called ComputeEF,

```
// computes electric field by differentiating potential
void computeEF(double dx, dvector &ef, const dvector &phi, bool
    second_order = true) {
    int ni = phi.size();        // number of mesh nodes

    // central difference on internal nodes
    for (int i=1;i<ni-1;i++)
        ef[i] = -(phi[i+1]-phi[i-1])/(2*dx);

    // one-sided first or second order difference on the boundaries
    if (second_order) {
        ef[0] = (3*phi[0]-4*phi[1]+phi[2])/(2*dx);
        ef[ni-1] = (-phi[ni-3]+4*phi[ni-2]-3*phi[ni-1])/(2*dx);
    }
    else {  // first order
        ef[0] = (phi[0]-phi[1])/dx;
        ef[ni-1] = (phi[ni-2]-phi[ni-1])/dx;
    }
}
```

The optional Boolean parameter switches between the first- and the second-order differencing. The two approximations are compared to the analytical solution from Equation 1.32 in Figure 1.7. A clear discrepancy is seen with the first order scheme.

1.4.5 Particle Motion

Now that we have computed potential and the electric field, we are ready to introduce our test particle. We define the particle by its mass, charge, position, and velocity. The particle motion is governed by the equations of motion, Equation 1.1 with force given by $F_x = qE_x$. Note that this formulation assumes that magnetic field is not present, $B = 0$. Integration of particle motion in a magnetic field is left for Chapter 7. Once again, we use the finite difference formulation to rewrite the time derivatives in terms of a finite time step Δt. The important distinction is that time flows only forward, and thus it is not possible to use the central difference method. Doing so would imply that results at time k depended on the state at time $k+1$. One possible integration scheme leads to

$$x^{k+1} = x^k + v^k \Delta t \qquad (1.65)$$

$$v^{k+1} = v^k + (q/m)E^k \Delta t \qquad (1.66)$$

This scheme is called Forward Euler method. It is an *explicit* method, since the new value at time $k+1$ depends only on data at time k. Therefore, an equation can be defined for each particle that explicitly provides the new position. This is in contrast with *implicit* methods, in which the value at $k + 1$ depends on other values also at $k + 1$. In general, implicit schemes lead to linear systems that require matrix solution. In the above formulation, $E^k = E(x^k)$ is the electric field sampled at particle's location at the time step k. We see how to interpolate the mesh data to the particle shortly. For now, assume that electric field is constant. The integration is demonstrated by the following code:

```
int main() {
  // load particle
  double m = ME;       // particle mass
  double q = -QE;      // particle charge
  double x = 0;        // initial position
  double v = 0;        // stationary

  // simulation parameters
  double dt = 1e-9;    // timestep
  double E = -100;     // electric field

  // open particle trace file
  ofstream out("trace.csv");
  if (!out) {cerr<<"Failed to open file"<<endl; return -1;}
  out<<"t,x,v\n";

  // particle loop
  for (int it=0;it<10;it++) {
    // write trace data
    out<<it*dt<<","<<x<<","<<v<<"\n";
```

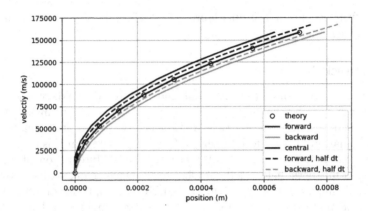

Figure 1.8: Particle velocity versus position for the forward, backward, and central methods.

```
  // advance velocity and position
  x += v*dt;
  v += (q/m)*E*dt;
}
  return 0;  // normal exit
}
```

We begin by defining a particle by its mass, charge, and initial position and velocity. We use a similar approach in our future codes, but instead of having just a single particle, we use many thousands of them. Next we set the integration time step Δt and also the fixed value of electric field. Then a file is opened to store particle positions and velocities at each time step. This output allows us to create a trace plot. We iterate through a prescribed number of time steps. At each step, the new position is computed from $x^{k+1} = x^k + v^k \Delta t$ using compound addition. It is important to notice that the code does not retain the past values of position and velocity. At any iteration, at most the single prior value is needed. Storing information for all previous time steps simply fills the computer memory with unnecessary data. This memory can be better utilized for more particles or a finer spatial gridding. However, for analysis, we may be interested in plotting traces for a small subset of particles. This is the purpose of including the trace file.

1.4.5.1 Forward Euler Method

Results from this integration are plotted in Figure 1.8 with the solid dark gray line. Instead of plotting position and velocity individually against time, we plot these two parameters against each other to obtain a phase space plot. This graph can be compared to the expected values. With constant E, velocity

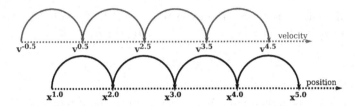

Figure 1.9: Visualization of the Leapfrog method. Positions and velocities are known at times offset by $0.5\Delta t$.

and position should follow

$$v = \frac{qE}{m}t + v_0 \tag{1.67}$$

$$x = \frac{qE}{2m}t^2 + v_0 t + x_0 \tag{1.68}$$

This trend is plotted by the open circles. Clearly, there is some discrepancy.

1.4.5.2 Backward Euler

Our integration scheme used the old velocity at time v^k to integrate the particle position from x^k to x^{k+1}. Alternatively, we could use the new velocity

$$x^{k+1} = x^k + v^{k+1}\Delta t \tag{1.69}$$

The only difference between these two schemes is the coefficient on the v term. Implementing this scheme simply involves switching the order of commands so that velocity is updated first,

```
v += (q/m)*E*dt;
x += v*dt;
```

We obtain the result shown by the light gray line. This scheme is known as the *Backward Euler Method*. Unfortunately, the solution is still wrong. It appears that one scheme over-estimates the solution, while the other under-estimates it. We can also check the influence of the finite time step. The two dashed lines show the results after 20 $\Delta t/2$ iterations. The number of steps was doubled so the same wall time is simulated. The integration schemes are *consistent*, since as the time step decreases, the solution approaches the expected analytical profile. In the limit of $\Delta t \rightarrow 0$, we expect both schemes to converge on the correct solution. Of course, this is not practical.

1.4.5.3 Leapfrog Method

Since during the integration, both v^k and v^{k+1} values are available, we could utilize their average. We thus have

$$x^{k+1} = x^k + \left(\frac{v^k + v^{k+1}}{2}\right)\Delta t \tag{1.70}$$

Finally, this scheme produces the expected result! It is plotted by the solid line. Utilizing the definition of $v^{k+1} = v^k + \Delta t(q/m)E$, we find

$$\frac{v^k + v^{k+1}}{2} = v^k + \frac{\Delta t}{2}\frac{q}{m}E \equiv v^{k+0.5} \tag{1.71}$$

Position needs to be integrated using velocities at the half step. Since velocities are needed only at these half-step times, it makes sense to let the initial particle velocity be defined at $k = -0.5$. The velocities are then simply integrated through a full step at each pass. This approach avoids the need to explicitly perform the averaging. While this may seem like a trivial optimization, given the millions of particles and hundreds of thousands of time steps in a typical PIC simulation, it is to our advantage to eliminate any inefficiencies. With this scheme, velocities are known at times $k = -0.5, 0.5, 1.5, \ldots$. Particle positions are known at times $k = 0, 1, 2, \ldots$. As shown in Figure 1.9, the two properties leapfrog over each other, giving this scheme its name. The *Leapfrog* particle integrator is demonstrated in the listing below. It is identical to the code we had for the Backward method, except that we perform a velocity rewind when injecting the particle. This rewind is performed by integrating velocity through $-\Delta t/2$ on line 18. Note that only the velocity is affected by the rewind.

```cpp
int main() {
    // load particle
    double m = ME;        // particle mass
    double q = -QE;       // particle charge
    double x = 0;         // initial position
    double v = 0;         // stationary

    // simulation parameters
    double dt = 1e-9;     // timestep
    double E = -100;      // electric field

    // open particle trace file
    ofstream out("trace.csv");
    if (!out) {cerr<<"Failed to open file"<<endl; return -1;}
    out<<"t,x,v\n";

    // velocity rewind
    v -= 0.5*(q/m)*E*dt;

    // particle loop
    for (int it=0; it<10; it++) {
        // write trace data
        out<<it*dt<<","<<x<<","<<v<<"\n";

        // compute future position and velocity
        v += (q/m)*E*dt;
        x += v*dt;
    }

    return 0;    // normal exit
}
```

1.4.5.4 Alternative Approaches

The Forward Euler method suffers from a number of limitations. Explicit schemes are simple to implement but tend to have a limited numerical stability. For some partial differential equations, schemes based on the Forward Euler method become *unconditionally unstable*, meaning that the solution always diverges regardless of the computational time step or cell spacing. For other schemes, there may be a region of *conditional stability* that involves excessively tiny time steps. Due to the finite computer math, a large number of tiny steps can result in the propagation of round-off errors. The Forward Euler method is also only first-order accurate. Integration errors scale linearly with Δt. With a second-order scheme, halving the simulation time step can be expected to produce a four-fold decrease in error. Backward Euler, on the other hand, is an example of an implicit scheme. Implicit schemes are generally unconditionally stable, meaning that error remains confined within some finite bounds. It is important to realize that stability does not automatically imply the solution is correct. Implementing a fully-implicit particle integrator involves computing velocity v^{k+1} using the electric field $E^{k+1} = E(x^{k+1})$. Since the electric field is a function of charge density, and hence particle positions, the future positions are needed to compute the acceleration that brings the particles to those positions. This chicken-and-egg problem leads to a linear system that is solved with the help of matrix math. For the huge number of particles used in PIC simulations, solving the system can be computationally prohibitive. For this reason, implicit PIC codes are not very common. An overview of a simple implicit algorithm can be found in [41].

The two Euler's methods discussed above are just a tiny sample of a huge family of explicit and implicit integration schemes for partial differential equations. There are many schemes that attempt to resolve the stability of an implicit method without requiring the matrix inversion. These methods typically fall in one of two classes. The first type are *predictor-corrector* methods. They consist of two steps, with the first one producing a test solution using a scheme similar to Forward Euler. This is then followed by a corrector step to reduce integration errors. Another class of methods is known as *multipoint*. The leapfrog scheme is one example since velocities at two different time steps are utilized to advance the position. Another multipoint method that you may be familiar with is the fourth-order accurate Runge-Kutta method (RK4). For a general partial differential equation

$$\frac{\partial f}{\partial t} = g(t, f) \tag{1.72}$$

the value at a future time step is given by

$$f^{k+1} = f^k + \frac{1}{6}\left(k_1 + 2k_2 + 2k_3 + k_4\right) \tag{1.73}$$

where

$$k_1 = \Delta t g \left(t^k, y^k \right) \tag{1.74}$$

$$k_2 = \Delta t g \left(t^k + \frac{\Delta t}{2}, y^k + \frac{k_1}{2} \right) \tag{1.75}$$

$$k_3 = \Delta t g \left(t^k + \frac{\Delta t}{2}, y^k + \frac{k_2}{2} \right) \tag{1.76}$$

$$k_4 = \Delta t g \left(t^k + \Delta t, y^k + k_3 \right) \tag{1.77}$$

The benefit of the RK4 method is that it allows for larger time steps to be used in the integration. This comes at the expense of the additional overhead associated with the four computations of the forcing term g and the subsequent averaging. Given that a typical PIC simulation contains millions of particles, these extra computations introduce a non-negligible overhead. In the case of PDE solvers used in fluid models, the additional overhead is made up by the ability to use larger integration time steps. In plasma simulations, the factor limiting the time step is generally not the convergence limit of the integrator. In fully-kinetic codes, we need to resolve electron oscillations at plasma frequency. PIC simulations also require that particles travel no more than one cell length per time step to prevent the electric field from becoming discontinuous as seen by the particles. For this reason, it is not customary to utilize higher order schemes such as RK4. All codes developed in this book implement the Leapfrog scheme.

1.4.6 Interpolation

We now have a particle integrator, but it uses a constant value for the electric field. Instead, we need some mechanism to evaluate the field from Section 1.4.4 at the particle position. The electric field is defined only on the mesh nodes. One possibility is to find the nearest grid node, and simply use the field from that node. This approach is not recommended. To see why, imagine a particle moving from left to right. For some interval, it occupies a region near a node i, and thus is accelerated by electric field E_i. As soon as the particle crosses the midpoint between nodes i and $i+1$, it suddenly experiences acceleration from E_{i+1}. Such discontinuities can induce numerical instabilities and lead to non-physical results. Instead, we prefer that the electric field is continuous. Throughout the majority of this book, we utilize a first-order scheme in which the field is interpolated linearly. Let's define a parameter $d_i \in [0,1]$ that specifies the fractional distance of a point located between nodes i and $i+1$. Linear interpolation tells us that a value of some function with values f_i when $d_i = 0$ and f_{i+1} when $d_i = 1$ is

$$f = f_i + d_i(f_{i+1} - f_i) \tag{1.78}$$

which can be rewritten as

$$f = (1 - d_i)f_i + (d_i)f_{i+1} \tag{1.79}$$

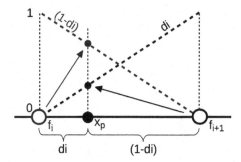

Figure 1.10: Visualization of the one-dimensional linear gather operation.

The operation given by the above expression is known as *gather*, since it gathers multiple mesh-based values to evaluate a quantity at a specified location. The opposite process is called *scatter*. It is used to deposit particle-based data to the mesh. We use the scatter operation in the next chapter to compute particle density. In Equation 1.79, the $(1 - d_i)$ and (d_i) terms are known as basis functions, or alternatively, shape factors. They are visualized in Figure 1.10 by the two dashed lines. Their magnitude at the particle position x_p is used to scale values at the two surrounding nodes, f_i and f_{i+1}.

Both gather and scatter algorithms require that the cell containing the particle is known. We also need the relative position of the particle within the cell. For a uniform mesh, this calculations is trivial. By inverting the map for node positions, we obtain $l_i = (x - x_0)/\Delta x$. Here l_i is the *logical coordinate*. It is simply a floating-point counterpart to the integer index i we used previously to compute node positions. The integer part of l_i is the node index, while the decimal part is the fractional distance to the next node d_i. We thus define a helper function

```
double XtoL(double x, double dx, double x0=0) {return (x−x0)/dx;}
```

to evaluate the logical coordinate. We use C++'s (**int**) cast to convert a floating point number to an integer. An integer cast discards the fractional part - there is no rounding. This integer value gives us the node index to the "left" of the particle (or below or in-front for higher dimensional codes). The fractional part d_i is obtained by subtracting the node index from the floating point logical coordinate. This is implemented in the gather function as follows,

```
double gather(double li, const dvector &field) {
    int i = (int)li;            // cast to integer
    double di = li−i;           // obtain fractional part
    return field[i]*(1−di) + field[i+1]*(di);
}
```

We use these functions as follows

```
dvector ef(ni);     // node−based electric field
```

```
for (int ts=0; ts <2000; ts++) { // iterate over time steps
    // sample electric field at particle position
    double li = XtoL(x,dx);
    double ef_p = gather(li ,ef);

    // integrate velocity and position
    v += (q/m)*ef_p*dt;
    x += v*dt;
}
```

1.4.7   Diagnostics

We now have implemented all pieces needed to complete the code. But as with any computer simulation, it is important to verify that results make physical sense. The particle oscillations arise from conservation of energy. Therefore, we should check that energy is indeed conserved. The total energy of a particle at time k is given by the sum of its potential and kinetic energy

$$\frac{1}{2}mv^2 + q(\phi - \phi_{max}) = C \qquad (1.80)$$

where C is some constant. Implementing this diagnostic check requires that velocities and positions are evaluated at the same physical time. With our leapfrog integration, these two parameters are actually offset by half a time step, with velocity trailing the position. One possibility is to bring the two parameters to the same physical time by tracking the old position, and sampling ϕ at $x^{k-0.5} = (x^k + x^{k-1})/2$. This is accomplished using the following code. The two energies are converted from Joules to electron volts by dividing by the elementary charge.

```
double phi_p = gather(XtoL((x+x_old)/2,dx),phi); // phi(x(k-0.5))
double ke = 0.5*m*v*v/QE;                         // KE in eV
double pe = q*(phi_p-phi_max)/QE;                 // PE in eV
```

The resulting code is given below in entirety.

```
#include <vector>
#include <iostream>
#include <fstream>
#include <math.h>

// constants
namespace Const{
    const double QE = 1.602176565e-19;    // C, elementary charge
    const double EPS_0 = 8.85418782e-12;  // C/V/m, vacuum perm.
    const double ME = 9.10938215e-31;     // kg, electron mass
};

using namespace std;
using namespace Const;
using dvector = vector<double>;

// function prototypes
```

```cpp
bool outputCSV(double x0, double dx, const dvector &phi, const
    dvector &rho, const dvector &ef);
void solvePotentialDirect(double dx, dvector &phi, const dvector
    &rho);
bool solvePotentialGS(double dx, dvector &phi, const dvector &rho,
    int max_it=5000);
void computeEF(double dx, dvector &ef, const dvector &phi, bool
    second_order=true);
double gather(double li, const dvector &field);

// converts physical position x to a logical coordinate l
double XtoL(double x, double dx, double x0=0) {return (x-x0)/dx;}

// main
int main() {
  const int ni = 21;          // number of nodes
  const double x0 = 0;        // origin
  const double xd = 0.1;      // opposite end
  double dx = (xd-x0)/(ni-1); // node spacing

  dvector rho(ni,QE*1e12);
  dvector ef(ni);
  dvector phi(ni);

  // solve potential
  solvePotentialGS(dx,phi,rho);
  // solvePotentialDirect(dx,phi,rho);   // alternate solver

  // compute electric field
  computeEF(dx, ef, phi, true);

  // generate a test electron
  double m = ME;
  double q = -QE;
  double x = 4*dx;      // four cells from left edge
  double v = 0;         // stationary

  double dt = 1e-10;  // timestep

  // velocity rewind
  double li = XtoL(x,dx);
  double ef_p = gather(li,ef);
  v -= 0.5*(q/m)*ef_p*dt;

  // save initial potential for PE calculation
  double phi_max = phi[0];
  for (int i=1;i<ni;i++)
    if (phi[i]>phi_max) phi_max = phi[i];

  // open file for particle trace
  ofstream out("trace.csv");
  if (!out) {cerr<<"Failed to open trace file"<<endl; return -1;}
  out<<"time,x,v,KE,PE\n";
  double x_old = x;

  // particle loop
  for (int ts=1; ts<=4000; ts++) {
```

```cpp
        // sample mesh data at particle position
        double li = XtoL(x,dx);
        double ef_p = gather(li,ef);

        // integrate velocity and position
        x_old = x;
        v += (q/m)*ef_p*dt;
        x += v*dt;

        double phi_p = gather(XtoL(0.5*(x+x_old),dx),phi);
        //phi(x(k-0.5))
        double ke = 0.5*m*v*v/QE;                    // KE in eV
        double pe = q*(phi_p-phi_max)/QE;            // PE in eV

        // write to a file
        out<<ts*dt<<","<<x<<","<<v<<","<<ke<<","<<pe<<"\n";

        if (ts==1 || ts%1000==0)                 // screen output every 1000
          timesteps
          cout<<"ts: "<<ts<<", x:"<<x<<", v:"<<v<<", KE:"<<ke<<",
            PE:"<<pe<<"\n";
    }

    // ouput results to a CSV file for plotting
    outputCSV(ni,dx,phi,rho,ef);

    return 0;   // normal exit
}

// outputs the given fields to a CSV file, returns true if success
bool outputCSV(double x0, double dx, const dvector &phi, const
    dvector &rho, const dvector &ef) {
    ofstream out("results.csv");          // open file for writing
    if (!out) {
      cerr<<"Could not open output file!"<<endl;
      return false;
    }

    out<<"x,phi,rho,ef\n";                 // write header
    for (size_t i=0;i<phi.size();i++) {
      out<<x0+i*dx;                        // write i-th position
      out<<","<<phi[i]<<","<< rho[i]<<","<<ef[i]; // write values
      out<<"\n";       // new line, not using endl to avoid buffer
        flush
    }

    return true; // file closed automatically here
}

// solves Poisson's equation with Dirichlet boundaries using the
  Thomas algorithm
void solvePotentialDirect(double dx, dvector &phi, const dvector
  &rho)
{
    int ni = phi.size();  // number of mesh nodes
    dvector a(ni);        // allocate memory for the matrix
      coefficients
```

```
    dvector b(ni);
    dvector c(ni);
    dvector d(ni);

    // set coefficients
    for (int i=0;i<ni;i++) {
      if (i==0 || i==ni-1) {   // Dirichlet boundary
        b[i] = 1;              // 1 on the diagonal
        d[i] = 0;              // RHS, 0 V
      }
      else {
        a[i] = 1/(dx*dx);      //
        b[i] = -2/(dx*dx);
        c[i] = 1/(dx*dx);
        d[i] = -rho[i]/EPS_0;
      }
    }

    // initialize
    c[0] = c[0]/b[0];
    d[0] = d[0]/b[0];

    // forward step
    for (int i=1;i<ni;i++) {
      if (i<ni-1)
        c[i] = c[i]/(b[i]-a[i]*c[i-1]);

      d[i] = (d[i] - a[i]*d[i-1])/(b[i] - a[i]*c[i-1]);
    }

    // backward substitution
    phi[ni-1] = d[ni-1];
    for (int i=ni-2;i>=0;i--)
      phi[i] = d[i] - c[i]*phi[i+1];
}
// solves potential using the Gauss Seidel Method
bool solvePotentialGS(double dx, dvector &phi, const dvector &rho,
    int max_it) {
    double L2;
    double dx2 = dx*dx;        // precompute dx*dx
    const double w = 1.4;
    int ni = phi.size();       // number of mesh nodes

    // solve potential*/
    for (int solver_it=0;solver_it<max_it;solver_it++) {
      phi[0] = 0;              // Dirichlet boundary on left
      phi[ni-1] = 0;           // Dirichlet boundary on right

      // Gauss Seidel method
      for (int i=1;i<ni-1;i++) {
        double g = 0.5*(phi[i-1] + phi[i+1] + dx2*rho[i]/EPS_0);
        phi[i] = phi[i] + w*(g-phi[i]);   // SOR
      }

      // check for convergence
      if (solver_it%50==0) {
```

```
      double sum = 0;

      // internal nodes, automatically satisfied on Dirichlet
         boundaries
      for (int i=1;i<ni-1;i++) {
        double R = -rho[i]/EPS_0 - (phi[i-1] - 2*phi[i] +
          phi[i+1])/dx2;
        sum+=R*R;
      }

      L2 = sqrt(sum/ni);
      if (L2<1e-6) {
        cout<<"Gauss-Seidel converged after "<<solver_it<<"
          iterations"<<endl;
        return false;
      }
    }
  }
  cout<<"Gauss-Seidel failed to converge, L2="<<L2<<endl;
  return true;
}

// computes electric field by differentiating potential
void computeEF(double dx, dvector &ef, const dvector &phi, bool
  second_order) {
  int ni = phi.size();  // number of mesh nodes

  // central difference on internal nodes
  for (int i=1;i<ni-1;i++)
    ef[i] = -(phi[i+1]-phi[i-1])/(2*dx);

  // boundaries
  if (second_order) {
    ef[0] = (3*phi[0]-4*phi[1]+phi[2])/(2*dx);
    ef[ni-1] = (-phi[ni-3]+4*phi[ni-2]-3*phi[ni-1])/(2*dx);
  }
  else {  // first order
    ef[0] = (phi[0]-phi[1])/dx;
    ef[ni-1] = (phi[ni-2]-phi[ni-1])/dx;
  }
}

// uses linear interpolation to evaluate f at li
double gather(double li, const dvector &f) {
  int i = (int)li;
  double di = li-i;
  return f[i]*(1-di) + f[i+1]*(di);
}
```

We compile the code and run it as

```
$ g++ ch1.cpp -o ch1
$ ./ch1
GS solver converged after 400 iterations
ts: 1, x:0.0200005, v:4773.91, KE:6.47884e-05, PE:8.14269
ts: 1000, x:0.0260696, v:-1.02448e+06, KE:2.98372, PE:5.19839
```

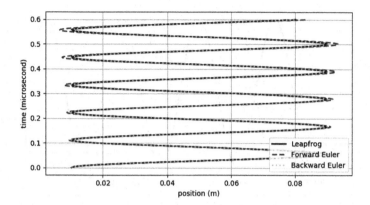

Figure 1.11: Particle position versus time. The forward Euler scheme results in a non-physical energy gain.

```
ts: 2000, x:0.0416597, v:-1.62704e+06, KE:7.52567, PE:0.668466
ts: 3000, x:0.0605681, v:-1.58226e+06, KE:7.11713, PE:1.05116
ts: 4000, x:0.0752719, v:-907960, KE:2.34359, PE:5.81261
```

For this simulation, we inject the particle 4 cell lengths $(4\Delta x)$ from the left boundary. The particle is initially stationary but as can be seen from the screen output above, it doesn't stay still for long.

Simulation results are further visualized in Figure 1.11. This image plots the position of the particle on the x axis. The y axis is time. The solid black line shows the results obtained with the leapfrog scheme. The particle continuously oscillates back and forth, as expected. The light gray line shows the path obtained with the forward Euler scheme attained by reversing the order of velocity and position integrations. A clear non-physical gain of energy is seen. The particle traverses increasingly longer distance. The backward Euler scheme, obtained by skipping the rewind, is also shown. This trajectory looks almost identical to the Leapfrog solution, but upon a close inspection, a small shift can be observed. But most importantly, the solution is bound. This demonstrates the stability inherent in implicit methods. Figure 1.12 plots the particle energies as a function of time. The energy is indeed conserved, as indicated by the gray horizontal line plotting the total energy.

1.5 SUMMARY

In this chapter we first reviewed different approaches for modeling plasmas and learned about a velocity distribution function. We then covered differences between fluid and kinetic simulation methods. The Electrostatic Particle in Cell (ES-PIC) method was then introduced. We saw how to implement its fundamental concepts: solving the Poisson's equation, computing the electric

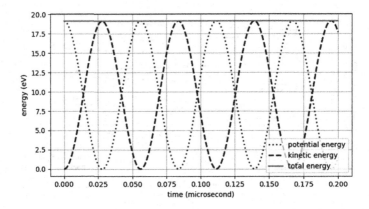

Figure 1.12: Plot of kinetic and potential energy. The total energy remains conserved.

field, integrating particle equations of motion, and gathering field data onto a particle position. We demonstrated these techniques by developing code simulating an electron oscillating in a potential well and verified that energy is indeed conserved.

EXERCISES

1.1 *Falling ball.* Imagine that a 50 cm diameter solid aliminum (density = 2700 kg/m^3) ball is dropped from rest at 10,000 m. Consider the force of gravity ($\vec{F} = -mg\hat{j}$) and numerically integrate particle position and velocity versus time. For simplicity, assume constant gravity ($a = 9.81$ m/s^2) Terminate the simulation when the ball hits the ground ($y = 0$). Plot position and velocity versus time with both gravity and drag included. Derive an analytical expression for the velocity as a function of height and compare to your numerical predictions.

1.2 *Drag.* Next, add the action of aerodynamic drag, $F_d = (1/2)\rho_g C_d A v_{rel}^2$. Use $C_D = 0.47$ and assume constant air density, $\rho_g = 1.225$ kg/m^3. Plot the height and velocity versus time on the same graph as above. You should observe that the ball reaches a certain *terminal velocity*. Compare the simulation value to theory.

1.3 *Surface interaction.* Add surface rebound. When the ball hits the ground, $y \leq 0$, compute the distance below the surface, $\Delta y = 0 - y$. Set the new height to this offset, and flip the velocity direction. To improve visualization, load the particle with a finite x velocity. Plot the particle trace, $x(t)$ versus $y(t)$. You should see the particle completing a series of consecutively shorter parabolas until it eventually comes to a rest. The decrease in the parabola height is due to the drag term. Verify this by running the simulation with only gravity included.

1.4 *Neumann boundary.* Modify the potential solver to solve a system with a zero Neumann boundary condition on the left edge, $\partial\phi/\partial x|_{x=0} = 0$. To do this, solve the expression used for computing the one-sided electric field for the phi[0] value. Compare the solution for the potential and the electric field to the analytical solution.

CHAPTER 2

Grounded Box

2.1 INTRODUCTION

W E ARE now ready to develop our first plasma simulation code. We start with a simple concept: plasma confined within a grounded box with reflective walls. The simulation utilizes the electrostatic particle in cell (ES-PIC) method and is *fully-kinetic*. This implies that both ions and electrons are treated as particles. This simple setup may not sound too exciting, but it actually leads to an interesting oscillation. It also allows us to develop the basic framework that is utilized in all subsequent codes. In the following chapters, we only make small modifications to simulate more complex phenomena. Along the way, we also cover some important C++ practices. We learn about class declaration, dynamic memory allocation, operator overloading, and template arguments.

2.2 SIMULATION SETUP

The simulation domain is shown in Figure 2.1. It is a box with 0.2 m sides spanning from $(-0.1, -0.1, 0)$ to $(0.1, 0.1, 0.2)$. The domain is discretized into a $21 \times 21 \times 21$ uniform Cartesian mesh. The simulation domain is initialized with uniform ion density $n_i = 10^{11}$ m^{-3}. The electron density is identical to the ions, but, and this is the important part, electrons occupy only the $[-0.1, 0) \times [-0.1, 0) \times [0, 0.1)$ octant. In other words,

$$n_e(\vec{x}) = \begin{cases} n_i & ; \ \vec{x} \in [\vec{x}_0, \vec{x}_c) \\ 0 & ; \ \text{otherwise} \end{cases} \tag{2.1}$$

This region is shown in the figure by the gray box. The setup is clearly unstable. The remaining seven octants are filled with a net positive charge. Electrons can be expected to move into these regions in an attempt to reduce the local charge separation. In doing so, they overshoot the ions. Given an infinitely large domain, the electrons eventually return as they are trapped in a potential well, similar to what was studied in Chapter 1. We approximate this behavior by making the walls *reflective*. Any particle leaving the simulation is

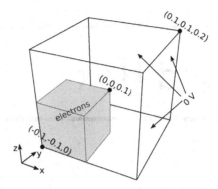

Figure 2.1: Computational domain for the plasma in a box example. Electrons initially occupy only the shaded region.

elastically reflected back. The 0 V Dirichlet boundary condition is set on all boundaries.

2.2.1 Pseudocode

The code is developed in pieces. I like to start with a skeletal application containing just the hooks for functions to be implemented later. These are then "fleshed-out" one by one. The starting code may look similar to the following:

```
int main(int argc, char *args[]) {
    World world(/*...*/);          // initialize computational domain

    Species ions(/*...*/);         // initialize particle species
    Species electrons(/*...*/);

    SolvePotential();              // get initial field
    ComputeElectricField();        // differentiate potential

    GenerateParticles();           // introduce particles

    // main loop
    for (ts=0; ts<num_ts; ts++) {
        ComputeChargeDensity();    // scatter particle positions

        SolvePotential();          // solve Poisson's equation
        ComputeElectricField();    // differentiate potential

        IntegrateVelocity();       // advance particle velocity
        IntegratePosition();       // advance particle position

        RunTimeDiagnostics();      // placeholder for diagnostics
    }

    OutputResults();               // save results to the disc
    return 0;                      // normal exit
```

}

We start by defining a variable of type `World` that encapsulates information about the computational domain. We then define objects to hold ion and electron species. This object of type `Species` stores information common to all particles, such as their mass or charge. It also stores positions and velocities of the individual particles. Next we compute the initial electric field. This field is needed to rewind particle velocities for the Leapfrog method. We then load particles. The code then enters the *main loop*. Generally, we run the simulation for some prescribed number of time steps, `num_ts`. Each step consists of first updating particle velocities from

$$\frac{d\vec{v}}{dt} = \frac{q}{m}\left(\vec{E} + \vec{v} \times \vec{B}\right) \tag{2.2}$$

which in the case of no magnetic field, reduces to

$$\frac{d\vec{v}}{dt} = \frac{q}{m}\vec{E} \tag{2.3}$$

Particle positions are integrated according to

$$\frac{d\vec{x}}{dt} = \vec{v} \tag{2.4}$$

The new positions are then used to compute charge density ρ by interpolating their positions to the grid. Poisson's equation is then solved,

$$\nabla^2 \phi = -\frac{\rho}{\epsilon_0} \tag{2.5}$$

and electric field is obtained by differentiating potential ϕ,

$$\vec{E} = -\nabla\phi \tag{2.6}$$

Finally, we include some screen and file diagnostics to provide an insight into how the simulation is progressing. And that's all! The loop iterates until the desired number of time steps is reached.

2.3 WORLD OBJECT

The first step is to construct containers for storing the mesh geometry and node-based values. The Cartesian mesh used in this example is shown in Figure 2.2. Due to the regular structure, the mesh can be fully described using just nine quantities: the three floating point values for the origin x_0, y_0, z_0; three values for cell spacing in the x, y, and z directions: Δx, Δy, and Δz; and the node counts, ni, nj, and nk. In vector form these are $\vec{x}_0$, $\vec{\Delta h}$, and $\vec{nn}$. This information is stored in an object of type `World`. A simple version is

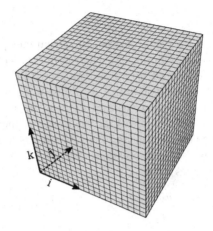

Figure 2.2: Cartesian mesh used in this example.

```
class World {
public:
    double x0[3];      // mesh origin
    double dh[3];      // cell spacing in x,y,z
    int nn[3];         // number of nodes in x,y,z
};
```

This data structure requires the external code to manipulate the arrays directly. For instance,

```
int main() {
    World world;
    world.nn[0] = 21;       // set node counts
    world.nn[1] = 21;
    world.nn[2] = 21;

    world.x0[0] = -0.1;   // set origin
    world.x0[1] = -0.1;
    world.x0[2] = 0;

    world.dh[0] = 0.01;    // set mesh spacing
    world.dh[1] = 0.01;
    world.dh[2] = 0.01;
}
```

2.3.1 Constructor

This "C"-style approach is not ideal as it does not offer any safety checks. There is nothing preventing the code in **main** from changing the variable holding the number of mesh nodes once the associated arrays have been initialized. This could lead to memory corruption. C++ offers two tools for controlling data access. First, we can define class members as **private** or **protected**, which prevents any code outside the class from accessing them. Second, we can define a special function called the *constructor* that is called automatically whenever the object is initialized. It is defined as a function sharing the class name and having no return type. A constructor allows us to initialize *constant* fields. These members can be read from outside the class, but their value cannot change once set. The second version looks like this:

```
// World.h
class World {
public:
  World(int ni, int nj, int nk);   // constructor

  // sets the mesh span, also recomputes cell spacing
  void setExtents(double x1, double y1, double z1, double x2,
    double y2, double z2);

  const int nn[3];      // number of nodes
  const int ni, nj, nk;  // number of nodes in individual variables

protected:
  double x0[3];     // mesh origin
  double dh[3];     // cell spacing
  double xm[3];     // mesh max bound
  double xc[3];     // domain centroid
};
```

Besides the constructor, we also add few additional variables. These include the "max bound". This is the point diagonally opposed from the origin, $\vec{x}_m = \vec{x}_0 + \Delta \vec{h} \cdot (\vec{nn} - 1)$. We also add the mesh centroid $\vec{x}_c = (\vec{x}_0 + \vec{x}_m)/2$. The non-constant data is moved to a **protected** block to prevent external access. The member function **setExtents** sets the mesh bounding box and also computes cell sizes. The call from **main** now looks as follows

```
int main() {
  World world(21,21,21);   // calls World constructor
  world.setExtents(-0.1,-0.1,-0.1,0.1,0.1,0.2);
}
```

2.3.2   Declaration and Implementation

So far we have only *declared* the class functions, but have not yet *implemented* them. This can be done during the declaration by providing the function body, and this is indeed the right choice for small functions that should be *inlined*. Inlining eliminates the overhead associated with function calls. But there are downsides to including the code directly in the class definition. The first, and perhaps a minor disadvantage is that it makes the code more cluttered. The second reason has to do with the build process. C++ generates the application in three steps. First, a *preprocessor* checks for special macros starting with the # sign. One of these, #**include**, allows us to import contents of another file. Next, a *compiler* generates the actual machine code. In order to call functions or to instantiate classes, the compiler needs to known the object *prototypes*. A prototype tells the compiler what kind of arguments the function expects and what is the return type. Similar information is provided by a class declaration such as the one on Page 49. It allows the compiler to construct and use an object of type World without actually knowing what the member functions do. In the final step, a *linker* searches for the function bodies and assures they are available as needed.

2.3.3 Multiple Files

Instead of storing the entire code in a single source file as was done in Chapter 1, we can break up the program into multiple files, such as Main.cpp, Output.cpp, Solver.cpp, Species.cpp, and World.cpp. These files can then be included in the build process through

```
$ g++ -O2 Main.cpp Output.cpp Solver.cpp Species.cpp
  World.cpp -o ch2
```

The optional -O2 flag instructs gcc to apply optimizations to make the code run faster. If these are the only .cpp files in the current directory, we can write

```
$ g++ *.cpp -o ch2
```

This command compiles and links the five files into an executable ch2. Alternatively, we can compile the files without linking them by including the -c flag,

```
$ g++ -c Main.cpp Output.cpp Solver.cpp Species.cpp World.cpp
```

or alternatively

```
$ g++ -c *.cpp
```

The output of this call will be five *object files*, Main.o, Output.o, Solver.o, Species.o, and World.o. We then link them as

```
$ g++ Main.o Output.o Solver.o Species.o World.o -o ch2
```

This is identical to the initial compile and link call, except that now the input files are the .o object files.

Let's say that we only make changes to the main loop, defined in Main.cpp. There is no need to recompile the other four files. We can rebuild the application using

```
$ g++ -c Main.cpp
$ g++ *.o -o ch2
```

While this may be a trivial time save for the small programs developed here, production applications can sometimes take hours to compile from scratch. The make tool is used to automate this process by automatically determining which files need to be rebuilt. By separating the function body from its definition, we reduce the amount of code that needs to be recompiled. We write the definition of World in the World.h *header file* and include the appropriate implementations in the World.cpp *source file*. Other files that need to use our *World* object include the definition using

```
#include "World.h"
```

These functions do not need to be recompiled when the code in World.cpp changes. The function implementation is specified by prepending the function name by the name of the class. For instance,

```
void World::setExtents(double x1, double y1, double z1, double x2,
    double y2, double z2) {
  /* ... */
}
```

There is another reason why we may need to separate function definitions from their implementations. Assume we have two classes A and B that each need to call a function from the other class,

```
class B;         // B is a class to be defined later

class A {
public:
  void run(B &b) {b.doSomething();}
  void doSomething() {/* ... */}
};

class B {
public:
  void run(A &a) {a.doSomething();}
  void doSomething() {/* ... */}
};

int main() {
  A a;
  B b;
  a.run(b);
  b.run(a);

  return 0;
}
```

This code will not build because when the compiler gets to run(B &b) function in class A, it does not yet know that B has a member method called doSomething(). This is the case despite instructing the compiler that B is a class using the forward declaration class B on line 1. Switching the order of classes does not help, since both classes depend on each other. But by separating the implementation from the definition, we can get the code to compile. We don't even need to use multiple files,

```
class B;         // B is a class to be defined later

class A {
```

```
public:
  void run(B &b);
  void doSomething() {/* ... */}
};

class B {
public:
  void run(A &a);
  void doSomething() {/* ... */}
};

// member function definitions
void A::run(B &b) {b.doSomething();}
void B::run(A &a) {a.doSomething();}

int main() {
  A a;
  B b;
  a.run(b);
  b.run(a);
}
```

While this example may seem rather contrived, it is in fact encountered in practice. In our example, a `World` reference is needed to instantiate a `Species` object. However, we also use `Species` in a `World` function to compute charge density. The body of this function had to be moved out of the header to avoid this circular reference.

2.3.4 Header Guards

Finally, since header files can include other headers, it is possible for the same file to be inadvertently included multiple times. This can lead to compilation errors. To prevent this, we wrap the contents within a *header guard*,

```
#ifndef _WORLD_H
#define _WORLD_H

class World {
  /* ... */
};

#endif
```

This construct uses the preprocessor to check if a macro `_WORLD_H` is already defined within the current compile unit. If not, we define it, and then include the file contents. Otherwise, nothing is included. This construct is used on all headers in the examples even if not shown in the printed snippets.

2.3.5 Implementation

With this in mind, let's continue implementing the World class. We add the following code to `World.cpp`

```
#include "World.h"

// constructor
World::World (int ni, int nj, int nk) : ni{ni}, nj{nj}, nk{nk},
  nn{ni,nj,nk} {}

// sets mesh extents and computes cell spacing
void World::setExtents(double x1, double y1, double z1,
  double x2, double y2, double z2) {
  //set origin (xmin)
  x0[0] = x1; x0[1] = y1; x0[2] = z1;

  // diagonally-opposite corner (xmax)
  xm[0] = x2; xm[1] = y2; xm[2] = z2;

  // compute spacing by dividing length by the number of cells
  for (int i=0; i<3; i++)
    dh[i] = (xm[i]-x0[i])/(nn[i]-1);

  // compute centroid
  for (int i=0; i<3; i++)
    xc[i] = 0.5*(x0[i]+xm[i]);
}
```

The constructor uses an *initializer list* to set the class constant members. The variable before the curly braces is the class member, while the pieces inside are the values assigned to it. These are often constructor arguments. It is acceptable to use the same name for both the class member and the argument. However, all classes contain a special pointer called **this** that can be used to explicitly refer to a class member. Using **this->ni=ni;** is an alternate way to set a class member to the value held by the local variable (or function argument) if both variables share the same name. Since the constructor does not do anything besides initializing data, we leave the body empty.

The **setExtents** function first sets values for the $\vec{x}_0$ and $\vec{x}_m$ vectors. This initialization could also be part of the constructor, however, it would lead to way too many function arguments. Separating the call makes the code a bit easier to read, at least in my opinion. Next, we compute mesh spacing from

$$\Delta x = \frac{x_m - x_0}{n_i - 1} \tag{2.7}$$

with similar expressions used for Δy and Δz. Here n_i is the number of nodes in the x direction and $n_i - 1$ is the corresponding cell count. We also set the mesh centroid, $\vec{x}_c = (\vec{x}_m + \vec{x}_0)/2$. The centroid is used solely for the purpose of loading the electron population. Finally, we call a function to compute node volumes. This function is yet to be implemented.

2.3.6 Mesh Resolution

Before continuing, we should note that cell spacing cannot be set completely arbitrarily. Plasma is non-neutral only on length scales smaller than the Debye length,

$$\lambda_D = \sqrt{\frac{\epsilon_0 k_B T_e}{n_e q_e^2}} \tag{2.8}$$

where k_B is the Boltzmann constant. This equation neglects the ion contribution, but since generally $T_i \ll T_e$, it is customary to ignore it. Since we are interested in simulating the mixing of electrons and ions, we need to resolve the local charge separation. We thus require that the cell volume is smaller than volume of the Debye sphere,

$$(\Delta x \Delta y \Delta z) < \frac{4}{3}\pi\lambda_D^3 \tag{2.9}$$

Alternatively, it is customary to require that

$$\max(\Delta x, \Delta y, \Delta z) < \lambda_D \tag{2.10}$$

In our example, computing the Debye length prior to the simulation is not trivial, since both ions and electrons are loaded cold, $T_e = T_i = 0$. However, as shown in Section 2.12, the total system kinetic energy of the electrons is bounded by $KE < 2 \times 10^{-11}$ J. Given that the system simulates 10^8 electrons, we can compute the per-electron kinetic energy to be 2×10^{-19} J. For the Maxwellian velocity distribution, we also have $KE = (3/2)kT$. Therefore, if we assume that electrons are Maxwellian, $T_e \approx 9,600$ K. It is customary to express temperature in terms of electron volts, with 1 eV $= (q/k_B) \approx$ 11604.5 K. Therefore, $T_e = 0.83$ eV. Using this value along with $n_e = 10^{11}$ m^{-3}, we obtain $\lambda_D = 0.0214$ m. The $20 \times 20 \times 20$ mesh resolution yields $\Delta x = (x_m - x_0)/n_i = (0.2\text{m})/20 = 0.01$ m. Our mesh is sufficiently fine to resolve the Debye length. If in doubt, perform mesh convergence studies by running the simulation at different mesh resolutions and comparing the results. Furthermore, in my experience, the Gauss-Seidel Poisson solver does not converge unless the mesh resolution requirement is satisfied. Divergence of the field solver is a good indicator that a finer mesh is required.

2.4 FIELD OBJECT

We have now fully specified the mesh geometry, but the mesh does not contain any data. In Chapter 1, we used a `std::vector` to store a one-dimensional array of ni entries. Unfortunately, there is no 3D counterpart to the C++ standard library `vector` and we need to create our own data container. We call this object `Field`.

The rest of this section discusses the implementation details. Some of what follows may be a bit dry so feel free to skip this section. If you do so, just know that `Field` stores 3D double precision data. We also implement a variant called `FieldI` for storing integers, and `Field3` for three-component vectors. Along the way, we also define a container for three-component floating point and integer arrays called `double3` and `int3`. These objects use operator overloading to enable support for common math operations. For instance, instead of saying

```
double a[3], b[3], c[3];
for (int i=0;i<3;i++)
  c[i] = a[i] + 5*b[i];      // compute each dimension in a loop
```

we can simply write

```
double3 a, b;
double3 c = a + 5*b;         // use overloaded * and + operators
```

Similarly, we can perform operations on entire 3D fields, such as

```
Field a, b;
Field c = a + 5*b; // every c[i][j][k] = a[i][j][k] + 5*b[i][j][k]
```

2.4.1 Memory Allocation

The primary purpose of the `Field` container is to store mesh node or cell-centered data. We want to be able to access $\phi_{i,j,k}$ with

```
Field phi(ni,nj,nk);    // initialize memory for ni*nj*nk values
phi[i][j][k] = some_value;
```

C++, unlike other languages such as Java, does not natively support allocation of multi-dimensional arrays. Such arrays are approximated by creating *arrays of pointers*. The code

```
double **v = new double*[5];
```

allocates an array of five *pointers to double precision data*. Each entry, such as `v[3]`, is a pointer that can be directed at some arbitrary memory location. We can let

```
v[3] = new double[10];
```

Now `v[3]` points to an array of 10 doubles. We access these values with `v[3][0]` through `v[3][9]`. Allocating three-dimensional data is similar. We start with an array of pointers-to-pointers. Then each entry is made to point to an array of pointers-to-doubles, each of which is assigned to point to an array of doubles. It looks like this:

```
// Field.h
class Field {
public:
  // constructor
  Field (int ni, int nj, int nk) : ni{ni}, nj{nj}, nk{nk} {
    data = new double**[ni];           // ni pointers-to-pointers
    for (int i=0;i<ni;i++) {
      data[i] = new double*[nj];       // nj pointers to doubles
      for (int j=0;j<nj;j++)
        data[i][j] = new double[nk];   // nk double
    }
  }
```

```
// destructor , frees memory in reverse order
~Field () {
  if (data==nullptr) return;   // return if unallocated
  for (int i=0; i<ni; i++) {   // release memory in reverse order
    for (int j=0; j<nj; j++)
      delete data[i][j];
    delete data[i];
  }

  delete[] data;
  data = nullptr;              // mark as free
}

const int ni,nj,nk;                    // number of nodes

protected:
  double ***data;
};
```

The *destructor* is another special function that is called automatically when the object goes out of scope. We use it to free the memory. We now write

```
int main () {
  Field phi(21,21,21);                 // calls Field constructor
  /* ... */
  return 0;
}   // Field destructor called here automatically
```

The resulting memory arrangement is shown in Figure 2.3. The highlighted block is referenced via [0][1][2] indexes. As can be seen, the allocated data does not reside in consecutive memory. Only the final nk entries are stored consecutively. This kind of memory fragmentation can lead to sub-optimal performance due to *cache misses*. High performance libraries may prefer to use an allocation strategy in which a single one dimensional $ni \times nj \times nk$ memory block is allocated first. The data is then accessed via clever pointer math. That approach is beyond the scope of this book, especially since care must be taken to keep the code platform independent.

2.4.2 Operator Overloading

As of now, there is no way to actually access the data stored in the Field object. One option is to move the internal **double ***data** class member to the **public** block. We would then access the data as **phi.data[i][j][k]**. This can be rather cumbersome. Luckily, C++ allows us define custom **operators** through *operator overloading*. Almost all operators can be overloaded, including the array access [] brackets,

```
class Field {
  // overload the array access operator []
  double** operator[] (int i) {return data[i];}
```

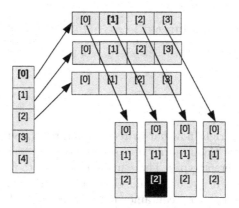

Figure 2.3: Memory allocation for a 3D data array.

```
protected:
  double ***data;
}
```

We can also overload the assignment operator to allow setting all field entries to a constant value. This is shown in the following snippet:

```
Field& operator= (double s) {          // assignment operator
  for (int i=0;i<ni;i++)
    for (int j=0;j<nj;j++)
      for (int k=0;k<nk;k++)
        data[i][j][k] = s;
  return *this;                        // return reference to self
}
```

This function simply iterates through all data entries and sets them to the provided scalar. For generality, we return a reference to the current **Field** instance by *dereferencing* the **this** pointer. Multiple operations can then be chained together. We can now write

```
Field phi(21,21,21);
phi = 0;                // initialize all values to zero
phi[3][4][5] = 1.0;     // use the overloaded operator to set data
```

The initial clearing of data is important, since C++ does not initialize variables when they are created. Since this step is needed for all dynamically allocated data, we add a call to this function to the constructor,

```
// constructor
Field (int ni, int nj, int nk) : ni{ni}, nj{nj}, nk{nk} {
    /* memory allocation code from above */
    (*this) = 0;      // explicitly use the asssignment operator
  }
```

Alternatively, we could call the operator directly. Operators are simply member functions with unusual names:

```
// constructor
Field (int ni, int nj, int nk) : ni{ni}, nj{nj}, nk{nk} {
    /* memory allocation code */
    operator=(0);    // call the overloaded operator= function
}
```

2.4.3 Templates

The `Field` object is right now hardwired to operate on double precision data. While this is true for the majority of cases, there are occasions where we need to store data of different types. These include integers to flag node types, and and (x, y, z) vectors for velocity and electromagnetic fields. C++ allows us to define `Field` as a generic container using *templates*. This generic version, renamed to `Field_`, is given below.

```
template <typename T>
class Field_ {
public:
    // constructor
    Field_ (int ni, int nj, int nk) : ni{ni}, nj{nj}, nk{nk} {
        data = new T**[ni];              // ni pointers-to-pointers of
            type T
        for (int i=0;i<ni;i++) {
            data[i] = new T*[nj];        // allocate nj pointers to T
            for (int j=0;j<nj;j++)
                data[i][j] = new T[nk];  // allocate nk objects of type T
        }
        (*this)=0;                        // clear data
    }

    // destructor, frees memory in reverse order
    ~Field_() {
        /* same code as before */
    }

    // data access operator
    T** operator[] (int i) { return data[i]; }

    // overload the assignment operator
    Field_<T>& operator= (T s) {
        for (int i=0;i<ni;i++)
            for (int j=0;j<nj;j++)
                for (int k=0;k<nk;k++)
                    data[i][j][k] = s;
        return *this;                     // return reference to self
    }

    const int ni,nj,nk;                   // number of nodes

protected:
    T ***data;                            // pointer of type T
};
```

The class definition is prepended with `template <typename T>` parameter.

This instructs the compiler to treat T as a generic type to be specified by the user whenever the object is being instantiated. All instances where a `double` type was hardcoded were replaced with the T type. We can now instantiate a 3D array of doubles as `Field_<double>`. An integer array can be similarly instantiated using `Field_<int>`. Constantly writing the template argument can become tiring. We get around this by defining "nicknames" for these types with the help of the `using` keyword,

```
using Field  = Field_<double>;  // field of doubles
using FieldI = Field_<int>;     // field of integers
```

We can now write `Field` and `FieldI` to refer to these two types. The template substitutions happen at compile time. Therefore, the compiler needs access to the entire code that depends on template arguments. In practical terms, this means that templated functions need to be fully implemented in the header files.

2.4.4 Move and Copy Constructors

For completeness, we should also define two special types of constructors that allow `Field` objects to be copied and moved. The first kind, the *copy constructor*, generates a deep copy by copying another Field element by element. A *move constructor* "steals" data from another object. It is used to return temporary objects from functions that otherwise would need to be copied. Code for these two constructors is listed below. The copy constructor allocates memory by calling the standard constructor via the initializer list. It then sets the items one by one. This step utilizes an overloaded () operator to read the other field's data. The [] operator supports read/write access to the data as it returns the *reference* to the object. It is not compatible with constant members, as the value could be changed via the reference. The () operator provides a read-only access to the data by returning the stored value. The move constructor on the other hand only sets the node counts, and "steals" the data by making our `data` pointer point to the data held by the other field. The `data` pointer in the other field is invalidated by setting it to a `nullptr`. This prevents the destructor from attempting to free the memory. The assignment operator is also overloaded to support moves. The && token indicates the `other` is a reference to a temporary object. These objects, also known as *r-values*, are encountered when returning local variables from a function.

```
class Field {
  // copy constructor
  Field_(const Field_ &other) :
  Field_{other.ni,other.nj,other.nk} {
    for (int i=0;i<ni;i++)
      for (int j=0;j<nj;j++)
        for (int k=0;k<nk;k++)
          data[i][j][k] = other(i,j,k);
  }
```

```
// move constructor
Field_(Field_ &&other):
  ni{other.ni},nj{other.nj},nk{other.nk} {
    if (data) ~Field_();      // deallocate own data
    data = other.data;        // steal the data
    other.data = nullptr;     // invalidate

// move assignment operator
Field_& operator = (Field_ &&f) {
  if (data) ~Field_();        // deallocate own data
  data=f.data;f.data=nullptr; return *this;}

// read-only access to data[i][j][k]
operator() (int i, int j, int k) const {return data[i][j][k];}
};
```

2.4.5 Additional Operators

We now define few additional functions that come in handy later on. We start with an element-wise division operator,

```
void operator /= (const Field_ &other) {
  for (int i=0;i<ni;i++)
    for (int j=0;j<nj;j++)
      for (int k=0;k<nk;k++) {
        if (other.data[i][j][k]!=0)
          data[i][j][k] /= other[i][j][k];
        else
          data[i][j][k] = 0;
      }
}
```

This operator allows us to write

```
Field a;
Field b/=a;       // sets b[i][j][k] = b[i][j][k]/a[i][j][k]
```

We also define a similar operator for the compound addition. The loop over $i - j - k$ is identical to the code above and is not repeated for brevity.

```
Field_& operator += (const Field_ &other) {
  /* loop over i-j-k from above */
      data[i][j][k] += other(i,j,k);
  return (*this);
}
```

Next, we write two operators for scaling a field by a scalar,

```
// compound multiplication
Field_& operator *= (double s) {
  /* loop over i-j-k from above */
      data[i][j][k] *= s;
  return (*this);
}
```

```
// multiplication operator, returns new Field set to f*s
```

```
friend Field_<T> operator*(double s, const Field_<T>&f) {
  Field_<T> r(f);
  return std::move(r*=s);    // force move
}
```

There are two versions defined. The first is the compound operator that multiplies values in place,

```
Field a;
a*=10.0;         // set a[i][j][k] = 10*a[i][j][k]
```

The second version is a defined as *friend function*. This separates the function from a particular instance of the `Field_` object. The operator creates a deep copy of the other object and uses the `*=` operator to scale values. It should be returned with the help of the move constructor, but just to help the compiler realize that `r` is a temporary data object, we wrap the return statement in `std::move`. We can now write

```
Field b = 10*a;
```

2.4.6 Output

In C++, data output is handled through an overloaded `<<` operator. We can also overload this operator so that we can write

```
Field phi(ni,nj,nk);
ofstream out("file.txt");      // open output file
out<<phi;                      // write the entire field to the file
```

to save the entire field to a file. This is done with the following function

```
// writes data to a file stream
template<typename T>
std::ostream& operator<<(std::ostream &out, Field_<T> &f) {
  for (int k=0;k<f.nk;k++,out<<"\n")  // new line after each "k"
    for (int j=0;j<f.nj;j++)
      for (int i=0;i<f.ni;i++) out<<f.data[i][j][k]<<" ";
  return out;
}
```

A new line is printed with `"\n"` instead of `endl` to avoid flushing the buffer to the disc prematurely.

2.4.7 Three-Component Vectors

Besides doubles and integers, we would also like to store three-component vectors to capture the electric field and flow velocities. To accomplish this, we define another custom type called `vec3`. This object is defined with the `struct` keyword. Historically in native C, structures could only hold data, while C++ classes contained functions. These days, the only differences between a `struct` and a `class` is that the members of the former are public by default. I prefer to use structs for small data objects that contain only a minimal code base. We define three constructors allowing us to initialize the vector either with three arguments or with a single array. Alternatively, an empty constructor creates a zero vector. We also override the array access operator []. Similarly we override the () operator to support read-only access. We use the **using** keyword to let **double3** refer to a **vec3<double>** and **int3** to **vec3<int>**.

```
template <typename T>
struct vec3 {
  vec3 (const T u, const T v, const T w) : d{u,v,w} {}
  vec3 (const T a[3]) : d{a[0],a[1],a[2]} {}
  vec3 (): d{0,0,0} {}
  T& operator[](int i) {return d[i];}
  T operator()(int i) const {return d[i];}
  vec3<T>& operator=(double s) {d[0]=s;d[1]=s;d[2]=s;return
    (*this);}
  vec3<T>& operator+=(vec3<T> o)
    {d[0]+=o[0];d[1]+=o[1];d[2]+=o[2];return(*this);}
  vec3<T>& operator-=(vec3<T> o)
    {d[0]-=o[0];d[1]-=o[1];d[2]-=o[2];return(*this);}

protected:
  T d[3];
};
using double3 = vec3<double>;  // assign new names
using int3 = vec3<int>;
```

We also need to define operators for some mathematical operations and data output. Some of these are called by the corresponding overloaded operators in **Field**. For instance, the **data[i][j][k]+=other(i,j,k)** call in **Field_<T>::operator+=** requires that **operator+=** is user defined if **T** is a custom type. We also add operators for supporting basic mathematical operations such as addition and multiplication.

```
// vec3-vec3 operations
template<typename T>   // addition of two vec3s
vec3<T> operator+(const vec3<T>& a, const vec3<T>& b) {
  return vec3<T> (a(0)+b(0),a(1)+b(1),a(2)+b(2));   }
template<typename T>   // subtraction of two vec3s
vec3<T> operator-(const vec3<T>& a, const vec3<T>& b) {
  return vec3<T> (a(0)-b(0),a(1)-b(1),a(2)-b(2));   }
template<typename T>   // element-wise multiplication of two vec3s
vec3<T> operator*(const vec3<T>& a, const vec3<T>& b) {
  return vec3<T> (a(0)*b(0),a(1)*b(1),a(2)*b(2));   }
template<typename T>   // element wise division of two vec3s
vec3<T> operator/(const vec3<T>& a, const vec3<T>& b) {
  return vec3<T> (a(0)/b(0),a(1)/b(1),a(2)/b(2));   }

// vec3 - scalar operations
template<typename T>      // scalar multiplication
vec3<T> operator*(const vec3<T> &a, T s) {
  return vec3<T>(a(0)*s, a(1)*s, a(2)*s);  }
template<typename T>      // scalar multiplication 2
vec3<T> operator*(T s, const vec3<T> &a) {
  return vec3<T>(a(0)*s, a(1)*s, a(2)*s);  }

// output
template<typename T>   // ostream output
```

```
std::ostream& operator<<(std::ostream &out, vec3<T>& v) {
  out<<v[0]<<" "<<v[1]<<" "<<v[2];
  return out;
}
```

The `double3` object allows us to perform some additional housekeeping. `World::setExtents (double x1, double y1, ...)` is one candidate. With the current implementation, it is hard to remember the order of arguments. Is the second value y_1 or x_2? Instead, by using `double3` we group the two points together inside of curly braces,

```
world.setExtents({-0.1,-0.1,0.0},{0.1,0.1,0.2});
```

The new function definition is

```
void World::setExtents(const double3 &_x0, const double3 &_xm) {
  x0 = _x0;        // set our copy of the origin
  xm = _xm;        // do the same for xmax
  /* ... */
}
```

The `const T&` syntax avoids the overhead of object duplication, while also allowing passing of temporary objects as is the case here.

2.4.8 Adding Fields to World

Now that we have a `Field` data type, we use it to add storage for several properties within the `World` object.

```
class World {
public:
  // constructor
  World (int ni, int nj, int nk);

  const int3 nn;           // number of nodes
  const int ni,nj,nk;      // number of nodes in individual variables

  Field phi;               // potential
  Field rho;               // charge density
  Field3 ef;               // electric field components
}
```

The objects are initialized using the World constructor initializer list:

```
World::World(int ni, int nj, int nk):
  ni{ni}, nj{nj}, nk{nk},  nn{ni,nj,nk},
  phi(ni,nj,nk),rho(ni,nj,nk),node_vol(ni,nj,nk),
  ef(ni,nj,nk)  { }
```

2.4.9 Initial Output

At this point, we would like to visualize the constructed mesh. We use Kitware's Paraview for this purpose. Paraview is a front end for Kitware's Visualization Toolkit (VTK), a powerful library for data processing and rendering.

VTK supports multiple types of input files, including "image data" for storing structured Cartesian meshes. The mesh geometry is described by specifying the origin, mesh spacing, and node counts. The file contents are stored using an .xml syntax. The function for outputting the mesh and the associated fields is given below. We define it as a member of *namespace* Output. This construct is analogous to using *static* classes you may be familiar with from Java.

```cpp
namespace Output { void fields(World &world); }   // in Output.h

// saves output in VTK format
void Output::fields(World &world) {
  stringstream name;        // build file name
  name<<"fields.vti";       // here we just set it to a given string

  // open output file
  ofstream out(name.str());
  if (!out.is_open()) {cerr<<"Could not open "<<name.str()
    <<endl;return;}

  // ImageData is a VTK format for structured Cartesian meshes
  out<<"<VTKFile type=\"ImageData\">\n";
  double3 x0 = world.getX0();
  double3 dh = world.getDh();
  out<<"<ImageData Origin=\""<<x0[0]<<" "<<x0[1]<<" "<<x0[2]<<"\"
    ";
  out<<"Spacing=\""<<dh[0]<<" "<<dh[1]<<" "<<dh[2]<<"\" ";
  out<<"WholeExtent=\"0 "<<world.ni-1<<" 0 "<<world.nj-1<<" 0
    "<<world.nk-1<<"\">\n";

  // output data stored on nodes (point data)
  out<<"<PointData>\n";

  // node volumes, scalar
  out<<"<DataArray Name=\"NodeVol\" NumberOfComponents=\"1\"
    format=\"ascii\" type=\"Float64\">\n";
  out<<world.node_vol;   // use the overloaded << operator
  out<<"</DataArray>\n";

  // potential, scalar
  /* ... */      // output world.phi

  // charge density, scalar
  /* ... */      // output world.rho

  // electric field, 3 component vector
  out<<"<DataArray Name=\"ef\" NumberOfComponents=\"3\"
    format=\"ascii\" type=\"Float64\">\n";
  out<<world.ef;       // uses overloaded << from Field_ and vec3
  out<<"</DataArray>\n";

  // close the tags
  out<<"</PointData>\n";
  out<<"</ImageData>\n";
  out<<"</VTKFile>\n";
}  // file closed here as 'out' goes out of scope
```

We take advantage of the overloaded << operators defined for the Field object to output the field data. We also utilize several *accessors* for World protected data storing mesh mesh origin and cell spacing. This step is needed only if you care about access control. Otherwise, these fields can be left in the public block.

```
class World {
    double3 getX0() const {return double3(x0);}
    double3 getDh() const {return double3(dh);}
};
```

The |const| keyword indicates that this function does not modify class variables. These accessors demonstrate another benefit of defining our own data object to hold the three component floating point values. C++ does not allow a function to return an array. Returning an object is allowed, however. Without using this construct, we would be forced to return data via function arguments,

```
class World {
  void getX0(double res[3]) {
    res[0]=x0[0]; res[1]=x0[1]; res[2]=x0[2]; }
};
```

which would be used as follows

```
double x0[3];
world.getX0(x0);    // store the origin in the x0 array
```

This construct is less clear than

```
double3 x0 = world.getX0();
```

and is also prone to errors if the order of input and output arrays is accidentally reversed in a function with multiple arguments. We add the call to the output function to main,

```
int main() {
  World world(21,21,21);
  world.setExtents({-0.1,-0.1,0.0},{0.1,0.1,0.2});

  Output::fields(world);
  return 0;
}
```

2.4.10 Visualization

The code written so far is located in the ch2/step2 folder. Compile and run it as

```
ch2/step2 $ g++ *.cpp -o step2
ch2/step2 $ ./step2
```

It generates a file called **fields.vti** with contents shown below. We can see that a VTK image file starts with a header specifying the mesh origin,

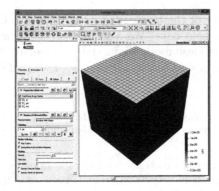

Figure 2.4: Visualization of the initial domain in Paraview.

spacing, and nodal ranges. It then contains several data arrays. These arrays are child elements of a `PointData` or a `CellData` element. We don't have any cell data yet but will add some in the future. Each data array contains a header specifying the field name, the type (integer or a double precision float), and the number of components. Fields with three components are automatically converted into vectors. The data can also be saved in a binary format, but we use the ASCII (text) convention throughout this book. We next open the file in Paraview to confirm that it indeed captures the domain of interest. We also confirm that all the data is present - although it is all zero at this point. This is shown in Figure 2.4.

```
<VTKFile type="ImageData">
<ImageData Origin="−0.1 −0.1 0" Spacing="0.01 0.01 0.01"
   WholeExtent="0 20 0 20 0 20">
<PointData>
<PointData>
<DataArray Name="NodeVol" NumberOfComponents="1" format="ascii"
   type="Float64">
0 0 0 0 0 0 0 0 0 ...
</DataArray>
...
</PointData>
</ImageData>
</VTKFile>
```

2.5 POTENTIAL SOLVER

After this long detour into the world of C++ data management, it is time to return to the Particle in Cell method. We start by writing the potential solver. This function solves the Poisson's equation using the Gauss-Seidel scheme introduced in Chapter 1. We start by discretizing the governing equation,

$$\nabla^2 \phi = -\rho/\epsilon_0 \qquad (2.11)$$

as

$$\frac{\phi_{i-1,j,k} - 2\phi_{i,j,k} + \phi_{i+1,j,k}}{\Delta^2 x} + \frac{\phi_{i,j-1,k} - 2\phi_{i,j,k} + \phi_{i,j+1,k}}{\Delta^2 y} +$$

$$\frac{\phi_{i,j,k-1} - 2\phi_{i,j,k} + \phi_{i,j,k+1}}{\Delta^2 z} = -\rho_i/\epsilon_0 \qquad (2.12)$$

The discretization is based on the second order central difference from Equation 1.36. Here $\Delta^2 x = (\Delta x)^2$. We next isolate all $\phi_{i,j,k}$ terms to the left hand side,

$$\left(\frac{2}{\Delta^2 x} + \frac{2}{\Delta^2 y} + \frac{2}{\Delta^2 z}\right)\phi_{i,j,k} = \rho_i/\epsilon_0 + \frac{\phi_{i-1,j,k} + \phi_{i+1,j,k}}{\Delta^2 x} +$$

$$\frac{\phi_{i,j-1,k} + \phi_{i,j+1,k}}{\Delta^2 y} + \frac{\phi_{i,j,k-1} + \phi_{i,j,k+1}}{\Delta^2 z} \qquad (2.13)$$

Finally, solving for $\phi_{i,j,k}$, we obtain an equation for the new estimate for potential at node i, j, k. Gauss-Seidel is commonly coupled with Successive Over Relaxation (SOR) to speed up convergence. The final set of equations becomes

$$\phi^*_{i,j,k} = \left(\rho_i/\epsilon_0 + \frac{\phi_{i-1,j,k} + \phi_{i+1,j,k}}{\Delta^2 x} + \frac{\phi_{i,j-1,k} + \phi_{i,j+1,k}}{\Delta^2 y} +\right.$$

$$\left.\frac{\phi_{i,j,k-1} + \phi_{i,j,k+1}}{\Delta^2 z}\right) / \left(\frac{2}{\Delta^2 x} + \frac{2}{\Delta^2 y} + \frac{2}{\Delta^2 z}\right) \qquad (2.14)$$

and

$$\phi_{i,j,k} \leftarrow \phi_{i,j,k} + w(\phi^*_{i,j,k} - \phi_{i,j,k}) \qquad (2.15)$$

Here the arrow indicates that we are overwriting the value of $\phi_{i,j,k}$ with the value on the right hand side. The above algorithm is valid only on the internal nodes on which the central difference can be evaluated. In a general case, we need additional equations to take control of boundaries. Since here the box walls are all assumed to be Dirichlet, we simply skip over them by limiting the loops to the internal mesh nodes, $i \in [1, ni - 2]$, $j \in [1, nj - 2]$, and $k \in [1, nk - 2]$. The solver is defined as a member function of a `PotentialSolver` object located in `PotentialSolver.h`.

```cpp
class PotentialSolver {
public:
    // constructor, sets members to given inputs
    PotentialSolver(World &world, int max_it, double tol):
        world(world), max_solver_it(max_it), tolerance(tol) { }

    // solves potential using Gauss-Seidel and SOR
    bool solve();

    // computes electric field = -gradient(phi)
    void computeEF();
```

```
protected:
  World &world;
  unsigned max_solver_it;    // maximum number of solver iterations
  double tolerance;          // solver tolerance
};
```

The constructor takes as an argument the reference to the World object, which we save as a class member. This reference is needed to access the phi field and the mesh geometry. We also set some solver parameters: the maximum number of iterations and the tolerance. The actual Gauss-Seidel algorithm is implemented in the solve function which can be found in PotentialSolver.cpp.

```
bool PotentialSolver::solve() {
  Field &phi = world.phi;  // references to avoid writing world.phi
  Field &rho = world.rho;

  // precompute 1/(dx^2)
  double3 dh = world.getDh();
  double idx2 = 1.0/(dh[0]*dh[0]);    // 1/dx^2
  double idy2 = 1.0/(dh[1]*dh[1]);    // 1/dy^2
  double idz2 = 1.0/(dh[2]*dh[2]);    // 1/dz^2

  double L2=0;                        // norm
  bool converged= false;

  // solve potential
  for (unsigned it=0;it<max_solver_it;it++) {
    for (int i=1;i<world.ni-1;i++)
      for (int j=1;j<world.nj-1;j++)
        for (int k=1;k<world.nk-1;k++) {
          // standard internal open node
          double phi_new = (rho[i][j][k]/Const::EPS_0 +
              idx2*(phi[i-1][j][k] + phi[i+1][j][k]) +
              idy2*(phi[i][j-1][k]+phi[i][j+1][k]) +
              idz2*(phi[i][j][k-1]+phi[i][j][k+1])) /
              (2*idx2+2*idy2+2*idz2);

          // SOR
          phi[i][j][k] = phi[i][j][k] + 1.4*(phi_new-phi[i][j][k]);
        }

    // check for convergence
    if (it%25==0) {
      double sum = 0;
      for (int i=1;i<world.ni-1;i++)
        for (int j=1;j<world.nj-1;j++)
          for (int k=1;k<world.nk-1;k++) {
            double R = -phi[i][j][k]*(2*idx2+2*idy2+2*idz2) +
                rho[i][j][k]/Const::EPS_0 +
                idx2*(phi[i-1][j][k] + phi[i+1][j][k]) +
                idy2*(phi[i][j-1][k]+phi[i][j+1][k]) +
                idz2*(phi[i][j][k-1]+phi[i][j][k+1]);

            sum += R*R;
          }
```

```
    L2 = sqrt(sum/(world.ni*world.nj*world.nk));
    if (L2<tolerance) {converged=true;break;}
  }
} // iteration loop

if (!converged) cerr<<"GS failed to converge, L2="<<L2<<endl;
return converged;
}
```

We start by setting local references to the phi and rho fields. This step is here for cosmetic reasons to avoid having to write world.phi every time we need to access potential. We also precompute $1/\Delta^2 x$ and so on. The solver runs for up to max_solver_it iterations. Every iteration consists of looping through the internal mesh nodes, and applying Equations 2.14 and 2.15. Every 25 iterations, we compute the residue vector

$$\vec{R} = \mathbf{A}\vec{x} - \vec{b} \qquad (2.16)$$

and compare the L2 norm

$$\sqrt{\frac{\sum_n R_n^2}{n}} \le \epsilon_{tol} \qquad (2.17)$$

to a tolerance. This convergence check skips over boundaries, since the Dirichlet condition is satisfied there naturally. If the solver fails to reach the desired tolerance, we print an error message and return false to indicate failure. We can then re-run the solver with an increased iteration limit, or as I often do, just ignore the non-convergence message if present only during the first few initial time steps. The value of the physical constant ϵ_0 (among with others) is provided via a namespace Const added to World.h,

```
namespace Const {
  const double EPS_0 = 8.8541878e−12;  // C/V/m, vac. permittivity
  const double QE = 1.602176565e−19;   // C, electron charge
  const double AMU = 1.660538921e−27;  // kg, atomic mass unit
  const double ME = 9.10938215e−31;    // kg, electron mass
  const double K = 1.380648e−23;       // J/K, Boltzmann constant
  const double PI = 3.141592653;       // pi
  const double EvToK = QE/K;           // 1eV in K ~ 11604
}
```

Since ρ is currently zero (we have not yet added any particles), the above code produces $\phi = 0$ everywhere. This is not very interesting. Just to verify the solver is indeed working, we let $\phi_{i=0} = 1$ and $\phi_{k=0} = 2$ by adding the following hack to main

```
int main() {
  // initialize domain
  World world(21,21,21);
  world.setExtents({−0.1,−0.1,0.0},{0.1,0.1,0.2});

  // set phi[i=0] = 1 for testing
  for (int j=0;j<world.nn[1];j++)
```

```
    for (int k=0;k<world.nn[2];k++)
      world.phi[0][j][k] = 1;   // phi[i=0] = 1

// set phi[k=0] = 2
for (int i=0;i<world.nn[0];i++)
  for (int j=0;j<world.nn[1];j++)
    world.phi[i][j][0] = 2;   // phi[k=0] = 2

// initialize and solve potential
PotentialSolver solver(world,5000,0);
solver.solve();

// save results
Output::fields(world);
return 0;
}
```

This resulting potential is shown in Figure 2.5.

2.6 ELECTRIC FIELD

In ES-PIC simulations, plasma potential is only a stepping stone for obtaining the electric field,

$$\vec{E} = -\nabla\phi \equiv -\left(\frac{\partial\phi}{\partial x}\hat{i} + \frac{\partial\phi}{\partial y}\hat{j} + \frac{\partial\phi}{\partial z}\hat{k}\right) \tag{2.18}$$

We again use the finite difference method to rewrite the derivatives. The central method, Equation 1.59, gives us

$$\vec{E} = \left(\frac{\phi_{i-1,j,k} - \phi_{i+1,j,k}}{2\Delta x}\right)\hat{i} + \left(\frac{\phi_{i,j-1,k} - \phi_{i,j+1,k}}{2\Delta y}\right)\hat{j} +$$
$$\left(\frac{\phi_{i,j,k-1} - \phi_{i,j,k+1}}{2\Delta z}\right)\hat{k} \tag{2.19}$$

This differencing is valid only on the internal nodes. On the boundaries, we use the one-sided second order accurate version we studied previously in Chapter 1,

$$(E_x)_{i=0} = \frac{3\phi_{0,j,k} + 4\phi_{1,j,k} - \phi_{2,j,k}}{2\Delta x} \tag{2.20}$$

$$(E_x)_{i=ni-1} = \frac{-\phi_{ni-3,j,k} + 4\phi_{ni-2,j,k} - \phi_{ni-1,j,k}}{2\Delta x} \tag{2.21}$$

with similar equations on the y and z faces. These relationships are implemented in the `computeEF` function,

```
// computes electric field = −gradient(phi)
void PotentialSolver::computeEF() {
  // reference to phi to avoid writing world.phi
  Field &phi = world.phi;
```

```
double3 dh = world.getDh();          //get cell spacing
double dx = dh[0];
double dy = dh[1];
double dz = dh[2];

for (int i=0;i<world.ni;i++)         // loop over nodes
  for (int j=0;j<world.nj;j++)
    for (int k=0;k<world.nk;k++) {
      double3 &ef = world.ef[i][j][k]; // ref to (i,j,k) ef vec3

      // x component, efx
      if (i==0)                      // forward difference
        ef[0] = -(-3*phi[i][j][k] + 4*phi[i+1][j][k] -
          phi[i+2][j][k])/(2*dx);
      else if (i==world.ni-1)  // backward difference
        ef[0] = -(phi[i-2][j][k] - 4*phi[i-1][j][k] +
          3*phi[i][j][k])/(2*dx);
      else                           // central difference
        ef[0] = -(phi[i+1][j][k] - phi[i-1][j][k])/(2*dx);

      // y component, efy
      if (j==0)
        ef[1] = -(-3*phi[i][j][k] +
          4*phi[i][j+1][k]-phi[i][j+2][k])/(2*dy);
      else if (j==world.nj-1)
        ef[1] = -(phi[i][j-2][k] - 4*phi[i][j-1][k] +
          3*phi[i][j][k])/(2*dy);
      else
        ef[1] = -(phi[i][j+1][k] - phi[i][j-1][k])/(2*dy);

      // z component, efz
      if (k==0)
        ef[2] = -(-3*phi[i][j][k] +
          4*phi[i][j][k+1]-phi[i][j][k+2])/(2*dz);
      else if (k==world.nk-1)
        ef[2] = -(phi[i][j][k-2] -
          4*phi[i][j][k-1]+3*phi[i][j][k])/(2*dz);
      else
        ef[2] = -(phi[i][j][k+1] - phi[i][j][k-1])/(2*dz);
    }
}
```

We again loop through all nodes, and on each, use the negative of Equations 2.19, 2.20, and 2.21 to compute the electric field. To simplify the syntax, we let a local variable **ef** be a reference to the $\vec{E}_{i,j,k}$ component of the electric field. This electric field be seen in Figure 2.5. Here we use Paraview's *clip* filter to show potential on the internal nodes, and a *stream tracer* to plot the electric field lines. This result was obtained with the "hacked" non-zero boundaries from the prior section.

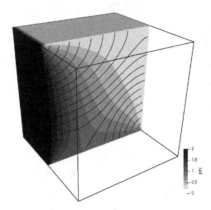

Figure 2.5: Plasma potential and electric field lines for non-zero boundaries.

2.7 PARTICLES

Next we need to write code for loading and moving particles. I like to group particles by gas species. This grouping eliminates the need to store common properties such as charge or mass for each individual particle. It also simplifies the computation of macroscopic flow properties, such as number density and the mean flow velocity. We start by defining a new data container for storing a single particle. This code is placed in `Species.h`.

```cpp
struct Particle {
    double pos[3];      // position
    double vel[3];      // velocity
    double mpw;         // macroparticle weight

    Particle(double x[3], double v[3], double mpw):    // constructor
    pos{x[0],x[1],x[2]}, vel{v[0],v[1],v[2]}, mpw{mpw} { }
};
```

This object stores just the minimum information needed to uniquely define a single particle. Specifically, we store the particle's position and velocity $\vec{x}$ and $\vec{v}$. We also store the macroparticle weight w_{mp}. Quite often, all particles share the same weight, and w_{mp} can be delegated to the species level. We keep it here for generality.

Next we define the `Species` object. The constructor takes as arguments a string name of the species, the mass and charge of each particle, and a reference to the World object. This object is used to initialize a local `Field` for storing the number density. We also declare several functions that will be implemented shortly. The actual particles are stored as a `vector<Particle>` array called `particles`.

```cpp
class Species {
public:
    // constructor
    Species(string name, double mass, double charge, World &world):
```

```
name(name), mass(mass), charge(charge),
  den(world.ni, world.nj, world.nk), world(world) { }

// returns the number of particles
size_t getNp() const {return particles.size();}

// moves all particles using electric field ef[]
void advance();

// compute number density
void computeNumberDensity();

// adds a new particle
void addParticle(double3 pos, double3 vel, double mpwt);

// loads num_mp particles in a x1-x2 box with num_den density
void loadParticlesBox(double3 x1, double3 x2, double num_den,
  int num_mp);

const string name;           // species name
const double mass;           // particle mass in kg
const double charge;         // particle charge in Coulomb

Field den;                   // number density
vector<Particle> particles;  // array for storing particles

protected:
  World &world;              // reference to the World object
};
```

The ion and electrons species are then added to the simulation by instantiating variables of type **Species**. We store these objects within another **vector**. This approach allows us to automate subsequent operations that involve looping over flying materials. The standard library **vector** supports automatic resizing. But resizing involves copying (or moving) the previously stored data, and therefore, if we know ahead of time how many items need to be stored, it is to our benefit to pre-allocate sufficient space using the **reserve** command,

```
#include "Species.h"
int main() {
  /* World initialization ... */

  // set up particle species
  vector<Species> species;
  species.reserve(2);   // pre-allocate space for two species
  species.push_back(Species("O+", 16*AMU, QE, world));
  species.push_back(Species("e-", ME, -1*QE, world));
```

Ions are given mass of 16 atomic mass units, and are singly charged. Similarly, mass m_e and charge $-e$ are assigned to the electrons.

2.7.1 Particle Loading

The populations are then initialized by loading the particles. Returning to our setup in Figure 2.1, we are interested in loading particles such that some specified number density is achieved within a specified bounding box. For ions, this bounding box has the same dimensions as the computational domain, while for the electrons, it constitutes only the one selected octant. Since number density is the number of particles per volume, a box with volume V and number density n contains $N_{real} = nV$ physical ions or electrons. This number of real particles is represented by N_{sim} simulation macroparticles, and hence the weight of each is $w_{mp} = N_{real}/N_{sim}$. Here N_{sim} (or M using our prior syntax) is a user input. This calculation is shown below. Once we have determined the weight, we use a loop to sample N_{sim} particle positions and velocities. Details of this sampling are for now ignored. Prior to adding particles, we allocate enough space for `num_mp` particles using a call to `reserve`. This avoids the need for the previously inserted data to be moved or copied when the `vector` needs to be resized.

```
void loadParticlesBox(double3 x1, double3 x2, double num_den, int
    num_mp) {
    double box_vol = (x2[0]-x1[0])*(x2[1]-x1[1])*(x2[2]-x1[2]);
    double num_real = num_den * box_vol;
    double mpw = num_sim/num_real;  // macroparticle weight

    // preallocate memory for particles
    particles.reserve(num_mp);

    for (int p=0; p<num_sim; p++) {
        /* sample position and velocity */
        /* add particle */
    }
}
```

2.7.2 Random Numbers

One obvious choice for sampling the position is to pick a random point inside the region of interest. We can generate a random number in $[x_1, x_2)$ range using

$$x = x_1 + \mathcal{R}(x_2 - x_1) \qquad \in [0, 1) \qquad (2.22)$$

where $\mathcal{R}$ is a random number. Random numbers are generated using functions called *random number generators* or RNGs. It is important to realize that RNGs do not generally produce true random numbers. Instead, they return consecutive values from a sequence large enough so that to a casual observer, numbers indeed appear to be random. The length of the sequence - the number of unique values that can be sampled before the numbers start repeating - is known as the *period*. It is crucial to use a generator with a large period. To see why, consider a one-dimensional simulation seeded by one million randomly positioned stationary particles. If the period is small, let's say only 10,000, the particles occupy only 10,000 unique positions. Each slot contains 100 particles stacked on top of each other. Although the program contains 100× more particles, the results are just as noisy as if only 10,000 particles have been used. This example may seem hypothetical, but it was actually frequently encountered in practice. The C standard library implemented only a single random generator through the *rand* function. The language standard did not specify the required period, and each compiler vendor was free to implement a different algorithm. Some compilers implemented generators with a period of only 32,768. This resulted in codes that ran fine on one platform producing erroneous (or at least noisy) results when compiled on a different system.

Prior to C++11, the solution was for programmers to implement their own generators. Numerical recipes [48] provided codes for several popular algorithms. Lucky for us, this is no longer needed as the new language extensions include a set of robust RNGs through the `<random>` header. These functions work by combining a generator with a distribution. One popular generator is the Mersenne Twister [44]. It is fast, and has a huge period of $2^{19337} - 1$. We combine it with a *uniform distribution* to sample numbers in the range of $R \in [0, 1)$. Note that 1.0 is not included in the range. This is desired. The values returned by the generator start with some initial *seed*. For a given seed, the generator produces the same sequence. This is useful for debugging but for production runs, we need to randomize the generator. Otherwise, running several independent simulations for ensemble averaging does not yield any benefit. One option is to seed the generator with the current time. C++11 provides an alternate mechanism via a `random_device`. This function attempts to provide a true random number based on various states of the operating system, such as the number of processes running, memory used, and so on. It is not optimized to be fast, and thus we only use it to set the initial seed. The following code illustrates how to implement the generator. We add this code to `World.h`, since this header is included by all other files.

```
// World.h
class Rnd {        // object for sampling random numbers
public:
  // constructor: set initial random seed and distribution limits
  Rnd(): mt_gen{std::random_device()()}, rnd_dist{0,1.0} {}
  double operator() () {return rnd_dist(mt_gen);}

protected:
  std::mt19937 mt_gen;                    // random number generator
  std::uniform_real_distribution<double> rnd_dist; // distribution
};

extern Rnd rnd;    // type Rnd object called rnd defined somewhere
```

The constructor sets the initial seed by creating an instance of the `random_device` object and sampling a value from it. If we wanted to repeat the simulation with the same set of random numbers, we would replace this initialization with `mt_gen{0}` (or some other fixed value). Then in `World.cpp`, we instantiate a `rnd` object of type `Rnd`,

```
Rnd rnd;         // create an instance of a Rnd object
```

A random number can then be sampled using `rnd()`. This call utilizes the overloaded () operator to return a random double precision number from a uniform $[0, 1)$ distribution. These types of custom objects that act as functions are called *functors*. We can now implement the body of the particle loading loop

```
void Species :: loadParticlesBox (double3 x1, double3 x2, double
    num_den , int num_mp) {
  /* same code as above */

  // load particles on an equally spaced grid
  for (int p=0;p<num_mp;p++) {
    // sample random position
    double3 pos;
    pos [0] = x1 [0] + rnd () *(x2 [0] − x1 [0]) ;
    pos [1] = x1 [1] + rnd () *(x2 [1] − x1 [1]) ;
    pos [2] = x1 [2] + rnd () *(x2 [2] − x1 [2]) ;

    // set initial velocity
    double3 vel {0 ,0 ,0};        // stationary particle
    addParticle (pos , vel ,mpw) ;   // add a new particle to the array
  }
}
```

The **addParticle** function simply adds a new entry to the **particle** vector,

```
void Species :: addParticle (double3 pos , double3 vel , double mpw) {
    particles . emplace_back (pos , vel ,mpw) ;
}
```

Finally, we add a call to the particle loader to **main**,

```
int main () {
  /* ... */

  // set up particle species
  vector<Species> species ;
  species . push_back ( Species ("O+", 16*AMU, QE, world )) ;
  species . push_back ( Species ("e−", ME, −1*QE, world )) ;

  int np_ions = 80000;     // number of simulation ions
  int np_eles = 10000;     // number of simulation electrons
  species [0]. loadParticlesBox ( world . getX0 () , world . getXm () ,1e11 ,
      np_ions ); // ions
  species [1]. loadParticlesBox ( world . getX0 () , world . getXc () ,1e11 ,
      np_eles ); // electrons

  // use range−style loop to print particle counts
  for (Species &sp: species )
    cout<<sp.name<<" has "<<sp.getNp ()<<" particle "<<endl ;
}
```

Note that ions are created in $[\vec{x}_0, \vec{x}_m)$, while electrons are created in the smaller $[\vec{x}_0, \vec{x}_c)$ region, where $\vec{x}_c$ is the centroid. We also print the particle counts to the screen. This code is called inside a *range style* **for** loop that is one of the more recent additions to the C++ language. This syntax uses *iterator* objects to advance the current index. It is a more concise way of writing

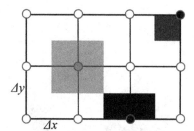

Figure 2.6: Visualization of node volumes on a 2D mesh.

```
for (vector<Species >::iterator iter=species.begin();
    iter!=species.end(); ++iter) {
    Species &sp = *iter;   // element at current iterator position
    /* use sp here */
}
```

This version of the code is found in **step3**. Running the code produces

```
O+ has 80000 particle
e- has 10000 particle
```

2.7.3 Node Volumes

At this point, there is no way to check whether the loading worked as expected. While we could add an output function to generate a scatter plot of particle positions, we leave this for the next chapter. Instead, we compute number density from particle data. Number density is the number of particles within a small region, divided by volume of that region. The cell size controls the level of spatial detail that we can resolve. Since we are interested in computing node-centered quantities, we define a cell-sized control volume centered on the nodes. This effectively means that each volume extends to the centroids of the neighbor cells.

In a Cartesian mesh, each cell has a volume $V = \Delta x \Delta y \Delta z$. Nodes have the same volume as cells, at least in the internal region away from boundaries. On boundary faces, the node volume is reduced in half. These nodes have one index that is either zero or $ni - 1$, $nj - 1$, or $nk - 1$. Nodes along edges with two zero or $ni - 1$, $nj - 1$, or $nk - 1$ indexes have the volume reduced by a factor of four. Finally, the node volume is reduced by eight on corners. In other words, the node volume reduces by a factor of two for each boundary index. This is visualized, although for a 2D grid, in Figure 2.6. This computation is performed by a function called **computeNodeVolumes** in **World** and it is called whenever the world extents are changed. The function declaration is added to the **protected** block of class **World**.

```
// computes node volumes, dx*dy*dz on internal nodes
void World::computeNodeVolumes() {
    for (int i=0;i<ni;i++)              // loop over nodes
```

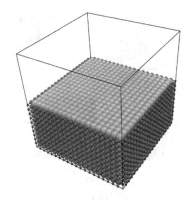

Figure 2.7: Node volumes visualized using glyphs.

```
for (int j=0;j<nj;j++)
    for (int k=0;k<nk;k++)    {
        double V = dh[0]*dh[1]*dh[2];    // standard volume
        if (i==0 || i==ni-1) V*=0.5;    // adjust on boundaries
        if (j==0 || j==nj-1) V*=0.5;
        if (k==0 || k==nk-1) V*=0.5;
        node_vol[i][j][k] = V;
    }
}
```

A call to this function is added to `setExtents`

```
void World::setExtents(double3 _x0, double3 _xm) {
    /* ... */
    computeNodeVolumes();
}
```

The `node_vol` field is visualized in Figure 2.7 using `clip` and `glyph` filters. The glyphs are scaled and colored according to the field value. As can be seen, the internal nodes all share the same volume and the smallest volume is found on the eight corners.

2.7.4 Scatter: Particle to Mesh Interpolation

One option to compute number density is to count the number of particles within each control volume. This approach is visualized in Figure 2.8. The number density on the central node is $n = 4w_{mp}/V$, assuming all four particles carry the same weight w_{mp}. This zeroth-order approach is not recommended as it does not lead to a smooth variation of densities as particles cross between cells. Instead, we implement a first order scheme visualized in Figure 2.9. This process is known as *scatter*. It is the reverse of the *gather* process introduced in Chapter 1. It allows us to interpolate particle-based data to the computational grid. The data is scattered to the eight (in 3D) nodes making up the cell containing the particle. Since nodes are shared between cells, this interpolation assures that even the surrounding cells are "aware" of the particle.

The first step of the scatter algorithm is determining which cell a particle belongs to. This is trivial for a Cartesian grid, since the equations for node positions can be inverted analytically. We have

$$l_i = (x - x_0)/\Delta x \tag{2.23}$$
$$l_j = (y - y_0)/\Delta y \tag{2.24}$$
$$l_k = (z - z_0)/\Delta z \tag{2.25}$$

These indexes are floating point numbers. The integer part corresponds to the node index, while the fractional part is the normalized distance to the neighbor node in the positive i, j, or k direction. We implement this calculation in a function called *XtoL*. The function is implemented directly in the World.h header as this allows the compiler to inline it. We also mark the function as const to signal that it does not modify class data. This annotation can lead to additional optimization.

```
class World {
   double3 XtoL(double3 x) const {
      double3 lc;
      lc[0] = (x[0]-x0[0])/dh[0];
      lc[1] = (x[1]-x0[1])/dh[1];
      lc[2] = (x[2]-x0[2])/dh[2];
      return lc;
   }
};
```

We then use the integer cast $\text{int}(l_i)$ to obtain the node index. Using particle position, a cell can be subdivided into eight octants, as shown in Figure 2.9. Their normalized volumes are used to determine the fraction of particle's data to deposit onto each of the eight nodes. The data is deposited in the "diagonally-opposed" direction. For instance, the volume of the shaded octant determines the contribution to the back-bottom-right node shown by the black circle. It may be beneficial to imagine that you can grab the particle and move it around. As the particle moves closer to this node, the shaded region grows in size. Eventually it occupies the entire cell once the particle coincides with

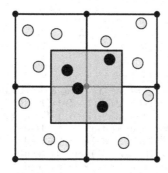

Figure 2.8: Zero-th order number density computation.

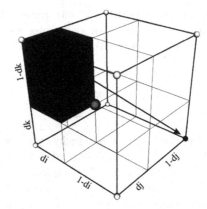

Figure 2.9: Visualization of the interpolation algorithm.

the node. Inversely, as the particle moves away, the shaded volume shrinks, as expected.

These fractional volumes can be computed from the logical coordinate. The fractional part $d_i = l_i - i$ is the normalized x distance between the particle and node i (and similarly for the j and k directions). The distance to the neighbor $i + 1$ node is $(1 - d_i)$. The normalized volume of the shaded region is thus $(d_i)(d_j)(1 - d_k)$. It contributes to node $(i + 1, j + 1, k)$. We can notice a pattern here. Sections with d_i, d_j, or d_k contribute to $i + 1$, $j + 1$, or $k + 1$ node, while $(1 - d_i)$, $(1 - d_j)$, and $(1 - d_k)$ distances map to i, j, and k. In the end, we add the following new member function to `Field`:

```
void Field::scatter(double3 lc, double value) {
    // make sure we are in domain
    if (lc[0]<0 || lc[0]>=nx-1 || lc[1]<0 || lc[1]>=ny-1 || lc[2]<0
        || lc[2]>=nz-1) return;

    // compute the cell index and the fractional distances
    int i = (int)lc[0];
    double di = lc[0]-i;

    int j = (int)lc[1];
    double dj = lc[1]-j;

    int k = (int)lc[2];
    double dk = lc[2]-k;

    // deposit fractional values to the 8 surrounding nodes
    data[i][j][k]       += value*(1-di)*(1-dj)*(1-dk);
    data[i+1][j][k]     += value*(di)*(1-dj)*(1-dk);
    data[i+1][j+1][k]   += value*(di)*(dj)*(1-dk);
    data[i][j+1][k]     += value*(1-di)*(dj)*(1-dk);
    data[i][j][k+1]     += value*(1-di)*(1-dj)*(dk);
    data[i+1][j][k+1]   += value*(di)*(1-dj)*(dk);
    data[i+1][j+1][k+1] += value*(di)*(dj)*(dk);
```

```
    data [ i ] [ j +1][k+1] += value*(1− di ) *( dj ) *(dk) ;
}
```

2.7.5   Number Density

We are now ready to add a `computeNumberDensity` function to the `Species` class:

```
void  Species :: computeNumberDensity () {
    den = 0;      // set all values to zero
    for ( Particle &part : particles )   {    // loop over particles
        double3 lc = world . XtoL ( part . pos ); // get logical coordinates
        den . scatter ( lc , part . mpwt ) ;      // deposit weight
    }

    den /= world . node_vol ;                 // divide by node volume
}
```

This function starts by clearing the data by setting all values to zero with the help of the overloaded assignment operator from Section 2.4.2. We then loop through all particles. We compute the logical coordinate, and use it to scatter the macroparticle weight to the mesh. The scatter is additive which is why the field needs to be cleared prior to this step. After we are done with the particle loop, the **den** object contains the *count* of real particles interpolated to the mesh nodes. But since we are interested in density, we need to divide these counts by the node volumes. Here we utilize the overloaded `/=` operator. If developing your code in a language that does not support operator overloading (or if you simply don't like them), you can define a function called `divideByField` to perform this element-by-element division.

We also modify the `Output::fields` function to receive as a second argument a reference to the species vector. We iterate through the members, and output density field as a `DataArrays` called `nd.O+` and `nd.e-`.

```
void Output :: fields (World &world , vector<Species> &species ) {
    /* ... */
    // species number densities
    for ( Species &sp : species ) {
        out<<"<DataArray Name=\"nd. "<<sp . name<<"\"
            NumberOfComponents=\"1\" format=\"ascii\"
            type=\"Float64\">\n";
        out<<sp . den ;
        out<<"</DataArray>\n" ;
    }
}
```

2.8   CHARGE DENSITY

Now that we have number densities, computing charge density

$$\rho = \sum_s q_s n_s \tag{2.26}$$

is trivial. This equation is implemented in the `computeChargeDensity` function added to `World`

```
void World :: computeChargeDensity ( std :: vector <Species> &species ) {
    rho = 0;
    for (Species &sp: species) {       // loop over species
        if (sp.charge==0) continue;    // don't bother with neutrals
        rho += sp.charge*sp.den;       // accumulate density scaled by
            charge
    }
}
```

This function takes as an argument a reference to a `vector` storing `Species` objects. While we may be tempted to inline this function due to its simplicity, doing so would require `World.h` to `#include "Species.h"`. This header in turns depends on `World.h`. The result is a circular dependence, as discussed in Section 2.3.3. We initialize ρ to zero, and then loop over the species. Each species adds a contribution equal to qn, where q is the charge carried by each particle, and n is that species' number density. The function skips over neutrals as they do not contribute to ρ due to having zero charge. We also add the appropriate calls to main,

```
int main () {
    /* set up particle species */

    /* load particles */

    // compute number density
    for (Species &sp: species)
        sp.computeNumberDensity();

    // compute charge density
    world.computeRho(species);

    // save results
    Output :: fields (world, species);
}
```

2.8.1 Numerical Noise

Compiling and running the code, we obtain the results visualized in Figure 2.10. This figure uses `slices` to plot electron and ion densities. A slice reduces the dimensionality of data. Paraview (and VTK) is based on the concept of a *visualization pipeline*. Different "filters" can be chained together. The slice that is used to visualize the electron density is sliced once again as shown by the dark horizontal line. Data along this line is then plotted as an XY graph on the right.

The simulation contained 80,000 ions and 10,000 electrons loaded into $20 \times 20 \times 20$ and $10 \times 10 \times 10$ cell domains, respectively. Therefore, there are, on average, 10 simulation particles per cell. While we can see that both ion and electron densities are approximately equal to the expected 10^{11} m^{-3}, there is

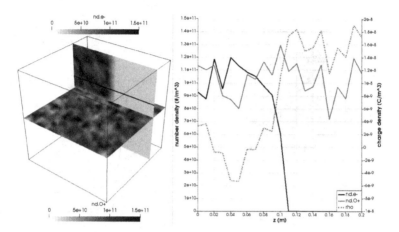

Figure 2.10: Number density with 80,000 ion and 10,000 electron macroparticles.

quite a lot of random variation about the mean. This variation is known as *numerical noise* and arises from the stochastic (random) loading of particles. Per our problem statement, we expect $n_e = n_i$ and hence $\rho = 0$ in the octant with the electrons. Instead, we obtain a finite, non-zero charge density. Each particle corresponds to $w_{mp} = (10^{11}\,\mathrm{m}^{-3}) \times (0.2\mathrm{m})^3/80000 = 10{,}000$ real ions. Due to the random loading, ion positions are not matched perfectly by the electrons. This results in each particle contributing charge density noise with magnitude $|\tilde{\rho}| \leq |q w_{mp}|$. Charge density is the forcing vector for the Poisson's equation. Non-zero ρ results in formation of a non-zero potential, and hence a finite electric field. Stationary particles start moving, even if they are loaded in a region that is supposed to be completely charge neutral. The effect of numerical noise is to introduce non-physical kinetic energy into the system. Collectively, such non-physical behavior is known as *numerical heating*. One way to reduce the noise is to increase the number of particles per cell. We can confirm this is indeed the case by re-running the simulation with the ion and electron macroparticle count increased to 8 and 1 million, respectively, by modifying the `int np_ions` and `int np_eles` lines in `main`. The new result is shown in Figure 2.11. While the noise has decreased significantly, it is still present. Utilizing higher particle counts may be impractical.

2.8.2 Quiet Start

Instead, we revisit the loading algorithm. Our goal is to minimize the non-physical non-neutrality that is found with randomly loaded particles. This can be accomplished by loading particles at uniform intervals. As long as the grid used for the electrons and the ions is the same, and if the two species have identical macroparticle weights, the contribution from one species is canceled

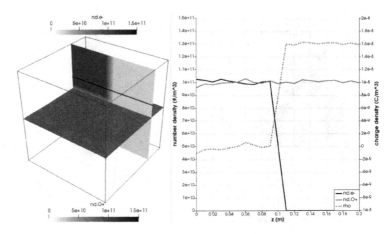

Figure 2.11: Number density with 8 million ion and 1 million electron macroparticles.

exactly by the other one. The function that performs this loading is listed below

```
void Species::loadParticlesBoxQS(double3 x1, double3 x2, double
  num_den, int3 num_sim) {
  double box_vol = (x2[0]−x1[0])*(x2[1]−x1[1])*(x2[2]−x1[2]);
  int num_sim_tot = (num_sim[0]−1)*(num_sim[1]−1)*(num_sim[2]−1);
  double num_real = num_den * box_vol;   // number of real particles
  double mpw = num_real/num_sim_tot;       // macroparticle weight

  // compute particle grid spacing
  double di = (x2[0]−x1[0])/(num_sim[0]−1);
  double dj = (x2[1]−x1[1])/(num_sim[1]−1);
  double dk = (x2[2]−x1[2])/(num_sim[2]−1);

  // preallocate memory for particles
  particles.reserve(num_sim_tot);

  // load particles on an equally spaced grid
  for (int i=0;i<num_sim[0];i++)
    for (int j=0;j<num_sim[1];j++)
      for (int k=0;k<num_sim[2];k++) {
        double pos[3];
        pos[0] = x1[0] + i*di;
        pos[1] = x1[1] + j*dj;
        pos[2] = x1[2] + k*dk;

        // shift particles on max faces back to the domain
        if (pos[0]==x2[0]) pos[0]−=1e-4*di;
        if (pos[1]==x2[1]) pos[1]−=1e-4*dj;
        if (pos[2]==x2[2]) pos[2]−=1e-4*dk;

        double w = 1;  // relative weight
        if (i==0 || i==num_sim[0]−1) w*=0.5;
```

```
    if (j==0 || j==num_sim[1]-1) w*=0.5;
    if (k==0 || k==num_sim[2]-1) w*=0.5;

    // add rewind
    double vel[3] = {0,0,0};  // particle is stationary

    addParticle(pos,vel,mpw*w);  // add to array
    }
}
```

Instead of providing the total number of particles, we call this function with the resolution of a three dimensional grid. A particle is loaded at each node position. The total number of simulation particles for the purpose of computing the macroparticle weight comes from the cell count, $N = (N_i - 1)(N_j - 1)(N_k - 1)$. We also give particles along the maximum bounds a tiny nudge back to the computational domain. This nudge is needed because the domain extends only up to, but not including, the max values, $\vec{x} \in [\vec{x}_0, \vec{x}_m)$. The weight of particles on domain boundaries is scaled to take into account fractional node volumes found here. This correction is technically required only on the computational mesh boundaries, but we apply it even on internal boundaries (for the electrons). This makes the sampling scheme consistent with the previous stochastic method. The function is used as follows:

```
int main() {
    // quiet start
    int3 np_ions_grid = {21,21,21};
    int3 np_eles_grid = {41,41,41};
    species[0].loadParticlesBoxQS(world.getX0(),world.getXm(),1e11,
        np_ions_grid);  // ions
    species[1].loadParticlesBoxQS(world.getX0(),world.getXc(),1e11,
        np_eles_grid);  // electrons
}
```

Results with this scheme are shown in Figure 2.12. There is no noise in the number density, and $\rho = 0$ in the octant occupied by electrons. This outcome is achieved despite using only 64,000 versus the 8 million ion macroparticles needed for the still noisy result in Figure 2.11. Unfortunately, quiet start is not a panacea. This perfect matching of densities is destroyed once particles start moving. Furthermore, the quiet start scheme is not conducive to many real-world engineering problems. Therefore, we should always perform convergence studies by running simulations with a varying number of macroparticles and with different computational mesh resolution. These studies help us separate physical behavior from a non-physical response arising from numerical noise.

2.9 PARTICLE PUSH

At this point, we have almost all the pieces of the final code. The major missing algorithm is the code for moving particles. As you remember from Chapter 1, moving particles involves two steps. We first update particle velocity from

$$\vec{v}^{k+0.5} = \vec{v}^{k-0.5} + (q/m)\vec{E}^k \Delta t \qquad (2.27)$$

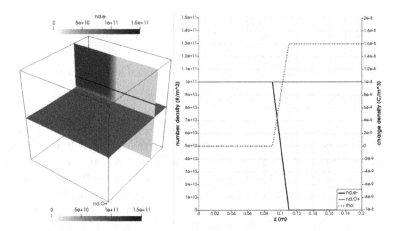

Figure 2.12: Number density obtained with the quiet start loading scheme.

Per the Leapfrog scheme, the new values correspond to velocity half time step in the future. This new velocity is then used to advance the particle position

$$\vec{x}^{k+1} = \vec{x}^k + \vec{v}^{k+0.5}\Delta t \tag{2.28}$$

In order to actually implement this method, we loop through the particles, and for each, we first obtain the electric field at the particle location. This step requires performing a *gather* operation.

2.9.1   Gather: Mesh to Particle Interpolation

Gather is the opposite of scatter, which was discussed in Section 2.7.4. While scatter is used to interpolate particle data to the grid, gather interpolates grid data to a particle. We saw the one-dimensional variant used in Section 1.4.6. Just as gather used normalized fractional volumes to determine how much of particle's data to deposit to each of the eight cell nodes, scatter uses these weights to collect the node data. The first order linear weighing scheme interpolates the value of some scalar V to the logical coordinate (l_i, l_j, l_k) with

$$
\begin{aligned}
V_{l_i,l_j,l_k} =& (1 - d_i)(1 - d_j)(1 - d_k)V_{i,j,k} + \\
& (d_i)(1 - d_j)(1 - d_k)V_{i+1,j,k} + \\
& (d_i)(d_j)(1 - d_k)V_{i+1,j+1,k} + \\
& (1 - d_i)(d_j)(1 - d_k)V_{i,j+1,k} + \\
& (1 - d_i)(1 - d_j)(d_k)V_{i,j,k+1} + \\
& (d_i)(1 - d_j)(d_k)V_{i+1,j,k+1} + \\
& (d_i)(d_j)(d_k)V_{i+1,j+1,k+1} + \\
& (1 - d_i)(d_j)(d_k)V_{i,j+1,k+1}
\end{aligned}
\tag{2.29}
$$

where

$$i = \text{int}(l_i) \tag{2.30}$$

$$di = l_i - i \tag{2.31}$$

$$dj = l_j - j \tag{2.32}$$

$$dk = l_l - k \tag{2.33}$$

$$\tag{2.34}$$

This function is added to the `Field` class,

```cpp
template <typename T>
class Field {
  // gathers field value at a logical coordinate lc
  T gather(double3 lc) {
    int i = (int)lc[0];
    double di = lc[0]-i;

    int j = (int)lc[1];
    double dj = lc[1]-j;

    int k = (int)lc[2];
    double dk = lc[2]-k;

    // interpolate data onto particle position
    T val = data[i][j][k]*(1-di)*(1-dj)*(1-dk)+
        data[i+1][j][k]*(di)*(1-dj)*(1-dk)+
        data[i+1][j+1][k]*(di)*(dj)*(1-dk)+
        data[i][j+1][k]*(1-di)*(dj)*(1-dk)+
        data[i][j][k+1]*(1-di)*(1-dj)*(dk)+
        data[i+1][j][k+1]*(di)*(1-dj)*(dk)+
        data[i+1][j+1][k+1]*(di)*(dj)*(dk)+
        data[i][j+1][k+1]*(1-di)*(dj)*(dk);

    return val;
  }
};
```

We use this function only with floating point scalars and arrays, in which the template argument `T` is either `double` or `double3`.

2.9.2 Velocity and Position Update

To actually push the particles, we add a function called `advance` to `Species`. The function is implemented as follows:

```cpp
void Species::advance() {
  // get the time step
  double dt = world.getDt();

  // get mesh bounds
  double3 x0 = world.getX0();
  double3 xm = world.getXm();

  // loop over all particles
```

```
for (Particle &part: particles) {
    // get logical coordinate of particle's position
    double3 lc = world.XtoL(part.pos);

    // electric field at particle position
    double3 ef_part = world.ef.gather(lc);

    // update velocity from F = qE
    part.vel += ef_part*(dt*charge/mass);

    // update position from v = dx/dt
    part.pos += part.vel*dt;
}
}
```

This code utilizes a `World::getDt()` accessor to retrieve the value of the simulation time step Δt. We then loop through the particles using the range-based `for` loop. The logical coordinate corresponding to the particle's position is then calculated. We evaluate the electric field at the particle position using the just described `gather` function. Velocity and position are then integrated via the Leapfrog scheme. This code takes advantage of the overloaded operators defined for the `double3` object. Alternatively, we could write

```
// update velocity
for (int i=0;i<3;i++)
    part.vel[i] += charge/mass*ef_part[i]*dt;

// update position
for (int i=0;i<3;i++)
    part.pos[i] += part.vel[i]*dt;
```

2.9.3 Particle Boundary Check

Right now, particles are free to leave the computational domain. Such particles will have logical coordinates falling outside the valid range. Attempting to scatter their weights results in memory corruption since `scatter` does not perform bounds checking for performance reasons (although such checking could be added to the debug version via the `assert` keyword). Similarly, performing a mesh gather would result in garbage values interpolated to the particle position. In order to avoid this, we need to implement a *particle boundary check*. Particles leaving the computational domain need to be either removed from the simulation, or reflected back, according to the problem setup. In this example we implement reflective walls. Along a single dimension, elastic reflection can be modeled by first computing the distance the particle has traveled outside the domain, and then adding or subtracting this distance from the domain boundary. On the left side we have

$$\Delta x_{out} = x_0 - x_p \qquad (2.35)$$

The new particle position is then given by

$$(x_p)_{new} = x_0 + \Delta x_{out} \tag{2.36}$$

$$= 2x_0 - x_p \tag{2.37}$$

We obtain a similar relationship on the right side,

$$\Delta x_{out} = x_p - x_d \tag{2.38}$$

The new particle position is then given by

$$(x_p)_{new} = x_d - \Delta x_{outside} \tag{2.39}$$

$$= 2x_d - x_p \tag{2.40}$$

The above algorithm only places the particle back in the computational domain. We also need to reflect the velocity. Specular reflection is modeled by multiplying the component in the wall normal direction by -1. This process is visualized in Figure 2.13. First, by checking the x direction, the particle is moved from the x_p position to the location marked by the medium gray circle. Then a check along the y direction moves the particle back to the computational domain. An alternate implementation of the boundary check involves determining the boundary the particle hit first and then performing a secondary particle push using a fractional time step. This algorithm is demonstrated in Chapter 4. The version with the simple boundary check is shown below,

```
//Species::advance() {

  // loop over all particles
  for (Particle &part: particles) {

    /* same code as above */

    // update position from v = dx/dt
    part.pos += part.vel*dt;

    // reflect particles leaving the domain
    for (int i=0;i<3;i++) {
      if (lc[i]<0) {
        part.pos[i]=2*x0[i]-part.pos[i];  part.vel[i]*=-1.0; }
      else if (lc[i]>=(world.nn[i]-1)) {
        part.pos[i]=2*xm[i]-part.pos[i];  part.vel[i]*=-1.0; }
    }
  }
}
```

2.9.4 Velocity Rewind

Implementing the Leapfrog scheme involves rewinding the velocity of injected particles backward through $-\Delta t/2$. Since all particles are injected via the addParticle member function, we add this logic there,

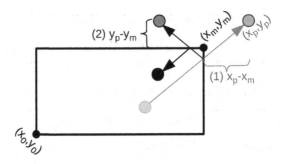

Figure 2.13: Elastic reflection of particles on the boundary.

```
void Species :: addParticle(double3 pos, double3 vel, double mpw) {
  // don't do anything if pos outside domain bounds [x0,xd)
  if (!world.inBounds(pos)) return;

  // get particle logical coordinate
  double3 lc = world.XtoL(pos);

  // evaluate electric field at particle position
    double3 ef_part = world.ef.gather(lc);

  // rewind velocity back by 0.5*dt*ef
  vel -= charge/mass*ef_part *(0.5*world.getDt());

  // add to list
  particles.emplace_back(pos,vel,mpw);
}
```

Here we also added boundary check using a new **inBounds** function added to **World**. The particle is added only if the position is within the computational bounds. Although this is not the case here, this check allows us to decouple the geometry of the sampling surface from the computational domain. In plume simulations, we may want to sample particles randomly across the entire exit plane even if the code simulates only a section of the plume. This function is listed below,

```
class World {
  bool inBounds(double3 pos) {
    for (int i=0;i <3; i++)
      if (pos[i]<x0[i] || pos[i]>=xm[i]) return false;
    return true;
  }
}
```

We return **false** as soon as an invalid position is encountered. If all three dimensions are within bounds, the function returns **true**.

2.9.5 Time Step

Since the PIC method involves time marching, we also need to define time parameters. Specifically, we need to tell the code what value to use for Δt, and how many time steps to simulate. We store this information in World. A new member function is used to set the values,

```cpp
class World {
public:
  // sets time step and number of time steps
  void setTime(double dt, int num_ts)
    {this->dt=dt; this->num_ts=num_ts;}

  // functions for accessing time information
  int getTs() const {return ts;}
  double getTime() const {return time;}
  double getDt() const {return dt;}
  bool isLastTimeStep() const {return ts==num_ts-1;}

  // advances to the next time step, returns true while time
     remains
  bool advanceTime() { time+=dt; ts++; return ts<=num_ts; }

protected:
  double dt;        // time step
  int num_ts;       // number of time steps
};
```

Here we added few additional functions for accessing time data. Furthermore, advanceTime increments the time step and updates the physical time simulated so far. This function returns false once the allotted number of steps has been exceeded. While this logic could be placed in main, delegating time advance to a function provides a convenient hook for placing any future algorithms that may need to run at every time step. We add the initialization to main,

```cpp
int main(int argc, char *args[]) {
  world.setTime(2e-10,10000);      // set dt and num_ts

  // main loop
  while(world.advanceTime()) { ... }
}
```

The simulation time step is set to $\Delta t = 2 \times 10^{-10}$ seconds and the number of time steps is 10,000. The total simulated time is thus $t = 2\mu s$. Just as there are requirements on the cell size per Section 2.3.6, the simulation time step also cannot be set arbitrarily. There are two constraints to consider. First, the time step needs to be small enough to resolve the fastest frequency of interest, $\Delta t \ll 1/\tau_{\max}$. This may be the *plasma frequency* in fully kinetic studies, or the *cyclotron frequency* in simulations involving motion of magnetized particles. Secondly, we require that particles do not travel more than a cell length per push. Otherwise, the electric field as seen by the particle is not continuous.

Therefore,

$$\Delta t < \Delta x / v_{max} \tag{2.41}$$

As can be seen from this equation, the time step is governed by the fastest moving particles. In the case of fully kinetic simulations, the fastest particles are the electrons. Due to their tiny mass, electrons move much faster than ions. Consider ions and electrons accelerated through the identical potential drop, and thus attaining equal kinetic energy, $(1/2)mv^2$. The electron velocity is related to the ion velocity by $v_e = v_i \sqrt{m_i/m_e}$. For hydrogen, $m_i = 1$ amu, this ratio evaluates to 42.6. For the much heavier 131.3 amu xenon encountered in plasma propulsion applications, the ratio increases to almost 500. A time step appropriate for the electrons is may not be acceptable if the process of interest is driven by ion dynamics. A large number integration steps need to be applied for an ion to cross a single cell. Not only is this inefficient, the repeated increment by tiny values can lead to a significant numerical error arising from the finite precision of computer math. Therefore, it is customary to modify the particle pushing algorithm to either eliminate electrons completely, or to use a different integration time step for each species. These alternate schemes, known as *hybrid methods* and *subcycling*, are discussed in Chapters 3 and 4.

2.10 DIAGNOSTICS

It is a good idea to include some run time diagnostics to monitor the simulation. I like to output minimal information, such as the current time step, and the number of simulation particles, to the screen. Additional information, such as the total system energy, is saved to a `.csv` (comma separated values) file for plotting. We add prototypes for two new functions to the `Output` namespace,

```cpp
namespace Output {
    void fields(World &world, std::vector<Species> &species);
    void screenOutput(World &world, std::vector<Species> &species);
    void diagOutput(World &world, std::vector<Species> &species);
}
```

The `screenOutput` function is implemented as follows:

```cpp
// writes information to the screen
void Output::screenOutput(World &world, vector<Species> &species){
    cout<<"ts: "<<world.getTs();
    for (Species &sp:species)
        cout<<setprecision(3)<<"\t "<<sp.name<<":"<<sp.getNp();
    cout<<endl;
}
```

The `diagOutput` function is slightly more complex,

```cpp
namespace Output {std::ofstream f_diag;}      // file handle

void Output::diagOutput(World &world, vector<Species> &species) {
    using namespace Output;            // to get access to f_diag
```

```
if (!f_diag.is_open()) {          // if file not open
  f_diag.open("runtime_diags.csv");
  f_diag<<"ts,time,wall_time";    // write header
  for (Species &sp:species)
    f_diag<<",mp_count."<<sp.name<<",real_count."<<sp.name
        <<",px."<<sp.name<<",py."<<sp.name<<",pz."<<sp.name
        <<",KE."<<sp.name;
  f_diag<<",PE,total_E"<<endl;
}

f_diag<<world.getTs()<<","<<world.getTime();
f_diag<<","<<world.getWallTime();

// write out species kinetic energy and momentum
double tot_KE = 0;
for (Species &sp:species) {
  double KE = sp.getKE();    // species kinetic energy
  tot_KE += KE;              // accumulate total kinetic energy
  double3 mom = sp.getMomentum();   // momentum

  f_diag<<","<<sp.getNp()<<","<<sp.getRealCount()
      <<","<<mom[0]<<","<<mom[1]<<","<<mom[2]<<","<<KE;
}

// write out total potential and kinetic energy
double PE = world.getPE();
f_diag<<","<<PE<<","<<(PE+tot_KE);

f_diag<<"\n";    // use \n to avoid flush to disc
if (world.getTs()%25==0) f_diag.flush();  // periodically write
}
```

Note that namespaces can also store variables, such as the `ofstream` file handle. This addition is made inside the `Output.cpp` file to keep it hidden from the rest of the code. On each call to `diagOutput`, we first check if the file is already open. If not, we create a new file called `runtime_diags.csv` and write out the header containing the column names. We then output relevant time data such as the current time step and the elapsed run time. We loop through the species, and for each output the number of simulation and real particles, and the species kinetic energy. The output is finalized with the system potential energy, and the total energy per

$$E = (PE) + \sum_s (KE)_s \qquad (2.42)$$

where the sum is over species. It is important to note that computation of these diagnostics imposes a finite overhead on the simulation. We should be judicious about how much diagnostic information to write, and how often. Here we write this information at every time step, but in production codes, it may be preferred to compute and output diagnostics only every 100 or so steps.

2.10.1 Elapsed Time

The functions for computing elapsed wall time, the number of real particles, and the kinetic and potential energies do not yet exist. C++11 includes a set of functions for performing high resolution timing implemented in <chrono>. We add a new variable to World to store the wall time at simulation start,

```
#include <chrono>
World () {
    double getWallTime () ;   // returns elapsed time in seconds

protected :
    std :: chrono :: time_point <std :: chrono :: high_resolution_clock >
        time_start ;   // time at simulation start
}
```

As can be seen, the type of this variable is rather long due to the need to specify the appropriate namespace and template arguments. Luckily, C++ now supports auto variables, which let the compiler deduct the type automatically. I have mixed feelings about using autos, as I feel they diminish the strongly-typed strength of the C++ language. They can however be a great space saver, especially when encountering complex declarations such as this one. We use the constructor to set the starting time

```
World :: World (int ni , int nj , int nk): ni{ni}, nj{nj}, nk{nk},
    nn{ni ,nj ,nk}, phi(ni ,nj ,nk) ,rho(ni ,nj ,nk) ,node_vol(ni ,nk ,nk) ,
    ef(ni ,nj ,nk)   {
        time_start =   chrono :: high_resolution_clock :: now () ;   // save
            starting time point
    }
```

The elapsed wall time is given by

```
double World :: getWallTime () {
    auto time_now = chrono :: high_resolution_clock :: now () ;
    chrono :: duration<double> time_delta = time_now − time_start ;
    return time_delta .count () ;
}
```

2.10.2 Particle Counts, Momentum, and Energy

We also need a function for obtaining the number of real particles. This is simply the sum of macroparticle weights of all particles, $N_{real} = \sum_p (w_{mp})_p$

```
double Species :: getRealCount () {
    double mpw_sum = 0;
    for (Particle &part : particles ) mpw_sum += part .mpw;
    return mpw_sum;
}
```

Similarly, the species s momentum is obtained from

$$P_s = m_s \sum_p (w_{mp})_p \vec{v}_p \tag{2.43}$$

where m_j is the mass of particles of species j.

```
double3 Species :: getMomentum () {
  double3 mom;
  for (Particle &part : particles ) mom += part.mpw*part.vel;
  return mass*mom;
}
```

Kinetic energy for species s is computed from

$$E_s = \frac{1}{2} m_s \sum_p (w_{mp})_p |\vec{v}_p|^2 \tag{2.44}$$

in function getKE,

```
double Species :: getKE () {
  double ke = 0;
  for (Particle &part : particles ) {
    double v2 = part.vel [0]*part.vel [0] + part.vel [1]*part.vel [1]
        + part.vel [2]*part.vel [2];
    ke += part.mpw*v2;
  }
  return 0.5*mass*ke;
}
```

We also need the total system potential energy,

$$PE = \frac{1}{2} \epsilon_0 \sum_n |E_n|^2 \tag{2.45}$$

where the sum is over all mesh nodes.

```
double World :: getPE () {
  double pe = 0;
  for (int i=0;i<ni;i++)
    for (int j=0;j<nj;j++)
      for (int k=0;k<nk;k++) {
        double3 efn = ef [i][j][k];    // el. field at this node
        double ef2 = efn [0]*efn [0]+efn [1]*efn [1]+efn [2]*efn [2];
        pe += ef2*node_vol [i][j][k];
      }
  return 0.5*Const :: EPS_0*pe;
}
```

2.10.3 Animation

We would like to animate the instantaneous field results at different simulation time steps. The previously developed Output::fields function writes to a file called fields.vti. Paraview automatically groups files following naming scheme prefix_xxxx.ext into an animation group. Therefore, we modify this function to include the time step number in the file name, such as fields_05700.vti. This is accomplished with the help of strstream data object and operators from iomanip,

```
void Output::fields(World &world, vector<Species> &species) {
  stringstream name;
  name<<"results/fields_"<<setfill('0')<<setw(5)<<world.getTs()<<
    ".vti";
    ofstream out(name.str());  // open output file
    if (!out.is_open()) {cerr<<"Could not open "<<name.str()
      <<endl;return;}
}
```

The files are written to a **results/** folder to keep our source directory from becoming cluttered. This is accomplished by including the forward slash directory separator in the file name. On Windows, you may need to modify this string to read `"results\\fields_"`.

2.11   MAIN LOOP

The entire **main** function is listed for completeness below. We start by performing initialization as described above. The simulation then enters the *main loop*,

```
int main(int argc, char *args[]) {
  // initialize domain
  World world(21,21,21);
  world.setExtents({-0.1,-0.1,0},{0.1,0.1,0.2});
  world.setTime(2e-10,10000);

  // set up particle species
  vector<Species> species;
  species.push_back(Species("O+", 16*AMU, QE, world));
  species.push_back(Species("e-", ME, -1*QE, world));

  // initialize potential solver and solve initial potential
  PotentialSolver solver(world,10000,1e-4);
  solver.solve();

  // obtain initial electric field*/
  solver.computeEF();

  // create particles
  int np_ions_grid[3] = {21,21,21};
  int np_eles_grid[3] = {41,41,41};
  species[0].loadParticlesBox(world.getX0(),world.getXm(),1e11,
    np_ions_grid);  // ions
  species[1].loadParticlesBox(world.getX0(),world.getXc(),1e11,
    np_eles_grid);  // electrons

  // main loop
  while(world.advanceTime()) {
    // compute charge density
    world.computeChargeDensity(species);

    // update potential
    solver.solve();

    // obtain electric field
```

```
solver.computeEF();

// move particles
for (Species &sp:species) {
    sp.advance();
    sp.computeNumberDensity();
}

// screen and file output
Output::screenOutput(world, species);
Output::diagOutput(world, species);

// periodically write out results
if (world.getTs()%100==0 || world.isLastTimeStep())
    Output::fields(world, species);
}

// output run time
cout<<"Simulation took "<<world.getWallTime()<<" seconds"<<endl;

return 0;      // indicate normal exit
}
```

We start by solving the "vacuum field" which is then used to rewind velocities of the injected particles. Each loop iteration starts by solving the potential and obtaining the electric field by calling the solve and computeEF methods of the PotentialSolver object. This step effectively yields $\vec{E}^k$. The charge density ρ used to evaluate ϕ is obtained by calling computeChargeDensity. We then advance particle velocities and positions to $v^{k+0.5}$ and x^{k+1}, respectively, by calling the advance function on each species. Next, we call the functions from the previous section to output diagnostic information. Every 100 time steps, we also save the instantaneous simulation results to a file using the Output::fields function. The loop iterates for the specified number of time steps. We end the simulation by displaying the run time.

2.12 RESULTS

Before we can run the simulation for the first time, we need to create the results subdirectory. This is needed, since platform-independent file system manipulation is only available starting with the C++17 language extensions.

```
$ mkdir results
$ g++ -O2 *.cpp -o ch2
$ ./ch2
ts: 0    O+:68921    e-:9261
ts: 1    O+:68921    e-:9261
ts: 2    O+:68921    e-:9261
ts: 3    O+:68921    e-:9261
. . .
```

The code continues to run for a few minutes. The simulation results can then be loaded into Paraview for plotting. Paraview contains a nice functionality to visualize field data using *volume rendering*. We can see several snapshots from an animation of electron density in Figure 2.14. The electrons, which are initially concentrated in one octant, start moving through the domain as the system attempts to reduce the space charge separation. Upon reaching the diagonally opposite boundary, the electrons are reflected. This process repeats for a while, although the structure eventually loses cohesion. We can also plot kinetic and potential energies from the `runtime_diags.csv` file. As can be seen, the total system energy remains conserved.

2.12.1 Convergence Check

The grid and particle number independence can be checked by running the simulation with the mesh resolution increased to $41 \times 41 \times 41$ nodes. Particle count is also increased by loading the particles on a finer spatial $81 \times 81 \times 81$ grid (for the ions). The relevant changes are shown below.

```
int main() {
  World world(41,41,41);
  int3 np_ions_grid = {81,81,81};
  int3 np_eles_grid = {41,41,41};
}
```

Results from this run in Figure 2.16 can be compared to the results on the coarse mesh in Figure 2.14. The new simulation produces an improved spatial resolution, as expected. However, the behavior is identical to what was observed previously. Therefore, we can conclude that the prior mesh and particle loading resolutions are satisfactory. For completeness, we should also check integration step independence, by running the simulation with a reduced value for Δt. This is left for an exercise.

2.13 SUMMARY

In this chapter we developed our first fully functional three dimensional plasma simulation code based on the electrostatic particle in cell (ES-PIC) method. The code simulates plasma confined in a reflecting box. The system is not charge neutral, resulting in electrons oscillating around the heavy ion core. Along the way, we learned how to solve potential, compute electric field, load and push particles, compute number density, and generate diagnostic output. In the next chapter, we reuse functions developed here to implement a simulation of plasma flowing past a charged sphere.

EXERCISES

2.1 *Potential Solver.* Derive the Gauss-Seidel finite difference scheme for the Poisson's equation. Then write your own solver that supports both Dirichlet and Neumann boundaries. You can use the first order scheme

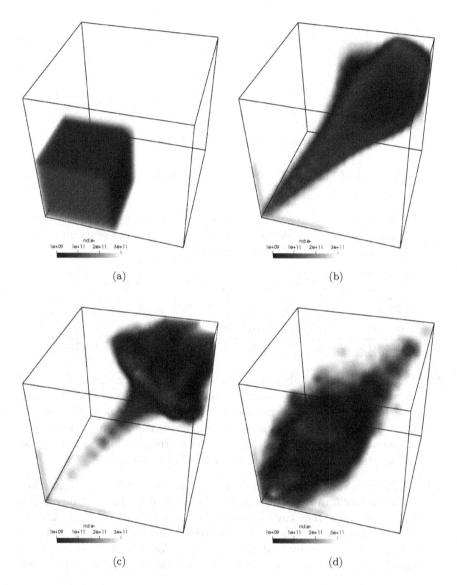

Figure 2.14: Animation of electron density on a $21 \times 21 \times 21$ grid with 8,000 electrons.

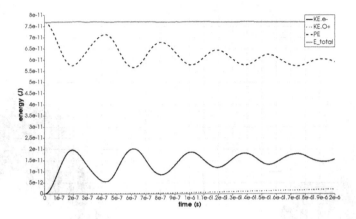

Figure 2.15: Plot of kinetic and potential energies. Total energy is conserved.

for $\partial\phi/\partial n = 0$. Use the code to solve potential on the domains given by the table below. Assume Neumann boundaries on the remaining faces. In all cases, $kT_{e,0} = 2$ eV and $\phi_0 = 0$ V.

Cells	Spacing (m)	n_0 (m^{-3})	Dirichlet Boundaries
$10 \times 10 \times 20$	$0.01, 0.01, 0.01$	0	ZMIN=0,ZMAX=10
$10 \times 10 \times 10$	$0.01, 0.01, 0.02$	0	ZMIN=0,ZMAX=10
$10 \times 10 \times 20$	$0.01, 0.01, 0.01$	10^{10}	ZMIN=0,ZMAX=10
$10 \times 10 \times 20$	$0.01, 0.01, 0.01$	10^{10}	ZMIN=0,ZMAX=10

2.2 *Particle Loading.* Modify the example code to inject identical ion and electron number density over the entire computational domain. Next run the simulation with the stochastic and quiet start loading schemes. If $\phi = 0$ V on all boundaries, we expect that stationary particles remain stationary, since $\rho = 0$ and there is no external forcing field. This will not be the case with the randomly loaded population.

2.3 *Fluid Ions.* Since ions are much more massive than the electrons, they do not respond rapidly to the induced electric field. In the time interval considered here, they can be assumed to remain stationary. Modify the code to simulate only the electron particles. Replace the ion population with a constant density field $n_i = 10^{10}$ m^{-3}. You should observe that this *hybrid* simulation runs faster due to the system tracking fewer particles.

2.4 *Convergence Study.* Check time step independence by running the simulation with a varying number of particles, different mesh resolution, and varying simulation time steps.

2.5 *Large Domain.* Instead of relying on the elastic boundaries, increase the

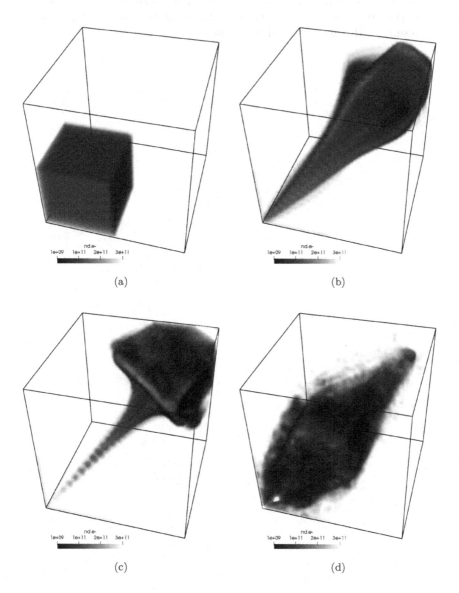

Figure 2.16: Animation of electron density on a 41 × 41 × 41 grid with 64,000 particles.

domain size without changing the extent of the ion and electron population. Your goal is to generate a $n_e = n_i = 0$ "moat" surrounding the ion
and electron box. Experiment with the domain size such that majority
of electrons are reflected back by electrostatic trapping. Compare these
results to the one obtained previously.

CHAPTER 3

Flow Around a Sphere

3.1 INTRODUCTION

I N THIS chapter, we develop a 3D Cartesian code for simulating the flow of plasma past a charged sphere. This simulation could be a crude approximation for an object orbiting a planet. Alternatively, it could model a spherical Langmuir probe embedded in a plasma beam. We actually already have most of the code implemented from the previous chapter, with the exception of the mechanism for including the sphere object. We also need to implement a continuous particle source. In order to speed up the simulation, we utilize a simple fluid model for the electrons, giving us a *hybrid-PIC* simulation. Along the way, we also learn about steady state, data averaging, matrix solver algorithms, and the quasi-neutral approximation.

3.2 SIMULATION DOMAIN

Our goal is to simulate the flow of stationary $n_i = 10^{10}$ m^{-3} ions past a $r = 0.05$ m sphere moving at $7,000$ m/s in the $-z$ direction. The sphere is assumed to be charged to a -100 V potential in respect to the ambient plasma. By changing the frame of reference, the sphere can be envisioned to be stationary with ions flowing towards it with $7,000$ m/s velocity in the $+z$ direction. This is the approach taken here.

The simulation domain is visualized in Figure 3.1. It is similar to what was used in Chapter 2, except that the domain length is extended to $z = 0.4$ m. The $r = 0.05$ m sphere is placed at $(0, 0, 0.15)$. It is modeled as a conductor at a constant potential. Therefore, the sphere introduces an internal Dirichlet boundary for the Poisson solver,

$$\phi(\vec{x}) = \phi_{sphere} \qquad \vec{x} \in \Omega_{sphere} \tag{3.1}$$

The outer boundaries are *open* to the ambient environment. We assume that they are sufficiently removed from the sphere so the Neumann boundary condition,

$$\nabla \phi \cdot \hat{n} = 0 \tag{3.2}$$

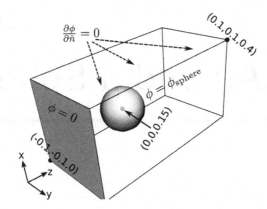

Figure 3.1: Computational domain for the flow past a charged sphere.

holds. Here $\hat{n}$ is the normal vector at the boundary face. The only exception is the z_{min} face. This plane represents the upstream bulk plasma. We set $\phi = 0$ V here to establish the potential drop. Oxygen ions are injected into the simulation domain from this inlet plane with the prescribed $v_z \equiv w = 7,000$ m/s velocity. The injection flow rate is based on the desired number density. The boundaries are open. Particles leaving the computational domain are removed from the simulation.

We implement the new domain extents in the World initialization in main. Since the domain length in z is doubled, we also double the node counts in this dimension to maintain cube-shaped cells. Simulation time step and number of time steps are updated to $\Delta t = 10^{-7}$ s and 400, respectively. The larger time step is acceptable since the simulation no longer tracks kinetic electrons. Finally, we also add hooks for adding the sphere and the $z = 0$ inlet.

```
int main(int argc, char *args[]) {
  // initialize domain
  World world(21,21,41);
  world.setExtents({-0.1,-0.1,0},{0.1,0.1,0.4});
  world.setTime(1e-7,400);

  // set objects
  double phi_sphere = -100;        // set default
  if (argc>1)
    phi_sphere = atof(args[1]);    // convert argument to float
  cout<<"Sphere potential: "<<phi_sphere<<" V"<<endl;
  world.addSphere({0,0,0.15},0.05,phi_sphere);
  world.addInlet();
  /* ... */
}
```

The sphere potential defaults to -100 V, however, here we illustrate how to use command line arguments to provide user inputs. The **args** C-style string array contains these arguments, with **args[0]** holding the executable path.

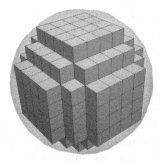

Figure 3.2: Sugarcubed representation of a sphere.

The first user input is in `args[1]`. `argc` contains the argument count. Once the code is built using

```
$ g++ *.cpp -o ch3
```

we can run a simulation for a $\phi_{\text{sphere}} = -25$ V using

```
$ ./ch3 -25
```

without having to recompile.

3.3 POTENTIAL SOLVER

3.3.1 Sugarcubing

The sphere acts as an internal Dirichlet boundary, and thus needs to be included in the potential solver. The computational domain is a uniform Cartesian mesh, composed of $\Delta x \times \Delta y \times \Delta z$ blocks. The sphere, on the other hand, is inherently smooth. The inability to resolve smooth surface is unfortunately a major limitation of a Cartesian mesh. In Chapter 6, we see how to use *unstructured mesh* that can resolve curved surfaces. But for now, we are stuck with what is known as *sugarcubing* or *staircasing*. With this approach, we simply flag the nodes (or cells) that are located within the object. This then results in a degenerate representation of the surface, similar to trying to create a sphere from sugarcubes. This is visualized in Figure 3.2.

We implement this model by introducing a new `object_id` integer data field to keep track of *fixed* (or Dirichlet) nodes. We let the value of 0 indicate a regular "gas domain" node, while positive values correspond to nodes internal to conductor objects. These nodes have a fixed potential and thus form a Dirichlet boundary condition. We add the variable to `World`

```
class World {
public:
    FieldI object_id;    // object id flag to flag fixed nodes
}
```

We also initialize it in the constructor,

```
World::World(int ni, int nj, int nk): ni{ni}, nj{nj}, nk{nk}, ...,
  object_id(ni,nj,nk) {
  /* ... */
}
```

The actual sugar cubing is performed within the addSphere function added to World. The input arguments include the sphere center, the sphere radius, and the sphere potential. We start by saving the geometry parameters as class members. We then loop through all mesh nodes, and for each, check if it is located inside the sphere. If it is, we simply set $(object\_id)_{i,j,k} = 1$. We also set $\phi_{i,j,k} = \phi_{sphere}$.

```
// sugarcubes a sphere centered at (x0,y0,z0)
void World::addSphere(double3 x0, double radius, double
  phi_sphere) {
  sphere_x0 = x0;                    // save sphere centroid
  sphere_rad2 = radius*radius;

  for (int i=0;i<ni;i++)             // loop over all nodes
    for (int j=0;j<nj;j++)
      for (int k=0;k<nk;k++) {
        double3 x = pos(i,j,k);      // node position
        if (inSphere(x)) {           // if point in sphere
          object_id[i][j][k] = 1;    // set object flag
          phi[i][j][k] = phi_sphere; // also set potential
        }
      }
}
```

The pos helper function is defined as shown below. This snippet also includes the new variables for storing sphere geometry parameters.

```
class World {
  // converts logical coordinate to physical position
  double3 pos(double3 lc) {
    double3 x = x0 + dh*lc;
    return x;
  }

  // another wrapper that takes 3 ints as inputs
  double3 pos(int i, int j, int k) {
    double3 x{(double)i,(double)j,(double)k};
    return pos(x);
  }
protected:
  double3 sphere_x0 {0,0,0};  // sphere centroid
  double sphere_rad2 = 0;     // sphere radius squared
};
```

The pos function simply evaluates $x = x_0 + l_i \Delta x$, with similar equations for y and z. It is in effect the inverse of the XtoL function we have seen previously. We implement two forms, one that takes as input a double3 and another one for which the inputs are three integers. This second form acts as a wrapper for an explicit cast of integers to doubles. This conversion is not allowed implicitly since on some architectures, a double precision floating point number may not be large enough to hold an integer. We also implement the `inSphere` function,

```
bool World::inSphere(double3 x) {
    double3 r = x - sphere_x0;        // ray to x, x0 + r = x
    double r_mag2 = (r[0]*r[0] + r[1]*r[1] + r[2]*r[2]);
    if (r_mag2<=sphere_rad2) return true;
    return false;
}
```

It simply computes the distance between the point and the sphere center, and compares it to the radius.

$$|\vec{x} - \vec{x}_{0,sphere}|^2 \leq r_{sphere}^2 \tag{3.3}$$

The comparison is made with the distance squared to avoid computing the square root. Calculating a square root is more computationally expensive than multiplying two numbers. The `if` statement could be combined with the `return` statement as `return r_mag2<=sphere_rad2`, but I prefer this more verbose, but perhaps less cryptic, formulation. We also fix potential on the $z = 0$ inlet. This is accomplished by iterating over all nodes with $k = 0$ and set $(\text{object\_id})_{i,j,k} = 2$ and $\phi_{i,j,k} = 0$. The function `addInlet` is defined as

```
void World::addInlet() {
    for (int i=0;i<ni;i++)         // loop over k=0 face
        for (int j=0;j<nj;j++) {
            object_id[i][j][0] = 2;   // set object flag
            phi[i][j][0] = 0;         // set potential
        }
}
```

3.3.2 Neumann Boundaries

The solver still retains the Dirichlet condition on the outer boundaries, per the example from the prior chapter. This is not the correct boundary for a domain that is open to the ambient environment. If the boundary is placed sufficiently far from the sphere, we can make the assumption that the local gradient in plasma density vanishes in the direction perpendicular to the boundary. Furthermore, we generally assume that the boundary is sufficiently removed for plasma neutrality to hold. Then

$$E_{\perp} \equiv \partial\phi/\partial\hat{n} = 0 \tag{3.4}$$

This is known as the zero Neumann boundary condition. For a Cartesian domain, we have $\partial\phi/\partial x = 0$, $\partial\phi/\partial y = 0$, and $\partial\phi/\partial z = 0$. We make the solver utilize this condition by default on all boundaries. It can be overriden with a non-zero `object_id` flag.

We apply the Neumann condition by applying first order differencing to Equation 3.4. On the $i = 0$ plane, we have

$$\frac{\phi_{1,j,k} - \phi_{0,j,k}}{\Delta x} = 0 \tag{3.5}$$

or $\phi_{0,j,k} = \phi_{1,j,k}$. On the opposite face where $i = ni - 1$, we obtain $\phi_{ni-1,j,k} = \phi_{ni-2,j,k}$. Similar expressions hold for the other two dimensions.

3.3.3 Boltzmann Electron Model

As discussed in Chapter 1, the kinetic treatment is only one of several options for simulating plasma constituents. For a species at the Maxwellian velocity distribution, we can derive expressions for conservation of mass, momentum, and energy. In unmagnetized plasma away from surfaces, it is generally safe to assume that electrons are thermalized, even if the density is low enough so that Knudsen number $Kn \gg 1$. This observation is tied to a phenomenon known as *Landau damping*, which is a collisionless decay of disturbances through electrostatic coupling. The momentum conservation equation for electrons in plasma containing only a single ion species can be written as

$$m_e n_e \left[\frac{\partial \vec{v}_e}{\partial t} + (\vec{v}_e \cdot \nabla) \vec{v}_e \right] = -e n_e \vec{E} - \nabla p_e - m_e n_e (\vec{v}_i - \vec{v}_e) \nu_{ei} \tag{3.6}$$

here p_e is the hydrodynamic pressure (in Pa), and ν_{ei} is the collision frequency (in s^{-1}) between the two populations moving at drift speed v_i and v_e.

Since electrons are much lighter than ions, they are able to respond almost instantaneously to any disturbance. This suggests that the left hand side can be ignored, since electrons are able to establish equilibrium between each ion move. The last term on the right hand side arises from collisional coupling between ions and electrons. This term can be rewritten in terms of current density, resulting in a form known as *Generalized Ohm's Law*. That formulation is discussed in Chapter 7. Instead, if we assume that mean ion and electron velocities are identical, this term drops out. We thus have

$$0 = -e n_e \vec{E} - \nabla p_e \tag{3.7}$$

Balance exists between the electrostatic and hydrodynamic pressure gradients. Next by substituting $\vec{E} = -\nabla \phi$ and using the ideal gas law $p_e = n_e k_B T_e$, we write

$$e n_e \nabla \phi = \nabla (n_e k_B T_e) \tag{3.8}$$

For an isothermal population, the temperature term can be taken outside the derivative. We can also integrate both sides of the equation along the same path,

$$e \int_{\phi_0}^{\phi} d\phi = k_B T_e \int_{n_0}^{n_e} \frac{1}{n_e} dn_e \tag{3.9}$$

to obtain

$$\frac{e(\phi - \phi_0)}{k_B T_e} = \ln \left(\frac{n_e}{n_0} \right) \tag{3.10}$$

or

$$n_e = n_0 \exp \left[\frac{e(\phi - \phi_0)}{k_B T_e} \right] \qquad T_e(K) \tag{3.11}$$

Equation 3.11 is known as the *Boltzmann relationship*. It provides an expression between electron density and plasma potential for a set of reference values. These properties correspond to plasma state sampled at some arbitrary, but identical, spatial location. In this expression, T_e is given in Kelvin. It is customary to express electron temperature in electronvolts. Given the conversion $1eV = e/k_B$, the Boltzmann equation can also be expressed as

$$n_e = n_0 \exp \left[\frac{\phi - \phi_0}{T_e} \right] \quad ; \ T_e(eV) \qquad (3.12)$$

This is the form we utilize in our simulation. In the derivation of the Boltzmann relationship, we assumed that the system does not contain magnetic field. If magnetic field is present, Equation 3.11 holds individually only along a single field line.

3.3.4 Non-Linear Poisson Solver

The equation for electron density. Equation 3.12 can be substituted into Poisson's equation, Equation 1.24,

$$\nabla^2 \phi = -\frac{e}{\epsilon_0}(Z_i n_i - n_e)$$

$$= -\frac{e}{\epsilon_0} \left[Z_i n_i - n_0 \exp \left(\frac{e(\phi - \phi_0)}{kT_e} \right) \right] \qquad (3.13)$$

Equation 3.13 is an example of a non-linear relationship, since the right-hand side forcing vector is a function of the unknown,

$$\nabla^2 \phi = b(\phi) \qquad (3.14)$$

In Section 3.6.4 we discuss a method for solving non-linear systems based on Newton-Rhapson linearization. However, for now we implement a "poor man's linearization" in our iterative Gauss-Seidel solver by assuming that the system is only weakly non-linear. We then evaluate the forcing vector using the solution from the previous solver iteration,

$$\nabla^2 \phi^k = b(\phi^{k-1}) \qquad (3.15)$$

As the solver approaches convergence, $|\phi^k - \phi^{k-1}| \to 0$ and hence $b(\phi^{k-1}) \to b(\phi^k)$. We thus modify the potential solver as follows,

```
bool PotentialSolver :: solve () {
    for (unsigned it=0;it<max_solver_it;it++) {
        for (int i=0;i<world.ni;i++)
            for (int j=0;j<world.nj;j++)
                for (int k=0;k<world.nk;k++) {
                    // skip over solid (fixed) nodes
                    if (world.object_id[i][j][k]>0) continue;
```

```
    if (i==0)
      phi[i][j][k] = phi[i+1][j][k];
    else if (i==world.ni-1)
      phi[i][j][k] = phi[i-1][j][k];
    else if (j==0)
      phi[i][j][k] = phi[i][j+1][k];
    else if (j==world.nj-1)
      phi[i][j][k] = phi[i][j-1][k];
    else if (k==0)
      phi[i][j][k] = phi[i][j][k+1];
    else if (k==world.nk-1)
      phi[i][j][k] = phi[i][j][k-1];
    else {  // standard internal open node
      // Boltzmann relationshp for electron density
      double ne = n0 * exp((phi[i][j][k]-phi0)/Te0);
      double phi_new =
            ((rho[i][j][k]-Const::QE*ne)/Const::EPS_0 +
             idx2*(phi[i-1][j][k] + phi[i+1][j][k]) +
             idy2*(phi[i][j-1][k]+phi[i][j+1][k]) +
             idz2*(phi[i][j][k-1]+phi[i][j][k+1])) /
             (2*idx2+2*idy2+2*idz2);

      // SOR
      phi[i][j][k]=phi[i][j][k]+1.4*(phi_new-phi[i][j][k]);
    }
  }

// check for convergence
if (it%25==0) {
  double sum = 0;
  for (int i=0;i<world.ni;i++)
    for (int j=0;j<world.nj;j++)
      for (int k=0;k<world.nk;k++) {
        // skip over solid (fixed) nodes
        if (world.object_id[i][j][k]>0) continue;

        double R = 0;
        if (i==0)
          R = phi[i][j][k] - phi[i+1][j][k];
        else if (i==world.ni-1)
          R = phi[i][j][k] - phi[i-1][j][k];
        else if (j==0)
          R = phi[i][j][k] - phi[i][j+1][k];
        else if (j==world.nj-1)
          R = phi[i][j][k] - phi[i][j-1][k];
        else if (k==0)
          R = phi[i][j][k] - phi[i][j][k+1];
        else if (k==world.nk-1)
          R = phi[i][j][k] - phi[i][j][k-1];
        else {
          // Boltzmann relationship for electrons
          double ne = n0 * exp((phi[i][j][k]-phi0)/Te0);
          R = -phi[i][j][k]*(2*idx2+2*idy2+2*idz2) +
              (rho[i][j][k]-Const::QE*ne)/Const::EPS_0 +
              idx2*(phi[i-1][j][k] + phi[i+1][j][k]) +
              idy2*(phi[i][j-1][k]+phi[i][j+1][k]) +
              idz2*(phi[i][j][k-1]+phi[i][j][k+1]);
```

```
        }
        sum += R*R;
     }
     L2 = sqrt(sum/(world.ni*world.nj*world.nk));
     if (L2<tolerance) {converged=true;break;}
  } // it%25

  if (!converged) cerr<<"GS failed to converge, L2="<<L2<<endl;
  return converged;
}
```

This version implements the internal fixed nodes per Section 3.3.1, Neumann boundaries per Section 3.3.2, and the Boltzmann relationship for the electrons per Section 3.3.3. As previously, we loop over all mesh nodes. On each, we first check for the presence of a non-zero `object_id` flag. If non-zero, we skip the node to keep the $\phi_{i,j,k}$ value set during the initialization. Otherwise, we check if the current node is on the domain boundary. Boundary nodes have at least one of the i, j, and k indexes equal to either zero or the maximum value, $(ni - 1, nj - 1, nk - 1)$. On these nodes we apply the Neumann condition. Finally, if neither of the above conditions are met, then we have a standard internal node on which the Poisson's equation $\nabla^2\phi = (\rho_i - en_e)/\epsilon_0$ holds. It is important to realize that the `rho` field object now holds only the ion contribution since electrons are no longer simulated directly. We evaluate electron density using Equation 3.12, and then include the en_e term in the Gauss-Seidel expression for $\phi_{i,j,k}$. Just as before, we compute the norm of the residue every 25 time steps by taking the modulus of the iteration index. The solver terminates once the norm reaches the prescribed tolerance.

The reference values are included in the `PotentialSolver` class found in `Solver.h`. We also include a function for setting them,

```
class PotentialSolver {
public:
  // sets reference values
  void setReferenceValues(double phi0, double Te0, double n0) {
    this->phi0 = phi0;
    this->Te0 = Te0;
    this->n0 = n0;
  }
protected:
  double phi0 = 0;    // reference plasma potential
  double n0 = 0;      // reference electron density
  double Te0 = 1;     // reference electron temperature in eV
};
```

Default values are included to avoid a compiler warning about the constructor leaving class members un-initialized. We add the call to this function to `main`,

```
int main() {
  /* ... */
  // initialize potential solver and solve initial potential
  PotentialSolver solver(world,10000,1e-4);
  solver.setReferenceValues(0,1.5,1e12);
  solver.solve();
```

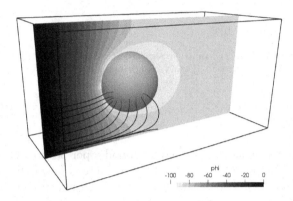

Figure 3.3: Initial potential in a system free of ions, $n_i = 0$. Electric field streamlines also shown.

```
    /* main loop */
}
```

The initial field obtained with $n_i = 0$ is shown in Figure 3.3. Electric field streamlines are also shown. In front of the sphere, the potential decreases almost linearly from $\phi = 0$ V on the inlet face to $\phi = -100$ V at the sphere. A lens configuration forms around the sphere which can be expected to focus ions into a region beyond the sphere.

3.4   PARTICLE INJECTION AND REMOVAL

Particles are injected into the simulation from the z_{min} face with constant velocity such that density at the injection plane is n_0. From mass conservation, we have

$$\frac{\partial n}{\partial t} = -\nabla \cdot (n\vec{v}) \tag{3.16}$$

Integrating both sides over the entire computational volume we have

$$\int_V \frac{\partial n}{\partial t} dV = -\int_V \nabla \cdot (n\vec{v}) dV \tag{3.17}$$

$$= -\oint_S n\vec{v} \cdot \hat{n} dA \tag{3.18}$$

where the term on the right hand side is rewritten using the Divergence Theorem. Then considering the initial state in which ions only flow into the domain, but there is no outflow, we have

$$\frac{dN}{dt} = nv_z A_{inlet} \tag{3.19}$$

This integration assumed that density and velocity are uniform along the entire integration surface for which $\hat{n} = -\hat{k}$. Here $dN/dt = \dot{N}$ is the number of real molecules entering the system per time through a surface with area A_{inlet} with normal (z direction) velocity v_z producing local density n. The number of real particles to inject per time step Δt is thus

$$N = \dot{N}\Delta t \qquad (3.20)$$

and the number of simulation macroparticles is

$$M_f = N/w_{mp} \qquad (3.21)$$

N is generally not evenly divisible by w_{mp} and M_f will be a floating point real number. We have two options. The first possibility is to first create $M = \text{int}(M_f)$ particles with weight w_{mp} and then produce one additional particle with a fractional weight $N - M w_{mp}$. Alternatively, we can take a stochastic approach and let

$$M = \text{int}(M_f + \mathcal{R}) \qquad (3.22)$$

where $\mathcal{R}$ is a uniformly distributed random number in $[0, 1)$. To prove to yourself this approach works, assume that $M_f = 10.2$. For a large number of samples, 80% of random values will be in $[0, 0.8)$ range. Since the int() cast merely truncates the floating point part, any of these values added to 10.2 lead to M being truncated to 10. For the remaining 20% cases, $\mathcal{R} \in [0.8, 1)$, which when added to 10.2, produces $M_f \geq 11$. Therefore $M = 0.8 \cdot 10 + 0.2 \cdot 11 = 10.2$.

3.4.1 Cold Beam Source

In this chapter, we implement what is known as a *cold beam source*. The name "cold beam" implies that the source injects mono-energetic particles. Temperature in the realm of gas kinetics corresponds to the random distribution of velocities about the mean, $v = \bar{v}_d + \tilde{v}_{th}$. Lower the temperature, lower the variation of velocities from particle to particle. A cold beam is the mathematical approximation of the $T \to 0$ limit, in which the thermal component $\tilde{v}_{th}$ vanishes. All particles share the same *drift velocity* $\bar{v}_d$. This injection source is implemented using a new class of type `ColdBeamSource` defined in `Source.h`.

```cpp
// simple monoenergetic source
class ColdBeamSource {
public:
    ColdBeamSource(Species &species, World &world, double v_drift,
        double den) : sp{species}, world{world}, v_drift{v_drift},
        den{den} { }
    void sample();    // generates particles
protected:
    Species &sp;      // reference to the injected species
    World &world;     // reference to world
    double v_drift;   // mean drift velocity
    double den;       // injection density
};
```

The class consists of two functions: the constructor which merely stores references to external objects of interest and a **sample** function which actually injects the particles. The source for the sample function is given below.

```
// samples monoenergetic particles
void ColdBeamSource :: sample () {
  double3 dh = world.getDh();
  double3 x0 = world.getX0();

  // area of the XY plane, A=Lx*Ly
  double Lx = dh[0]*(world.ni-1);
  double Ly = dh[1]*(world.nj-1);
  double A = Lx*Ly;

  // number of real particles to generate: N = n*v*A*dt
  double num_real = den*v_drift*A*world.getDt();

  // number of simulation particles
  int num_sim = (int)(num_real/sp.mpw0+rnd());

  // inject particles
  for (int i=0;i<num_sim;i++) {
    double3 pos {x0[0]+rnd()*Lx, x0[1]+rnd()*Ly, x0[2]};
    double3 vel {0,0,v_drift};
    sp.addParticle(pos,vel,sp.mpw0);
  }
}
```

We start by computing the area of the injection surface, which corresponds to the z_{min} domain face. The area is $A_{inlet} = L_x L_y = (ni - 1)\Delta x \times (nj - 1)\Delta y$. We next compute the number of real ions to generate in a time step Δt using Equation 3.20. The number of macroparticles is obtained using the stochastic approach per Equation 3.22. We inject this many particles at random locations on the injection plane. The position is sampled from

$$x = x_0 + \mathcal{R}_1 L_x \qquad (3.23)$$
$$y = y_0 + \mathcal{R}_2 L_y \qquad (3.24)$$
$$z = z_0 \qquad (3.25)$$

where $\mathcal{R}_1$ and $\mathcal{R}_2$ are two random numbers. Particle velocity is set to $\vec{v} = (0, 0, v_d)$.

This function utilizes a new scalar added to the Species class to store the default macroparticle weight, $w_{mp,0}$.

```
class Species {
public:
  Species (std :: string name, double mass, double charge, double
    mpw0, World &world) : name(name), mass(mass), charge(charge),
    mpw0(mpw0), den(world.ni,world.nj,world.nk), world(world) {
  }

  const double mpw0;      // default macroparticle weight
}
```

We modify the call in **main** to initialize the ion species with $w_{mp} = 10^4$. We then create a new **vector** to store the particle injection sources and populate it with one instance of **ColdBeamSource** sampling 10^{12} m^{-3} ions with 7 km/s drift velocity,

```
int main() {
    // set up particle species
    vector<Species> species;
    species.push_back(Species("O+", 16*AMU, QE, 1e4, world));

    // setup injection sources
    vector<ColdBeamSource> sources;
    sources.push_back(ColdBeamSource(species[0], world, 7000, 1e12));
}
```

3.4.2 Particle Removal

Particles leaving the computational domain need to be removed from the simulation. This is also true for particles impacting the sphere. The approach taken here is to kill these particles by setting their weight to zero, $w_{mp} = 0$. We subsequently perform a removal sweep to actually erase them from the data container. The code is implemented in **Species.cpp** as shown below

```
void Species::advance() {
    for (Particle &part: particles) {   // loop over particles
        // update position from v=dx/dt
        part.pos += part.vel*dt;

        // inside the sphere or outside the domain?
        if (world.inSphere(part.pos) || !world.inBounds(part.pos)) {
            part.mpw = 0;  // kill the particle by setting weight to zero
            continue;      // go to the next particle
        }
    }
}
```

We use the previously defined **World::inSphere** function to check if the final position of the particle is inside the sphere. This approach has a serious flaw in that it misses particles that completely cross the sphere during a single push. Given that the sphere spans many cells, this scenario is not likely for the majority of particles, as that would indicate that the time step is too large. But, even with a reasonable time step, there is still the possibility of a particle skimming the surface being missed. A more robust algorithm is discussed in Chapter 4.

Now, in order to actually remove the particles, we need a data storage object that supports not only adding, but also removing, items as needed. We now review some possibilities. The C++ **vector** is a wrapper for a contiguous memory array, **Particle *particles = new Particle[max_size]**. Here **max_size** is the actual allocated memory size, while **np** is the number of particles actually stored in the array. Whenever we add a new entry using **push_back** or **emplace_back**, the container first checks if space remains in

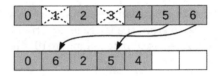

Figure 3.4: Illustration of particle removal using a vector.

the allocated block, np<max_size-1. If not, a new, larger array is allocated, and the previously stored objects are moved or copied into it using the appropriate constructors. A **vector** also supports removal of objects. However, removing any internal item [p] involves a memory copy to shift the subsequent [p+1,np-1] items. This is computationally inefficient.

Consider the particle array in Figure 3.4. Assume that the third and the fifth particles have left the computational domain. These ions leave holes in the particle array. We could simply leave these holes in place, and modify the particle push algorithm to skip over any dead particles with $w_{mp} = 0$. This is clearly inefficient from both computational and memory storage perspective. After a while, the memory becomes exhausted by an array storing predominantly dead particles. To avoid this, new particles could be added into the empty slots. This is also inefficient, as adding a particle now involves searching through a list containing possibly millions of entries for an empty slot. Therefore, the ideal approach is to simply close the holes. Since in PIC simulations the particle ordering is inconsequential, we can fill the holes with data from the end,

particles [p] = particles [np−1]

We then reduce the index of the last valid item, **--np**. In our example, this means that the particle [6] is moved to the slot previously occupied by particle [1] and [5] becomes the index of the last valid item. This data is then used to fill the hole at [3] and the final index is decreased to [4]. Unfortunately, the **vector** container does not implement functionality for directly setting the element count. Therefore, we keep track of it manually to avoid having to call the **erase** function individually for every hole that is being filled. We then perform a single call to **erase** using a range of values to trim the vector. This would involve removing the [5 : 6] elements in our example.

```
void Species :: advance () {
    // particle removal step
    size_t np = particles . size ();
    for (size_t p=0;p<np;p++) {
        if (particles [p].mpw>0) continue;   // ignore live particles
        particles [p] = particles [np−1];    // fill the hole
        np−−;      // reduce the count of valid array element
        p−−;      // decrement p so this position gets checked again
    }
    // now delete particles [np:end]
    particles . erase (particles . begin ()+np, particles . end ());
}
```

Figure 3.5: Illustration of particle removal using a linked list.

3.4.3 Alternate Storage Schemes

Instead of utilizing a vector, we could store the particles in a *linked list*. This is a collection of data blocks that are randomly scattered through memory and are linked to their neighbor through pointers. A singly-linked list only knows its next neighbor, while a doubly-linked list also links to its predecessor. The major benefit of a linked list is the ease of removal and addition of entries. In order to remove the element [1] in a [0]-[1]-[2] chain, we simply make [0]'s next neighbor be [2] and set [2]'s predecessor to [0]. We then deallocate the memory space occupied by [1]. This is visualized in Figure 3.5. We can also easily add new entries by allocating memory for a new particle, and adding it to the front of the chain. But linked lists also suffer from serious disadvantages. First, due to the scattered memory layout, data stored in a linked list is not as cache friendly as the contiguous vector storage. Secondly, a linked lists does not support direct retrieval. In order to access the p-th object, the previous $p - 1$ objects need to be traversed. This limitation is typically not important for plasma simulations as we rarely need to retrieve a particle by its array index. Implementing your own linked list is quite trivial, but the C++ standard library already includes an implementation in the form of `list`.

In order to use a linked list, we first change the container type from `vector` to `list` in `Species.h`. We also need to `#include <list>`.

```
class Species {
  std::list<Particle> particles;  // linked list for particles
};
```

The particle removal step in **advance** is then modified to call **erase** for every dead particle. Since erasing entries from a container invalidates the iterator, we split the range-based loop into two steps. We start by initializing a list iterator to point to the first element. We then continue looping until the iterator reaches the end. The particle at the current position is obtained by *dereferencing* the iterator in a fashion similar to dereferencing a pointer. When a live particle is encountered, we advance the iterator and continue on to the next particle. Otherwise, the item is deleted from the list and a new iterator is obtained. This iterator is not advanced, as it already points to the next particle that needs to be checked.

```
void Species::advance() {
  // perform a particle removal step
  std::list<Particle>::iterator it = particles.begin();
  while(it!=particles.end()) {
```

```
    Particle &part = *it;            // dereference the iterator
    if (part.mpw>0) {++it;continue;}  // ignore live particles
    it = particles.erase(it);    // delete the item and get a new
        iterator to the current position
  }
}
```

This version of the code is found in `ch3/list`. On my computer, it takes 203 seconds versus 177 seconds for the `vector` version. To further isolate the container performance, we can comment out calls to `solver.solve()` and `Output::fields`. This version takes 27 seconds versus 8 seconds for a similarly modified `vector` version. Clearly, while the linked list offers a simpler interface for removing particles, it comes at a noticeable performance hit. Switching to a linked list resulted in a particle push time increased by a factor of three.

Finally, instead of using a linked list to link individual particles, we can use it to link fixed size particle arrays. This is the approach I used in my thesis [21], where I called this data structure `PartBlock`. This data container provides contiguous memory storage for the majority of particles, while easily supporting resizing without needing to copy data. If more particle are needed, we simply create a new `PartBlock`. Each block may store somewhere between 10,000 to 100,000 particles. The downside of this container is that adding a new particle requires a traverse through a small number of blocks until one is found with an empty slot. This is generally a tiny performance hit, especially if the number of blocks is small.

3.5 STEADY STATE AND DATA AVERAGING

Believe it or not, the above sections cover all changes needed to adapt the code from Chapter 2 to simulate the flow past a sphere. The entire code is found in `ch3/ver1`. Compile and run it as usual,

```
$ ch3/ver1: g++ -O2 *.cpp -o ch3
$ ch3/ver1: ./ch3
Sphere potential: -100 V
ts: 0    0+:2800
ts: 1    0+:5600
...
ts: 399   0+:330439
ts: 400   0+:330412
Simulation took 177 seconds
```

Figure 3.6 shows ion density after several different numbers of time steps. The sphere acts as a particle sink and a wake forms behind the sphere. The region of increased ion density is due to the lensing effect seen in Figure 3.3. Particle based plasma simulations are inherently noisy due to their dependence on random numbers. This noise is clearly apparent in these plots. The last two plots at time steps 160 and 400 seem identical, but upon a closer inspection,

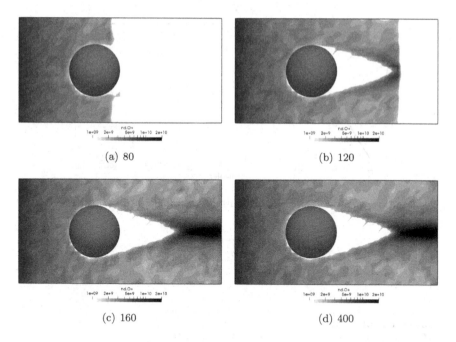

(a) 80　　　　　　　　　　　　　　　(b) 120

(c) 160　　　　　　　　　　　　　　　(d) 400

Figure 3.6: Ion density after the specified number of time steps.

we notice small differences due to noise. PIC simulations are ideally suited to cases in which a *steady state* is achieved. The results can then be averaged over a large number of time steps to reduce the numerical noise.

The first step is determining when the steady state is achieved. At steady state, the system mass, momentum, energy, and charge become time invariant. This check can be performed visually by inspecting data in the `runtime_diags.csv` file. The normalized values are plotted in Figure 3.7. We can then rerun the simulation with a flag indicating that averaging should begin around time step 150. Alternatively, this check can be performed automatically. We simply compare the normalized change in magnitude of properties of interest to some defined limit. The following function, added to `World`, sets a steady state flag to true once the time step to time step change in system mass, momentum, and energy becomes less than 0.1% of the current value.

```
bool World :: steadyState (vector<Species> &species) {
    // do not do anything if already at steady state
    if (steady_state) return true;

    double tot_mass = 0;
    double tot_mom = 0;
    double tot_en = getPE();
    for (Species &sp: species) {
        tot_mass += sp.getRealCount();   // number of real molecules
```

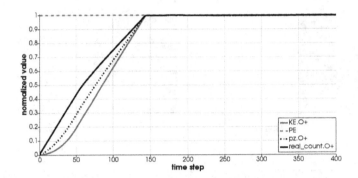

Figure 3.7: Evolution of system mass, momentum, and energy.

```
    double3 mom = sp.getMomentum();
    tot_mom += mag(mom);              // momentum magnitude
    tot_en += sp.getKE();            // kinetic energy
}

// compute relative change from prior values
const double tol = 1e−3;
if (abs((tot_mass−last_mass)/tot_mass)<tol &&
    abs((tot_mom−last_mom)/tot_mom)<tol &&
    abs((tot_en−last_en)/tot_en)<tol) {
    steady_state = true;
    cout<<"Steady state reached at time step "<<ts<<endl;
}

// update prior values
last_mass = tot_mass;
last_mom = tot_mom;
last_en = tot_en;
return steady_state;
}
```

Four new variables are added to the World class,

```
class World {
protected:
    bool steady_state = false;  // set to true once at steady state
    double last_mass = 0;       // mass at the prior time step
    double last_mom = 0;        // momentum at the prior time step
    double last_en = 0;         // energy at the prior time step
};
```

We then start averaging the density data. The approach taken here is to define a new field within the Species class to hold the average density,

```
class Species {
    Field den_ave;    // averaged number density
};
```

The Output::fields function is also modified to include the output of this field into a scalar nd-ave.(species_name). The values are set by computing the running average,

$$\bar{A}_k = \frac{A_k + (k-1)\bar{A}_{k-1}}{k} \tag{3.26}$$

Here $\bar{A}_k$ is the average of some field A after k samples. A_k is the latest instantaneous sample. This code is implemented by modifying the `Field` class,

```
class Field_ {
  // incorporates new instantaneous values into running average
  void updateAverage(const Field_ &I) {
    ++ave_samples;              // increment number of samples
    for (int i=0;i<ni;i++)      // loop over nodes
      for (int j=0;j<nj;j++)
        for (int k=0;k<nk;k++)
          data[i][j][k] = (I(i,j,k) +
            (ave_samples-1)*data[i][j][k])/ave_samples;
  }

protected:
  int ave_samples = 0;   // number of samples in the average
};
```

This algorithm is included in the `updateAverages` wrapper added to `Species`,

```
class Species {
  void updateAverages() {den_ave.updateAverage(den);}
};
```

which is in turn called from `main` once we reach the steady state as

```
// Main.cpp main loop
while(world.advanceTime()) {
  // particle push, etc...
  if (world.steadyState(species)) {
    for (Species &sp:species)
      sp.updateAverages();     // update averages at steady state
  }
}
```

This version of the code is found in the `ver2` folder. The code reports that steady state is reached at time step 145. This agrees with our visual observation in Figure 3.7. The averaged value of number density obtained by averaging instantaneous results for time steps 145 to 400 is then shown in Figure 3.8. Comparing this picture to Figure 3.6, we can clearly see the reduction in numerical noise.

3.6   POTENTIAL SOLVER REVISITED

In Section 3.4.3, we observed that particle push contributed only 8 of the 177 seconds required to complete the simulation. Including the file output, the run time increases only by one second to 9 seconds. Therefore, 168 seconds, or almost 95% of the wall time, is spent solving potential. Clearly, some improvements can be made here.

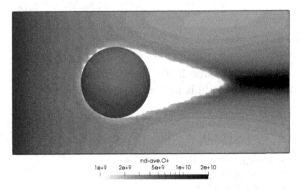

nd-ave.O+

1e+9   2e+9   5e+9   1e+10   2e+10

Figure 3.8: Ion density averaged over multiple time steps.

3.6.1   Matrix Representation

The Gauss-Seidel scheme is simple to implement, but is also slow to converge. Some faster alternatives include *multigrid*, and *alternate direction implicit* (ADI). Other matrix solving algorithms can be found in high-performance numerical libraries, such as LAPACK [1]. Full description of these methods is beyond the scope of this book, but if interested, you will find an example multigrid code on my blog [2]. A "Dynamic ADI" implementation is described in [27].

These solvers operate on a generic $\mathbf{A}\vec{x} = \vec{b}$ system, in which the coefficient matrix may be subjected to some requirements. For instance, it may need to be *banded* (as is the case with the Poisson's equation) or *positive definite*, implying that the magnitude of the term on the diagonal is greater than or equal to the sum of magnitudes of the remaining coefficients on the same row, $|a_{ii}| \geq \sum_j |a_{ij}(1 - \delta_{ij})|$. Here the first index is the row, and the second is the column. δ_{ij} is the Kronecker delta with value of 1 if $i = j$, and zero otherwise.

In order to use these advanced solvers, we need to recast our potential solver into a matrix system. Our prior Gauss-Seidel implementation operated directly on values in the three-dimensional ϕ field and applied the appropriate coefficients based on the node type. Alternatively, we can initialize the solver by building the coefficient matrix $\mathbf{A}$. With this approach, the solver becomes oblivious to the particular equation being solved.

Figure 3.9 shows the *computational stencil* for the Finite Difference discretization of the Laplacian operator, $\nabla^2 \phi$. The solution on node $\phi_{i,j,k}$ depends on values of the six neighbors, $\phi_{i-1,j,k}$, $\phi_{i+1,j,k}$, $\phi_{i,j-1,k}$, $\phi_{i,j+1,k}$, $\phi_{i,j,k-1}$, $\phi_{i,j,k+1}$. Instead of using this three-dimensional indexing, we can *flatten* the system into a one dimensional array containing $nu = ni \times nj \times nk$ unknowns. We map from the 3D representation to the 1D one using consecutive node numbering,

$$u = k \cdot ni \cdot nj + j \cdot ni + i \qquad (3.27)$$

where u is the *unknown index*. Here we use the term "unknown" loosely as this

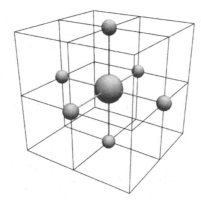

Figure 3.9: Computational stencil for the 3D Poisson equation.

indexing also includes the known Dirichlet values. With this scheme, the node at origin has unknown index 0. The neighbor in the $+x$ direction has index 1. This numbering increases until we reach the final node in the x direction, with index $u = ni - 1$. We then wrap around to $\phi_{0,1,0}$, which becomes node ni. Finally, once the entire $k = 0$ two dimensional plane is completed, we move onto node $\phi_{0,0,1}$, which becomes node $u = ni \cdot nj$. This flattening and inflating is accomplished using two new functions added to namespace vec in PotentialSolver.cpp,

```cpp
namespace vec {
    // converts a 3D field to a 1D vector
    dvector deflate(Field &f3) {
        dvector r(f3.ni*f3.nj*f3.nk);
        for (int i=0;i<f3.ni;i++)
            for (int j=0;j<f3.nj;j++)
                for (int k=0;k<f3.nk;k++)
                    r[f3.U(i,j,k)] = f3[i][j][k];
        return r;
    }

    // converts a 1D vector to a 3D field
    void inflate(dvector &d1, Field& f3) {
        for (int i=0;i<f3.ni;i++)
            for (int j=0;j<f3.nj;j++)
                for (int k=0;k<f3.nk;k++)
                    f3[i][j][k] = d1[f3.U(i,j,k)];
    }
};
```

Now let $u = U(i, j, k)$ be the index of node i, j, k. With the help of this function, we find that the stencil neighbors are separated by 1, ni, and $ni \cdot nj$ nodes in the x, y, and z direction, respectively. The discretized Poisson's equation can thus be written in a form operating on a one-dimensional vector,

$$a_x \left(\phi_{u-1} - 2\phi_u + \phi_{u+1}\right) + a_y \left(\phi_{u-n_i} - 2\phi_u + \phi_{u+n_i}\right) +$$
$$a_z \left(\phi_{u-n_i n_j} - 2\phi_u + \phi_{u+n_i n_j}\right) = -\frac{\rho_u}{\epsilon_0} \qquad (3.28)$$

Here $a_x = 1/\Delta^2 x$, $a_y = 1/\Delta^2 y$, and $a_z = 1/\Delta^2 z$. Combining the unknowns, we have

$$a_z \phi_{u-n_i n_j} + a_y \phi_{u-n_i} + a_x \phi_{u-1} - 2 \left(a_x + a_y + a_z\right) \phi_u$$
$$+ a_x \phi_{u+1} + a_y \phi_{u+n_i} + a_z \phi_{u+n_i n_j} = -\frac{\rho_u}{\epsilon_0} \qquad (3.29)$$

This is clearly a linear system that can be written in a matrix form. However, before getting started with the matrix version, it is important to realize that the size of the coefficient matrix is n_u^2. Given that PIC simulations often involve meshes with one million or more nodes, storing the *full matrix* is not only inefficient, it may actually require more memory than available on the system. This is completely unnecessary since every row has, at most, only seven non-zero entries. The coefficient matrix is thus *sparse*. It is also *banded*, since the non-zero coefficients are limited to 3 (in 1D), 5 (2D), or 7 (3D) bands centered on the main diagonal. Some storage schemes take advantage of the banded structure, but we do not do so here.

3.6.2   Sparse Matrix

Instead, we implement a simple data structure to store a general sparse matrix containing at most some pre-defined number of non-zero entries per row. We call this container **Matrix**. It is implemented as follows.

```
class Matrix {        // matrix with up to seven non zero diagonals
public:
    Matrix(int nr):nu{nr} {rows=new Row<nvals>[nr];}
    Matrix(const Matrix &o):Matrix(o.nu) {  // copy constructor
        for (int r=0;r<nu;r++) rows[r] = o.rows[r];
    };
    ~Matrix() {if (rows) delete[] rows;} // destructor
    dvector operator*(dvector &v);   // matrix-vector multiplication

    double& operator() (int r, int c); // A[r,c] in full matrix
    void clearRow(int r) {rows[r]=Row<nvals>();} // deletes row data
    Matrix diagSubtract(dvector &P);    // sub. vector from the diag
    Matrix invDiagonal();   // inverse of the diagonal in matrix form
    double multRow(int r, dvector &x); // multiply row r with vec x

    static constexpr int nvals = 7;     // maximum 7 non-zero values
    const int nu;                       // number of rows (unknowns)

protected:
    Row<nvals> *rows;                   // row data
};
```

At the heart of this container is a dynamically allocated array of Row<nvals> objects. As the name indicates, each Row stores data for a single row,

```
template <int S>       // template argument to set max row size
struct Row {            // structure to hold data for a single row
   Row() {for (int i=0;i<S;i++) {a[i]=0;col[i]=-1;}}
   void operator= (const Row &o) {
      for (int i=0;i<S;i++) {a[i] = o.a[i]; col[i]=o.col[i];}
   }
   double a[S];         // coefficients
   int col[S];          // full matrix columns, or -1 if not set
};
```

This container takes advantage of a template argument. Instead of being used to specify a data type as is the case with Field, the argument is used to pass in, at compile time, the number of non-zero entries. This allows us to allocate the storage for the coefficients and column indexes statically.

We store the data for each row as a combination of coefficients a, and the corresponding column indexes col. Negative indexes indicate unused slots. To illustrate this concept, let say that we want to set $A_{i,j} = 0.1$ in a previously empty matrix. This involves setting A.rows[i].a[0]=0.1 and A.rows[i].col[0]=j. Now, if we would like to set another entry on this row, such as $A_{i,j-1} = -2.3$, we end up with A.rows[i].a[1]=-2.3; and A.rows[i].col[1]=j-1. This access is provided by the overloaded () operator,

```
// returns reference to A[r,c] element in the full matrix
double& Matrix::operator()(int r, int c) {
   auto &row = rows[r]; int v;
   for (v=0;v<nvals;v++) {        // find this entry
      if (row.col[v]==c) break;   // if found
      if (row.col[v]<0) {row.col[v]=c; break} // set
   }
   assert(v!=nvals);  // check for overflow in debug mode
   return row.a[v];
}
```

We start by grabbing the appropriate row data. We then search through the stored column indexes for one matching the operator argument. If it is found, we return a *reference* to the data. Otherwise, if we encounter an empty slot, we initialize it to the specified column. A debug-only assertion check is included to make sure the calling algorithm does not attempt to set more values that we have space for. Returning the reference allows us to use the operator on both sides of an assignment, i.e.

```
A(i,j) = 0.1;
double s = A(i,j-1);  // valid even if A[i,j-1] not set explicitly
```

The Matrix class also overloads the * operator to support matrix-vector multiplication,

```
dvector Matrix::operator*(dvector &v) {
   dvector r(nu);
```

```
for (int u=0;u<nu;u++) {   // loop over rows
  auto &row = rows[u];
  r[u] = 0;
  for (int i=0;i<nvals;i++) {
    if (row.col[i]>=0) r[u]+=row.a[i]*v[row.col[i]];
    else break;            // end at the first -1
  }
}
return r;
}
```

We also override some additional operators for `dvector`, which is our nickname for `vector<double>`. Only the code for subtraction is listed below for the sake of brevity,

```
dvector operator-(const dvector &a, const dvector &b) {
  size_t nu = a.size();
  dvector r(nu);
  for (size_t u=0;u<nu;u++) r[u] = a[u]-b[u];
  return r;
}
```

These two overloaded operators allow us to compute the residue as

```
dvector R = A*x-b;
```

where `A` is a `Matrix` and `x` and `b` are of type `dvector`. With a bit of up-front work, C++ code can read just like the mathematical formula it is computing! The coefficient matrix is constructed during the initialization of the `Solver` object using a function called `buildMatrix`. Besides setting the finite difference coefficients, the function also initializes a `vector<NodeType> node_type` vector, where `NodeType` is defined as

```
enum NodeType {REG,NEUMANN,DIRICHLET};
```

This array lets us flag Neumann and Dirichlet boundary nodes to support a slightly different treatment of the right hand side vector per Equation 3.43. Portion of the function is below,

```
void PotentialSolver::buildMatrix() {
  node_type.reserve(nu);   // reserve space for node types

  // solve potential
  for (int k=0;k<nk;k++)
    for (int j=0;j<nj;j++)
      for (int i=0;i<ni;i++) {
        int u = world.U(i,j,k);
        A.clearRow(u);

        if (world.object_id[i][j][k]>0) {  // Dirichlet node?
          A(u,u)=1;                        // set 1 on the diagonal
          node_type[u] = DIRICHLET;        // flag as a fixed node
          continue;                        // go to the next node
        }

        node_type[u] = NEUMANN;            // set Neumann by default
        if (i==0) {A(u,u)=idx;A(u,u+1)=-idx;}
```

```
else if (i==ni−1) {A(u,u)=idx;A(u,u−1)=−idx;}
else if (j==0) {A(u,u)=idy;A(u,u+ni)=−idy;}
else if (j==nj−1) {A(u,u)=idy;A(u,u−ni)=−idy;}
else if (k==0) {A(u,u)=idz;A(u,u+ni*nj)=−idz;}
else if (k==nk−1) {A(u,u)=idz;A(u,u−ni*nj)=−idz;}
else {
  // standard internal stencil
  A(u,u−ni*nj) = idz2;              // coeff for i,j,k−1
  A(u,u−ni) = idy2;                 // coeff for i,j−1,k
  A(u,u−1) = idx2;                  // coeff for i−1,j,k
  A(u,u) = −2.0*(idx2+idy2+idz2);   // coeff for i,j,k
  A(u,u+1) = idx2;                  // coeff for i+1,j,k
  A(u,u+ni) = idy2;                 // coeff for i,j+1,k
  A(u,u+ni*nj) = idz2;              // coeff for i,j,k+1
  node_type[u] = REG;               // regular internal node
}
}
}
```

3.6.3 Preconditioned Conjugate Gradient Solver

To illustrate the benefit of utilizing a different algorithm, we now implement a matrix solver based on the *preconditioned conjugate gradient* (PCG) method. Conjugate Gradient (CG) is type of a *Krylov space* solver. These solvers mathematically observe the direction in which the error reduces most rapidly and propagate the solution along that vector. Convergence is further improved by *preconditioning*, or multiplying both sides of the system by a helper matrix. A simple example is the Jacobi preconditioner, which consists of only the diagonal entries of the original matrix,

$$P_{i,i} = A_{i,i} \qquad (3.30)$$

The advantage of this preconditioner is that its inverse can be computed trivially,

$$M_{i,i} = (\mathbf{P}^{-1})_{i,i} = 1/A_{i,i} \qquad (3.31)$$

Conjugate Gradient methods are only applicable to *positive definite* matrices. Poisson's equation for the three-dimensional Cartesian system satisfies this requirement. There are several different variants for the PCG algorithm for a general $\mathbf{A}\vec{x} = \vec{b}$ system using a preconditioner matrix $\mathbf{P}$. We follow the formulation in [23], which starts with the following initialization:

$$\vec{g} = \mathbf{A}\vec{x} - \vec{b}$$
$$\mathbf{M} = \mathbf{P}^{-1}$$
$$\vec{s} = \mathbf{M}\vec{g}$$
$$\vec{d} = -\vec{s} \qquad (3.32)$$

The solver then starts iterating, with each iteration consisting of the following steps:

$$\vec{z} = \mathbf{A}\vec{d}$$
$$\alpha = \vec{g} \cdot \vec{s}$$
$$\beta = \vec{d} \cdot \vec{z}$$
$$\vec{x} \leftarrow \vec{x} + (\alpha/\beta)\vec{d}$$
$$\vec{g} \leftarrow \vec{g} + (\alpha/\beta)\vec{z}$$
$$\vec{s} = \mathbf{M}\vec{g}$$
$$\beta = \alpha$$
$$\alpha = \vec{g} \cdot \vec{s}$$
$$\vec{d} \leftarrow (\alpha/\beta)\vec{d} - \vec{s} \tag{3.33}$$

The iterations continue until the norm of $\vec{g}$ reaches a prescribed tolerance. Thanks to our overloaded operators and a dot product command implemented in **namespace vec**, the PCG code reads quite similarly to the mathematical formulation above,

```cpp
bool PotentialSolver::solvePCGLinear(Matrix &A, dvector &x,
   dvector &b) {
  bool converged = false;
  double l2 = 0;
  Matrix M = A.invDiagonal(); // inverse of Jacobi preconditioner

  // initialization
  dvector g = A*x-b;
  dvector s = M*g;
  dvector d = -1*s;

  for (unsigned it=0;it<max_solver_it;it++) {
    dvector z = A*d;
    double alpha = vec::dot(g,s);
    double beta = vec::dot(d,z);

    x = x+(alpha/beta)*d;
    g = g+(alpha/beta)*z;
    s = M*g;

    beta = alpha;
    alpha = vec::dot(g,s);

    d = (alpha/beta)*d-s;
    l2 = vec::norm(g);
    if (l2<tolerance) {converged=true;break;}
  }

  if (!converged)  cerr<<"PCG failed to converge, norm(g) =
    "<<l2<<endl;
  return converged;
}
```

3.6.4 Newton-Raphson Linearization

The PCG solver solves the linear $\mathbf{A}\vec{x} = \vec{b}$ system. It is not easily non-linearized using the "poor man's" approach used with Gauss-Seidel. Instead, we use the Newton-Raphson algorithm to capture the non-linearity arising from the Boltzmann electron term. The NR algorithm is an extension of the popular Newton's root finding method to a system of equations. For a single equation, the root of a function $f(x) = 0$ can be found with Taylor series,

$$f(x) = f(\bar{x}) + (x - \bar{x})f'(\bar{x}) + O(2) \tag{3.34}$$

Since the left hand side vanishes when x is the root, we have

$$x^{k+1} = x^k - \frac{f(x^k)}{f'(x^k)} \tag{3.35}$$

Here x^k is the current guess, and x^{k+1} is the new estimate. This process repeats until convergence. The matrix form for

$$\mathbf{A}\vec{\phi} - b(\vec{\phi}) \equiv F(\vec{\phi}) = 0 \tag{3.36}$$

is

$$\vec{\phi}^{k+1} = \vec{\phi}^k - \mathbf{J}^{-1}F(\vec{\phi}^k) \tag{3.37}$$

where k is the iteration counter. The $\mathbf{J}$ matrix is known as the *Jacobian*, and has values consisting of derivatives of all equations against all unknowns,

$$J_{i,j} = \frac{\partial f_i}{\partial x_j} \tag{3.38}$$

Fortunately, the Jacobian is quite trivial for our problem. Each u-th term of the $\vec{b}$ vector is dependent only on the u-th value of the unknown $\vec{\phi}$. Furthermore, the Jacobian of the linear part is just the constant coefficient matrix. We thus write

$$\mathbf{J} = \mathbf{A} - \text{diag}(\vec{Q}) \tag{3.39}$$

Here diag() is an operator that diagonalizes the vector $\vec{Q}$ into a square matrix with zero values everywhere except on the diagonal. Components of the $\vec{Q}$ vector are the derivatives of the non-linear Boltzmann relationship, Equation 3.11,

$$Q_i = \frac{en_0}{\epsilon_0 kT_{e,0}} \exp\left(\frac{e(\phi_i - \phi_0)}{kT_{e,0}}\right) \tag{3.40}$$

Computing a matrix inverse is computationally expensive and is not recommended. Instead of computing $\mathbf{J}^{-1}$, we solve the linear system

$$\mathbf{J}(\vec{\phi})\vec{y} = F(\vec{\phi}) \tag{3.41}$$

The solution is then updated from

$$\vec{\phi}^{k+1} = \vec{\phi}^k - \vec{y} \tag{3.42}$$

The PCG linear solver developed in the previous section is used to solve Equation 3.41. Alternatively, we could use the linear Gauss-Seidel algorithm modified to operate on arbitrary coefficient matrix. The resulting Newton-Raphson solver is

```cpp
bool PotentialSolver::solveNRPCG() {
    const int NR_MAX_IT=20;      // maximum number of NR iterations
    const double NR_TOL = 1e-3;
    int nu = A.nu;
    Matrix J(nu);
    dvector Q(nu);
    dvector y(nu);
    dvector x = vec::deflate(world.phi);    // 3D field to 1D vector
    dvector b = vec::deflate(world.rho);

    // set RHS according to node type
    for (int u=0;u<nu;u++) {
        if (node_type[u]==NEUMANN) b[u] = 0;           // dpi_dn = 0
        else if (node_type[u]==DIRICHLET) b[u] = x[u];  // given val
        else b[u] = -b[u]/EPS_0;               // regular node, -rho/eps0
    }

    double norm;
    bool converged=false;
    for(int it=0;it<NR_MAX_IT;it++) { // non-linear solver loop
        // compute F by first subtracting the linear term
        dvector F = A*x-b;

        // subtract b(x) on regular nodes
        for (int n=0;n<nu;n++)
            if (node_type[n]==REG)            // regular nodes
                F[n] -= QE*n0*exp((x[n]-phi0)/Te0)/EPS_0;

        // Compute Q, which is the diagonal of d(bx)/dphi
        for (int n=0;n<nu;n++) {
            if (node_type[n]==REG)
                Q[n] = n0*QE/(EPS_0*Te0)*exp((x[n]-phi0)/Te0);
        }

        // Compute J = A-diag(Q)
        Matrix J = A.diagSubtract(Q);

        // solve Jy=F, using GS if PCG fails
        if (!solvePCGLinear(J,y,F))
            solveGSLinear(J,y,F);

        // clear numerical noise on (constant) Dirichlet nodes
        for (int u=0;u<nu;u++)
            if (node_type[u]==DIRICHLET) y[u]=0;

        // update x
        x = x-y;

        // test for convergence
        norm=vec::norm(y);
        if (norm<NR_TOL) { converged=true; break; }
```

```
}

if (!converged)
  cout<<"NR+PCG failed to converge, norm = "<<norm<<endl;

// convert to 3d data
vec::inflate(x,world.phi);
return converged;
}
```

We start by *deflating* the three dimensional potential and charge density field into one-dimensional vectors. This deflation represents unnecessary overhead that should be streamlined in production code by utilizing the same data packing (3D or 1D) everywhere. We use the previously described `node_type` array to set the appropriate values on the $\vec{b}$ vector,

$$b[u] = \begin{cases} -(e(n_i)[u])/\epsilon_0 & u \in \text{regular node} \\ \phi[u] & u \in \text{Dirichlet node} \\ 0 & u \in \text{Neumann node} \end{cases} \quad (3.43)$$

We then apply the Newton-Raphson algorithm. Occasionally, the PCG solver may fail to converge. If that happens, we recompute the solution using the linear Gauss-Seidel solver. Using our pre-computed matrix coefficients, this version of the solver is much shorter - and cleaner - than the version developed in Chapter 2.

```
bool PotentialSolver::solveGSLinear(Matrix &A, dvector &x, dvector
    &b) {
  double L2=0;                   // norm
  bool converged= false;

  // solver loop
  for (unsigned it=0;it<max_solver_it;it++) {
    for (int u=0;u<A.nu;u++) {
      // multiplication and sum of non-diagonal terms
      double S = A.multRow(u,x)-A(u,u)*x[u];
      double phi_new = (b[u]- S)/A(u,u);
      x[u] = x[u] + 1.*(phi_new-x[u]);   // SOR
    }

    if (it%25==0) {  // convergene test
      dvector R = A*x-b;
      L2 = vec::norm(R);
      if (L2<tolerance) {converged=true;break;}
    }
  }
  if (!converged) cerr<<"GS failed to converge, L2="<<L2<<endl;
  return converged;
}
```

This code version is found in the `ver2` subdirectory. The simulation now takes only 36 seconds to complete the 400 time steps. This is almost a five-fold improvement over the 176 seconds needed with GS. Even further improvement is possible through optimization. For instance, the NR solver computes the

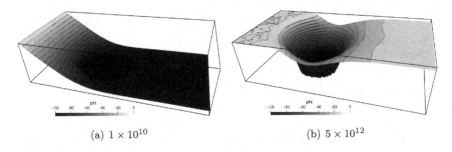

(a) 1×10^{10} (b) 5×10^{12}

Figure 3.10: Final potential at two different values of upstream plasma density.

Jacobi preconditioner at each solver iteration. This is unnecessary. This initialization could be moved to the constructor, and the result stored as a class member variable.

3.6.5 Quasi-Neutral Approximation

Yet sometimes an even greater improvement in performance is possible by changing the fundamental physical model. Figure 3.10 compares the final potential at two different values of upstream plasma density. These results are obtained by modifying the particle source parameters and also the reference value for the potential solver. We observe that while in the first case the *plasma sheath* surrounding the sphere extends all the way to the inlet, the higher density case in (b) results in sheath compression. In fact, ions injected from the inlet are not even aware of the sphere until only a short distance away from it. We can imagine that as plasma density increases even further, more of the domain becomes occupied by the neutral plasma, and the role of the sheath becomes diminishingly small.

Computing potential in the charge neutral region does not require solving the Poisson's equation. In fact, solving it is often prohibitively expensive. Poisson solvers fail to converge when the cell size is greater than the Debye length. Obtaining the second solution in Figure 3.10 required using a finer $41 \times 41 \times 81$ grid. Higher densities demand an even a greater increase in mesh resolution, until the simulation simply becomes infeasible using standard computational resources.

In deriving the Boltzmann relationship, Equation 3.11, we came across the intermediate result in Equation 3.10. Since outside the sheath the plasma is quasi-neutral, we can let $n_e = n_i$ to obtain

$$\phi = \phi_0 + \frac{k_B T_e}{e} \ln\left(\frac{n_i}{n_0}\right) \tag{3.44}$$

This equation provides a direct relationship between ion density, which is obtained from particle data, and plasma potential. Unlike the Poisson's equation which involves iterating through the unknowns multiple times, Equation 3.44

is applied to each node independently to evaluate potential using only local data. This quasi-neutral (QN) solver requires only n_u computations to set the potential. It is implemented as follows,

```
bool PotentialSolver::solveQN() {
    Field& phi = world.phi;          // references to simplify syntax
    Field& rhoi = world.rho;
    double rho0 = n0*QE;
    double rho_ratio_min = 1e-6;  // set potential floor

    // loop over all nodes
    for (int i=0;i<world.ni;i++)
        for (int j=0;j<world.nj;j++)
            for (int k=0;k<world.nk;k++) {
                if (world.object_id[i][j][k]>0) continue; // skip sphere
                double rho_ratio = rhoi[i][j][k]/rho0;
                if (rho_ratio<rho_ratio_min) rho_ratio=rho_ratio_min;
                phi[i][j][k] = phi0 + Te0*log(rho_ratio);
            }
    return true;
}
```

Instead of comparing the ratio n_i/n_0, we compare ρ_i/ρ_0 since ρ_i includes the entire positive carrier contribution in simulations containing multiple ion species. Also, since a log of zero is undefined, Equation 3.44 can be applied only when $\rho_i > 0$. In practical terms, we require a density *floor*. Here we set $\min(\rho_i/\rho_0) = 10^{-6}$. This floor also helps prevent the solver from producing excessively large negative potentials.

To simplify switching between different solvers, we add a new **enum** type to **Solver.h** to indicate which algorithm to use,

```
class enum SolverType {GS, PCG, QN};
```

The type is set in the call to the constructor in **Main.cpp**

```
PotentialSolver solver(world, SolverType::PCG,1000,1e-4);
solver.setReferenceValues(0,1.5,1e10);
solver.solve();
```

The **solve** function is modified to act a traffic cop,

```
bool PotentialSolver::solve() {
    switch(solver_type) {
        case SolverType::GS: return solveGS();
        case SolverType::PCG: return solveNRPCG();
        case SolverType::QN: return solveQN();
        default: return false;
    }
}
```

Potential obtained with the QN method for $n_0 = 5 \times 10^{12}$ m^{-3} is shown in Figure 3.11. Clearly a discrepancy exists, however, you can imagine that as the density increases, the solution from Figure 3.10 begins to approach this result. Since this method lacks the smoothing inherent in the Laplace operator, the results are noisy. Therefore, it may be a good idea to use averaged densities

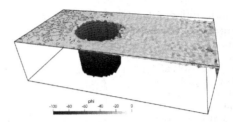

Figure 3.11: Potential for $n_0 = 5 \times 10^{12}$ m^{-3} computed using the QN method.

with the QN solver. This solver is frequently encountered in simulations of plasma thruster plumes. The densities in the core are sufficiently high to assume quasi-neutrality. We can then use the QN solver to fix the potential in the dense region, and use the Poisson solver to back fill the low density wake. More information on this "QN-switch" method can be found in [50].

3.7 SUMMARY

In this chapter we modified the code developed in Chapter 2 to simulate the flow of plasma around a charged sphere. In doing so, we learned how to develop a hybrid code, in which electrons are treated as fluid. We also learned about particle removal, steady state, and data averaging. In the second half of the chapter, we modified the code to improve performance by utilizing a Newton-Raphson backed Preconditioned Conjugate Gradient solver to obtain potential. We also learned how to compute potential in high density cases in which plasma can be assumed to be charge neutral. In the next chapter, we modify this code by improving the surface interaction physics, and by incorporating collisions.

EXERCISES

3.1 *Multiple Species.* Include additional species, such as multiply-charged ions, in the simulation. This involves adding new items into the vector<Species> and vector<Source> lists.

3.2 *Geometry.* Next modify the code to simulate a different geometry. Start by adding two new spheres centered on $(-0.1, 0, 0.2)$ and $(0.1, 0, 0.2)$. This is a possible example of a *periodic system*, if we assume this setup repeats over an infinitely long distance.

3.3 *Cylinder.* Modify the inSphere algorithm to model an infinitely long cylinder in the x direction. Instead of computing $|\vec{x} - \vec{x}_0| \leq r_{sph}$, compute $|(y, z) - (y_0, z_0)| \leq r_{cyl}$. Compare several x slices. You should observe that they are all identical, within the limits of numerical noise. This is an example of a *symmetric* problem.

3.4 *Larger Domain.* Increase the size of the computational domain to instance $(-0.2, -0.2, -0.2) - (0.2, 0.2, 0.8)$. Compare the results to the smaller version. Any discrepancy in the overlapping region is due to boundary conditions.

3.5 *Particle Trace.* At steady state, inject one or more test particles and, at every time step, output their positions and velocities to a file. You may want to use a comma-separated (`.csv`) format in which each line consists of `x,y,z,u,v,w,ts`. The resulting trace can be visualized using Paraview's `Table to Structured Grid` filter.

3.6 *Averaging.* Instead of waiting until steady state, start averaging at the beginning of the simulation. However, after every file save, reset `num_samples` to zero. With this scheme, every file contains results averaged over the prior 20 (the frequency of data output) time steps. This approach is preferred when modeling dynamic systems that do not attain a real steady-state.

CHAPTER 4

Material Interactions

4.1 INTRODUCTION

W E NOW extend the sphere code by including material interactions. We begin by letting ions neutralize on surface impact. We then incorporate Monte Carlo (MCC) and Direct Simulation Monte Carlo (DSMC) collisions to model charge exchange, momentum transfer, and ionization. We also learn about sampling the Maxwellian velocity distribution, and extend the output capabilities to include stream velocity, temperature, and particle phase plots. Finally, we briefly review steps needed to adapt the simulation to an arbitrary geometry described by a surface mesh.

4.2 MULTIPLE SPECIES

The electron - single ion species model considered so far is quite simplistic. Plasma generally consists of a soup of different chemical compounds at various ionization levels. Adding additional species into our simulation is quite trivial. We simply populate the **species** vector in **main** with new entries, and include corresponding injection sources. Let's assume that the upstream region is only 0.1% ionized, and that 20% of the positive charge density comes from doubly charged ions. We have $n_i = 10^{-3}n_a$, $n_{i1} = 0.8n_i$, and $n_{i2} = 0.1n_i$. The 80%-10% breakdown is due to the double-charged ions contributing $2e$ to the charge density. We add these new species by including additional entries in the **Species** and **Sources** vectors in **main**,

```
// particle species
vector<Species> species;
species.push_back(Species("O+", 16*AMU, QE, 1e2, world));
species.push_back(Species("O++", 16*AMU, 2*QE, 5e1, world));
species.push_back(Species("O", 16*AMU, 0, 1e5, world));

// injection sources
const double nda = 1e13;        // neutral density
const double ndi = 1e10;        // mean ion density
vector<ColdBeamSource> sources;
sources.emplace_back(species[0],world,7000,0.8*ndi); // O+
```

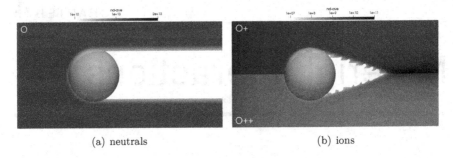

(a) neutrals   (b) ions

Figure 4.1: Initial number density of ions (left), and singly and doubly charged ions (right). Re-emission of neutrals is not yet included.

```
sources.emplace_back(species[1],world,7000,0.1*ndi);  // O++
sources.emplace_back(species[2],world,7000,nda);      // O
```

The rest of the code, including the push algorithms and the output subroutines, automatically accounts for these new species thanks to the use of the `vector` iterators. Running the program, we obtain the output shown in Figure 4.1. Clearly, while neutrals and double charged ions are included, the physics is not quite right. This is especially true for the neutrals. From our background in gas dynamics, we know that on impact molecules temporarily stick to the surface. The time spent on the surface is known as *residence time* and scales with surface temperature and material activation energy [46]. For light gases, residence time is measured in picoseconds, except for cryogenic surfaces. Therefore, we can safely assume that impacting neutrals bounce off immediately. This is clearly not captured in Figure 4.1. The sphere acts as a sink for neutrals.

4.3 SURFACE INTERACTIONS

Therefore, instead of removing neutrals, we need to re-inject them back into the computational domain. The logic is similar to the handler for a reflecting boundary in Chapter 2. We first determine the impact location and the fraction of Δt that was spent traveling to it. We then push the particle to the surface, and update its velocity to point away from the sphere. We complete the push by advancing the particle through the remaining Δt.

4.3.1 Rebound Algorithm

In a single push, a particle moves from point $\vec{x}_1$ to $\vec{x}_2$ as shown in Figure 4.2. Any point along the way, $\vec{x}$, can be described in terms of a parametric parameter t,

$$\vec{x}(t) = \vec{x}_1 + t(\vec{x}_2 - \vec{x}_1) \qquad t \in [0, 1] \tag{4.1}$$

Since $\vec{x}_2 = \vec{x}_1 + \vec{v}\Delta t$, this equation is identical to $\vec{x} = \vec{x}_1 + t(\vec{v}\Delta t)$. For $t = 0$, $\vec{x} = \vec{x}_1$, and for $t = 1$, $\vec{x} = \vec{x}_2$. Particles are not created inside the sphere, and thus we can be certain that the initial position is outside, $|\vec{x}_1 - \vec{x}_{0,sp}| > r_{sp}$. Here $\vec{x}_{0,sp}$ and r_{sp} are the sphere center and radius. If the final position is inside the sphere, then there must be a point $\vec{x}_p$ for which the radial distance equals r_{sp}. This is the location of the surface impact. Substituting the parametric equation 4.1 for $\vec{x}_p$, we have

$$|\vec{x}_1 + t_p(\vec{x}_2 - \vec{x}_1) - \vec{x}_0| = r_{sp} \tag{4.2}$$

We can solve this equation for t_p, which is the fractional distance of the particle trajectory up to the impact point. With a bit of algebra, we obtain a quadratic relationship $at^2 + bt + c = 0$, with

$$a = (\vec{x}_2 - \vec{x}_1)^2 \tag{4.3}$$
$$b = 2(\vec{x}_1 - \vec{x}_0) \cdot (\vec{x}_2 - \vec{x}_1) \tag{4.4}$$
$$c = (\vec{x}_1 - \vec{x}_0)^2 - r_{sp}^2 \tag{4.5}$$

Roots of a quadratic equation are given by,

$$t_p = \frac{-b \pm \sqrt{b^2 - 4ac}}{2a} \tag{4.6}$$

This algorithm is implemented in a new function added to `World.cpp`,

```
double World::lineSphereIntersect(const double3 &x1, const double3
    &x2) {
    double3 B = x2-x1;
    double3 A = x1-sphere_x0;
    double a = dot(B,B);
    double b = 2*dot(A,B);
    double c = dot(A,A)-sphere_rad2;
    double det = b*b-4*a*c;
    if (det<0) return 0.5;

    double tp = (-b + sqrt(det))/(2*a);
    if (tp<0 || tp>1.0)     {          // if invalid range
      tp = (-b - sqrt(det))/(2*a);  // try negative sign
      if (tp<0 || tp>1.0) {           // if still invalid
        cerr<<"Failed to find a line-sphere intersection!"<<endl;
        tp=0.5;                       // set as midpoint
      }
    }
    return tp;
}
```

This function first evaluates the quadratic equation with the positive sign for the root term. If the result is not in the valid range $t_p \in [0, 1]$, we try again with the negative sign. If this fails as well, we print an error message, and set $t_p = 0.5$. This check is included for testing purposes and is not actually triggered during the simulation. The calling function uses t_p to evaluate the

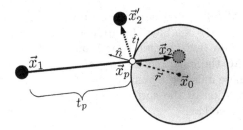

Figure 4.2: Particle intersection with a sphere.

impact location $\vec{x}_p = \vec{x}_1 + t_p(\vec{x}_2 - \vec{x}_1)$. Here we utilize a `dot` function added to `vec3` in `Field.h`. For completeness, we also add functions to evaluate vector magnitude, a unit vector, and a vector cross product. The `friend` keyword creates a function that is not associated with any particular instance of `vec3` but operates on this type of data.

```
template <typename T>
struct vec3 {
    // dot product of two 3-component vectors
    friend T dot(const vec3<T> &v1, const vec3<T> &v2) {
        T s=0;
        for (int i=0;i<3;i++) s+=v1(i)*v2(i);
        return s;
    }

    // vector magnitude
    friend T mag(const vec3<T> &v) {return sqrt(dot(v,v));}

    // unit vector, undefined for <0,0,0>
    friend vec3<T> unit(const vec3<T> &v) {return vec3(v)/mag(v);}

    // cross product
    friend vec3<T> cross(const vec3<T> &a, const vec3<T> &b) {
        return {a(1)*b(2)-a(2)*b(1), a(2)*b(0)-a(0)*b(2),
        a(0)*b(1)-a(1)*b(0)};
    }
};
```

The t_p parameter corresponds to the fraction of Δt that was spent reaching the sphere. After moving the particle position to the surface impact location, we still need to push it for the remaining $(1 - t_p)\Delta t$ to catch it up with the rest of the simulation. For generality, it is possible for the simulation to contain multiple objects or complex geometries with small crevices, leading to the particle undergoing multiple bounces per time step. Therefore, we wrap the push algorithm with a `while` loop that continues until the particle moves through the entire Δt. A local variable t_{rem} corresponds to the remaining fraction of Δt that the particle still needs to move through. After each push, we decrement this variable by the normalized distance the particle was actually able to traverse. The loop continues as long as $t_{rem} > 0$ (or some small epsilon) and the particle is alive, $w_{mp} > 0$. It is a good idea to include a secondary exit in the form of a maximum number of bounces as particles can sometimes become "stuck" due to numerical issues when using discretized surfaces (Section 4.10). We update the particle push from

```
part.pos += part.vel*dt;
if (world.inSphere(part.pos) || !world.inBounds(part.pos)) {
    part.mpw = 0;
}
```

to

```
double t_rem = 1;      // push through the entire time step
int n_bounces = 0;     // surface hit counter

// iterate while time remains and the particle is alive
while (t_rem>0 && part.mpw>0) {
    // kill stuck particles
    if (++n_bounces>20) {cerr<<"Stuck particle!"<<endl;part.mpw=0;}

    double3 pos_old = part.pos;
    part.pos += part.vel*t_rem*dt;

    // did this particle leave the domain?
    if (!world.inBounds(part.pos))    {
        part.mpw = 0;      // kill the particle
    }
    else if (world.inSphere(part.pos)) {
        double tp = world.lineSphereIntersect(pos_old,part.pos);
        part.pos = pos_old+tp*(part.pos-pos_old);
        part.vel = sampleReflectedVelocity(part.pos,mag(part.vel));
        t_rem *= (1-tp);  // update remaining time
        continue;          // skip the rest of the loop
    }

    // this particle finished the whole step
    t_rem = 0;
}
```

4.3.2 Diffuse Reflection

The above particle push code includes a call to `sampleReflectedVelocity`. This function implements the *diffuse reflection* model to produce a random post-impact velocity vector. In a diffuse reflection, the particle forgets its pre-impact direction, and bounces off in a random direction with the angle from the surface normal following the *cosine law*. Diffuse reflection also generally implies *thermal accommodation*, in which the incident energy is lost, and the particle re-emits with velocity corresponding to the surface temperature. We allow for partial accommodation by including a *coefficient of thermal accommodation* α_{th},

$$v_2 = v_1 + \alpha_{th}(v_s - v_1) \tag{4.7}$$

Here v_1 is the particle speed before the surface impact, v_s is a random velocity corresponding to the surface temperature. In the limit of $\alpha_{th} = 1$, the entire pre-impact kinetic energy of the particle is absorbed by the surface.

We use a model of Bird [12] to sample a random vector that follows the cosine law. Any probability function can be sampled by evaluating the normalized *cumulative distribution function* for a random number picked from the uniform distribution. For $P = \cos\theta$, we have

$$\mathcal{R}_1 = \int \cos\theta \equiv \sin\theta \qquad (4.8)$$

where $\mathcal{R}_1$ is a random number. There is no need to evaluate θ, since only the sine and cosine are needed. $\cos\theta$ is obtained from the identity,

$$\cos\theta = \sqrt{1 - \sin^2\theta} \qquad (4.9)$$

The rotation about the normal is obtained by sampling uniform distribution in $[0, 2\pi]$,

$$\psi = 2\pi\mathcal{R}_2 \qquad (4.10)$$

The reflected velocity vector is then built by combining contribution along the surface normal vector $\hat{n}$ and two surface tangents, $\hat{t}_1$ and $\hat{t}_2$.

$$\vec{v} = \cos\theta\hat{n} + \sin\theta\cos\psi\hat{t}_1 + \sin\theta\sin\psi\hat{t}_2 \qquad (4.11)$$

These direction vectors are evaluated at the surface impact location. For the analytical sphere centered at $\vec{x}_0$, the normal at point $\vec{x}_p$ is the unit vector

$$\hat{n} = \frac{\vec{x}_p - \vec{x}_{0,sp}}{|\vec{x}_p - \vec{x}_{0,sp}|} \equiv \text{unit}(\vec{x}_p - \vec{x}_{0,sp}) \qquad (4.12)$$

The orientation of the tangents is not important; we just need them to be perpendicular to each other and to the surface normal. Since the cross-product of any two vectors produces a mutually perpendicular vector, we can use one of the coordinate axes to construct the first tangent,

$$\hat{t}_1 = \hat{n} \times \hat{i} \qquad (4.13)$$

This algorithm fails if $\hat{n} \parallel \hat{i}$, in which case we use $\hat{j}$. Once we have the first tangent, obtaining the second one is trivial

$$\hat{t}_2 = \hat{t}_1 \times \hat{n} \qquad (4.14)$$

The resulting code is implemented as a member function of `World`,

```
double3 World::sphereDiffuseVector(const double3 &x) {
    // pick angles, theta = off normal, psi = azimuthal rotation
    double sin_theta = rnd();
    double cos_theta = sqrt(1-sin_theta*sin_theta);
    double psi = 2*Const::PI*rnd();

    double3 n = unit(x-sphere_x0);   // normal vector
    double3 t1; // create the first tangent
    if (dot(n,{1,0,0})!=0) t1 = cross(n,{1,0,0});
    else t1 = cross(n,{0,1,0});
    double3 t2 = cross(n,t1); // second tangent

    return sin_theta*cos(psi)*t1+sin_theta*sin(psi)*t2+cos_theta*n;
}
```

4.3.3 Reflection Velocity

The above algorithm samples only the direction of the emitted molecule. This direction needs to be scaled by the appropriate reflected speed v_2. We let $v_s = v_s(T)$ be some function that samples the speed distribution of emitted molecules given surface temperature T. The appropriate model may be problem specific. For instance, in the case of thermionic electron emission, we need to consider the surface material work function. The inherent potential well prevents low energy electrons from emerging, resulting in a distribution that is cut off and also shifted.

Here we assume that emitted particles follow the Maxwellian speed distribution function

$$\hat{g}(v) = 4\pi(m/2\pi k_B T)^{3/2}v^2 \exp(-v^2/v_{th}^2) \tag{4.15}$$

This function is simply the scalar magnitude of a vector with components following the one-dimensional distribution function at isotropic temperature

$$\hat{f}_m = (m/2\pi k_B T)^{1/2} \exp(-v^2/v_{th}^2) \tag{4.16}$$

The *thermal velocity* term v_{th} is given by $v_{th} = \sqrt{2k_B T/m}$. The approach used above to sample from the cosine distribution is not readily applied here as Equation 4.15 is not easily inverted to yield v. Birdsall [15] provides a simple algorithm to sample from a distribution given by

$$f(v) \sim \exp\left(-\frac{v^2}{2(v_{th}^2)_B}\right) \tag{4.17}$$

as

$$v_M = (v_{th})_B \left(\sum_{i=1}^{M} \mathcal{R}_i - \frac{M}{2}\right)\left(\frac{M}{12}\right)^{-1/2} \tag{4.18}$$

Comparing Equations 4.16 and 4.17, we observe that $(v_{th})_B = \sqrt{1/2}v_{th}$. In Equation 4.18, $\mathcal{R}_i$ is an i-th random number. A common simplification is to use only three values so that each velocity component is

$$v_M = v_{th}(\mathcal{R}_1 + \mathcal{R}_2 + \mathcal{R}_3 - 1.5) \tag{4.19}$$

This model is implemented in the **sampleVth** function added to Species,

```
double Species::sampleVth(double T) {
    double v_th = sqrt(2*Const::K*T/mass);   // thermal velocity
    double v1 = v_th*(rnd()+rnd()+rnd()-1.5); // x component
    double v2 = v_th*(rnd()+rnd()+rnd()-1.5); // y component
    double v3 = v_th*(rnd()+rnd()+rnd()-1.5); // z component
    return sqrt(v1*v1+v2*v2+v3*v3);           // magnitude
}
```

Putting everything together, we have

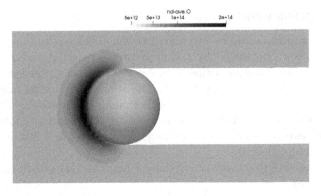

Figure 4.3: Neutral density with diffuse reflection implemented.

```
double3 Species :: sampleReflectedVelocity (double3 pos , double
    v_mag1) {
    double v_th = sampleVth(300);  // assume T_sphere = 300K
    const double a_th = 1;          // thermal accommodation coeff
    double v_mag2 = v_mag1 + a_th*(v_th−v_mag1);
    return v_mag2*world.sphereDiffuseVector(pos);  // new velocity
}
```

With these changes, we obtain the more physically-sound result shown in Figure 4.3. A region of increased neutral density is observed in front of the sphere due to reflection of impinging molecules. Note that we have not yet implemented any inter-molecular interactions and hence this result models the free molecular flow condition.

4.3.4 Species Change

Right now, the reflected particles retain their original species. While this is acceptable for neutrals, impacting ions recombine on the surface, and should be re-emitted as neutral atoms. Numerically, implementing a species change involves destroying the source particle and injecting some number of new particles of the new species. The difficulty arises when the two species have different macroparticle weights. Since our code stores macroparticle weights individually for each particle, we could eliminate this mass balance issue by always creating a single atom particle for each incident ion with $w_{mp,atom} = w_{mp,ion}$. This approach is not recommended. First, it can lead to an excessive number of particles and hence slow run times. Neutrals move much slower than ions. After a while, the simulation domain becomes saturated by neutrals that linger around the sphere. Second, the Direct Simulation Monte Carlo (DSMC) scheme introduced later in this chapter is applicable, at least in its native form, only to collisions between particles of equal weights. Algorithms have been devised to support variable weight, but they typically still require a constant weight per species [18]. Therefore, we maintain mass balance using a stochastic approach. We start by computing the fractional number of particles to re-emit as

$$M_f = w_{mp,p}/w_{mp,s} \tag{4.20}$$

where $w_{mp,p}$ is the weight of the product (post impact) species, and $w_{mp,s}$ corresponds to the source (incident) particle. This value may be greater or small than one. The integer number of particles to generate is obtained per

$$M = \text{int}\,(M_f + \mathcal{R}) \tag{4.21}$$

where $\mathcal{R}$ is a random number.

4.3.5 Sputtering and Surface Emission

Additional physics can also be included to model secondary electron emission or sputtering by ion bombardment. The ability to capture these non-linear processes directly for each impinging particle is one of the great strengths of the kinetic method. A detailed discussion of these processes is out of scope of this book, but just to illustrate the process, let's assume that the sphere is made of a hypothetical "spherium" material. Let's also assume that any incident ion with speed greater than 5 km/s sputters, on average, 0.1 native spherium atoms. Our hypothetical yield is thus given by

$$\gamma = \begin{cases} 0.1 & v_1 > 5000 \\ 0 & \text{otherwise} \end{cases} \tag{4.22}$$

The actual number of generated atoms is $\gamma w_{mp,s}$, where $w_{mp,s}$ is the macroparticle weight of the incident source species.

The resulting code that takes into account both neutralization of ions and surface sputtering is given below. Since this algorithm requires access to the neutral and "spherium" species, we pass these as arguments to the **advance** function. For any particle terminating inside the sphere, we first check the particle charge. Neutrals are characterized by $q = 0$ and are simply reflected back diffusely. Otherwise, for charged particles we compute the number of neutrals to generate. The source particle is then killed, and the corresponding number of new particles is added to the provided **neutrals** population. This hard-coded approach should be generalized to use a pre-defined species-species interactions table. The code as written here produces incorrect results when non oxygen ions turn into oxygen atoms. We also check for sputtering. We again use the stochastic approach to compute the number of spherium macroparticles and then inject them. A very important feature to notice is that a new injection velocity is sampled for each injected particle. Without this step, lowering the macroparticle weight for the new species would not lead to a decrease in numerical noise, since all particles would follow the same trajectory,

```
void Species :: advance(Species &neutrals , Species &spherium) {
/* ... */
```

```
else if (world.inSphere(part.pos)) {
  double tp = world.lineSphereIntersect(pos_old, part.pos);
  part.pos =     pos_old+tp*(part.pos-pos_old);
  double v_mag1 = mag(part.vel);    // pre-impact speed
  if (charge==0)   // neutrals
    part.vel = sampleReflectedVelocity(part.pos,v_mag1);
  else {   // ions
    double mpw_ratio = this->mpw0/neutrals.mpw0;
    part.mpw = 0;   // kill source particle

    // inject neutrals
    int mp_create = (int)(mpw_ratio+rnd());
    for (int i=0;i<mp_create;i++) {
      double3 vel = sampleReflectedVelocity(part.pos,v_mag1);
      neutrals.addParticle(part.pos,vel);
    }

    // emit sputtered material using a simple yield model
    double sput_yield = (v_mag1>5000)?0.1:0;
    double sput_mpw_ratio  = sput_yield*this->mpw0/spherium.mpw0;
    int sput_mp_create = (int)(sput_mpw_ratio+rnd());
    for (int i=0;i<sput_mp_create;i++) {
      double3 vel = sampleReflectedVelocity(part.pos,v_mag1);
      spherium.addParticle(part.pos,vel);
    }
  }
}
```

The new materials are added to **main**. This is also where we set the appropriate references to utilize in the call to **Species::advance**,

```
vector<Species> species;
species.push_back(Species("O+", 16*AMU, QE, 1e2, world));
species.push_back(Species("O++", 16*AMU, 2*QE, 5e1, world));
species.push_back(Species("O", 16*AMU, 0, 1e5, world));
species.push_back(Species("Sph", 100*AMU, 0, 2e2, world));
species.push_back(Species("Sph+", 100*AMU, 1, 2e2, world));
Species &neutrals = species[2];   // create named references
Species &spherium = species[3];
```

We do not include any injection sources for the new materials. Yet, running the simulation, we confirm that "spherium" is indeed generated once ions start reaching the sphere. At this point we do not have any spherium ions. These are added in Section 4.6.3 by incorporating the ionization reaction.

ts:	141	O+:318080	O++:79520	O:397600	Sph:0	Sph+:0
ts:	142	O+:320320	O++:80080	O:400400	Sph:0	Sph+:0
ts:	143	O+:322544	O++:80634	O:403200	Sph:1	Sph+:0
ts:	144	O+:324758	O++:81187	O:406000	Sph:3	Sph+:0
ts:	145	O+:326956	O++:81736	O:408800	Sph:6	Sph+:0

4.4   WARM BEAM

All real molecules have some finite temperature, which manifests itself as a random thermal velocity component. Now that we have a function to sample speed from the Maxwellian distribution, we define a new source of type WarmBeamSource,

```
class WarmBeamSource: public Source {
public:
    WarmBeamSource(Species &species, World &world, double v_drift,
        double den, double T) : sp{species}, world{world},
        v_drift{v_drift}, den{den}, T{T} {}

    void sample();        // generates particles

protected:
    Species &sp;          // reference to the injected species
    World &world;         // reference to world
    double v_drift;       // mean drift velocity
    double den;           // injection density
    double T;             // temperature
};
```

This code is identical to the one for cold beam, but instead of setting the velocity of all particles to $\vec{v} = (0, 0, v_d)$, we start by sampling a random value from the speed distribution function. We then create a random isotropic velocity direction vector $\vec{d}$. It is important to realize that it would be incorrect to sample $\vec{d} = \text{unit}(-1+2\mathcal{R}_1, -1+2\mathcal{R}_2, -1+2\mathcal{R}_3)$ as this approach effectively samples a random point on the surface of a $[-1:1] \times [-1:1] \times [-1:1]$ cube. Instead, we want to pick a random point from a uniform distribution on a $r = 1$ sphere. The algorithm to do this is [4]

$$\theta = 2\mathcal{R}_1 \tag{4.23}$$

$$d_0 = -1 + 2\mathcal{R}_2 \tag{4.24}$$

$$d_1 = \cos\theta\sqrt{1 - d_0^2} \tag{4.25}$$

$$d_2 = \sin\theta\sqrt{1 - d_0^2} \tag{4.26}$$

It is implemented as follows,

```
void WarmBeamSource::sample() {
    /* ... */
    // inject particles
    for (int i=0;i<num_sim;i++) {
        // sample random position and speed
        double3 pos {x0[0]+rnd()*Lx, x0[1]+rnd()*Ly, x0[2]};
        double v_th = sp.sampleVth(T);

        // sample random isotropic direction
        double theta = 2*Const::PI*rnd();
        double r = -1.0+2*rnd();  // random direction for d[0]
        double a = sqrt(1-r*r);   // scaling for unity magnitude

        double3 d;
        d[0] = r;
        d[1] = cos(theta)*a;
        d[2] = sin(theta)*a;
```

```
        double v_th = sp.sampleVth(T);
        double3 vel = v_th *d;
        vel[2] += v_drift; // add drift component
        sp.addParticle(pos, vel);
    }
}
```

For $T = 0$, this source reduces to the previously defined `ColdBeamSource`. That source could thus be eliminated and simply replaced by this new class. However, retaining both versions allows us to demonstrate *class inheritance*. The ability of classes to inherit features from other classes is a great strength of *objected-oriented programming*. Specifically, we can use this functionality to introduce a *base class* to acts as an interface with access to the general functionality implemented by each derived class. Specifically, we declare a base class of type `Source` with the following signature,

```
class Source {
public:
    virtual void sample() = 0; // pure virtual function
    virtual ~Source() {};      // to allow destruction through base
};
```

We then let both source classes inherit from it using

```
class WarmBeamSource: public Source { ... };
class ColdBeamSource: public Source { ... };
```

The `public` keyword controls access right to the parent class members.

The `Source` class contains a *virtual function* called `sample`. Virtual functions allow us to access an implementation in a derived class through a reference (or a pointer) to the base class. These functions can be declared as *pure virtual* using the `= 0` syntax. The presence of pure functions makes `Source` an abstract class. We cannot instantiate an object of this type, but we can use it as a generic pointer or a reference to any concrete inherited implementation. For instance,

```
Source *source1 = new ColdBeamSource(...);
WarmBeamSource wb(...);
Source &source2 = wb;      // reference to a WarmBeamSource object
source1->sample();         // calls ColdBeamSource::sample
source2.sample();          // calls WarmBeamSource::sample
delete source1;            // calls ~ColdBeamSource
// ~WarmBeamSource called when wb goes out of scope
```

In order to support deletion of objects through the base class, it is important that the base class destructor is also `virtual`. The abstraction provided by the `Source` interface then allows us to generalize the particle sampling code as

```
vector<Source *> sources;
for (Source *source: sources)
    source->sample();
```

The injection algorithm is now oblivious to the type of injection models in use. It simply requires that any particle source is derived from the base `Source` class and that it provides a concrete implementation of the `sample` function.

C++ requires that we use references or pointers when accessing classes through their base type. If `Source` were not an abstract class, the following code

```
WarmBeamSource wb(...);
Source source2 = wb;       // object slicing
source2.sample();          // calls Source::sample, incorrect
```

would compile but would not produce the desired behavior. Unlike in Java, where the above code would actually work due to objects being treated as references by default, in C++, the code on line 2 performs a copy of the `wb` object into `source2`. Since the two variables are of different types, only the members found in the base class are copied. This behavior is known as *object slicing*. It is also for this reason that we cannot declare a standard library vector to serve as a container for various derived types stored by value

```
vector<Source> sources;    // incorrect
```

It is also not possible to allocate an array of references. Therefore, we are left having to store pointers,

```
vector<Source*> sources;   // one option
```

In modern C++, it is not recommended to use *naked pointers* [53]. We can instead wrap them in a data type called `unique_ptr` to automatically delete the dynamically allocated data when the wrapper goes out of scope. We thus modify the `sources` storage container to store `unique_ptr<Source>` objects,

```
const double nda = 1e13;        // neutral density
const double ndi = 1e10;        // mean ion density
vector<unique_ptr<Source>> sources;
// add sources for neutrals, and singly and doubly charged ions
sources.emplace_back(new WarmBeamSource(neutrals,world,7000,nda,
    1000));
sources.emplace_back(new ColdBeamSource(ions,world,7000,0.8*ndi));
sources.emplace_back(new ColdBeamSource(ions2,world,7000,0.1*ndi));
```

The `unique_ptr` object overrides the `*` and `->` operators to provide direct access to the stored object.

4.4.1 Velocity Sampling

The velocity sampling scheme from Equation 4.18 is simple to implement but is also quite limited. The fastest one-dimensional velocity that can be sampled corresponds to all $\mathcal{R} = 1$, which for the case with $M = 3$ yields $v_{max} = 1.5v_{th}$. The fastest speed is $\sqrt{(1.5^2 + 1.5^2 + 1.5^2)v_{th}^2} \approx 2.27v_{th}$. I put together a short program to test the algorithm. It is found in `vdf/maxw_birdsall.cpp`. The code picks 10 million random velocities. The velocities are then binned into a histogram spanning $\pm 6v_{th}$ and containing 2000 bins. We also compute the time needed per sample by dividing the total run time of the sampling loop by the number of samples.

```cpp
int main(int num_args, char* args[]) {
  const int NUMS = 10000000;     // number of samples
  const int NUM_BINS = 2000;     // number of uniques we would like
  int M = 3;                     // parameter for Birdsall's method

  if (num_args>1) M = atoi(args[1]);
  if (M<1) M=1; else if (M>12) M=12;
  cout<<"Running with M = "<<M<<endl;

  double vth = 1e5;                  // thermal velocity

  dvector bins(NUM_BINS);

  double v_min = -6*vth;
  double v_max = 6*vth;
  double dv = (v_max-v_min)/(NUM_BINS);

  chrono::time_point<chrono::high_resolution_clock> time_start =
    chrono::high_resolution_clock::now();   // save time at start

  // sample the Maxwellian VDF using Birdsall's method
  for (int s=0;s<NUMS;s++) {
    double Rsum = 0;
    for (int i=0;i<M;i++)  Rsum += rnd();

    double v = sqrt(0.5)*vth*(Rsum-M/2.0)/sqrt(M/12.0);

    // bin result, add to nearest
    int bin = (int)((v-v_min)/dv+0.5);
    bins[bin]++;
  }

  auto time_now = chrono::high_resolution_clock::now();
  chrono::duration<double, std::nano> time_delta =
    time_now-time_start;
  cout<<"Time per sample: "<<time_delta.count()/NUMS<<" ns"<<endl;

  // normalize bins by the max value
  double max_val = 0;
  for (int i=0;i<NUM_BINS;i++)
    if (bins[i]>max_val) max_val=bins[i];

  for (int i=0;i<NUM_BINS;i++) bins[i]/=max_val;

  // write to a file
  ofstream out("maxw-birdsall"+to_string(M)+".csv");
  out<<"vel/vth,f_num,f_th\n";
  for (int i=0;i<NUM_BINS;i++) {
    double v = v_min+i*dv;
    double f_th = exp(-v*v/(vth*vth));
    out<<v/vth<<","<<bins[i]<<","<<f_th<<"\n";
  }
  return 0;
}
```

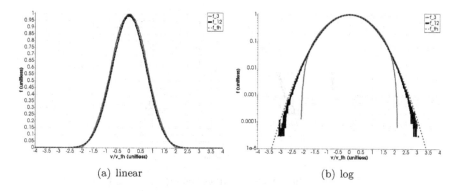

(a) linear   (b) log

Figure 4.4: Maxwellian VDF sampled using Birdsall's method for $M = 3$ and 12 plotted on a linear and log scale.

The code runs with $M = 3$ by default, but this value can be overridden through the command line argument. On my system, I obtain

```
$ ./maxw-birdsall
Running with M = 3
Time per sample: 47.377 ns
$ ./maxw-birdsall 12
Running with M = 12
Time per sample: 153.724 ns
```

Using 12 random numbers increased the sampling time by a factor of 3.2×. We can next visualize the results. Figure 4.4 compares the sampled velocity distribution function for $M = 3$ and $M = 12$ as solid gray and black lines, respectively. While the view on linear scale may look acceptable even for $M = 3$, the semilog plot in 4.4 clearly indicates the artificial cut off of high velocity particles. With $M = 12$, the velocity range is increased, but we are still limited to approximately $\pm 3 v_{th}$.

Clearly, some alternatives are needed. One such scheme is given in Appendix A of [19]. The approach I prefer to use is to sample a random velocity from a uniform distribution in $\tilde{v} \in [v_{min}, v_{max}]$ and then use it to evaluate the distribution function. We continue this sampling until $f(\tilde{v}) \geq \mathcal{R}$, where $\mathcal{R}$ is a random number. The distribution needs to be normalized such that $\max(f) \leq 1$. This approach is analogous to inverting the cumulative distribution function, but is applicable to cases where $f(v)$ cannot be easily solved for v. The example code is given in `maxw-fun.cpp`. The relevant change is listed below,

```
for (int s=0;s<NUMS; s++){
    double v;
    // sample by evaluating distribution function at uniformly
        sampled velocity
    while(true) {
```

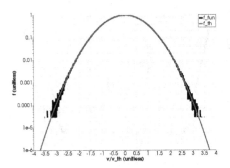

Figure 4.5: Maxwellian VDF sampled using function evaluation.

```
    // pick random velocity between bin_min and bin_max
    v = v_min + rnd()*(v_max−v_min);

    // compare against distribution function, max(fm)<=1
    double fm = exp(−v*v/(vth*vth));
    if (fm>rnd()) break;
  }

  // bin result, add to nearest
  int bin = (int)((v−v_min)/dv+0.5);
  bins[bin]++;
}
```

Running it, we obtain

```
$ ./maxw-fun
Time per sample: 322.038 ns
```

This method is over $2\times$ slower than the $M = 12$ approach of Birdsall, but as shown in Figure 4.5, the agreement with the theoretical model is greatly improved in the high velocity region.

Yet, despite using 10 million particles, neither approach was able to sample velocities $v > 3v_{th}$. Let's assume that we would like to generate particles up to $4v_{th}$. Here the normalized distribution function evaluates to $O(-7)$. Therefore, for each high energy particle, we need to sample approximately 10^7 low energy particles with $v < 4v_{th}$. This approach is not practical for simulations in which we need resolve the high velocity tail with a statistically sufficient number of particles. This is where the *variable weight* comes in. Instead of using the probability function to determine which velocities to sample, we instead divide the entire velocity space into a number of bins and in each, generate some prescribed number of particles with w_{mp} selected to reproduce the distribution function. This example is given in maxw_mpw.cpp. The relevant part is shown below,

```
const int NUM_BINS = 41;      // number of uniques we would like
double PARTS_PER_BIN = 10;    // number of particles to load per bin
double MPW0 = 1;              // nominal particle weight
```

```
double vth = 1e5;              // thermal velocity
const double pi = acos(-1.0);

dvector bins(NUM_BINS);
double v_min = -6*vth;
double v_max = 6*vth;
double dv = (v_max-v_min)/(NUM_BINS);

chrono::time_point<chrono::high_resolution_clock> time_start =
  chrono::high_resolution_clock::now();   // save the starting time

double a = 1/(sqrt(pi)*vth);

// loop over bins
for (int i=0;i<NUM_BINS;i++) {
  // evaluate normalized distribution function
  double v = v_min + (i)*dv;
  double fm = a*exp(-v*v/(vth*vth));
  // particle weight
  double mpw = MPW0*fm*dv/PARTS_PER_BIN;

  for (int p=0;p<PARTS_PER_BIN;p++) {
    // pick random velocity in this bin
    v = v_min+i*dv + rnd()*dv;

    // bin result, add to nearest
    int bin = (int)((v-v_min)/dv);
    bins[bin] += mpw;
  }
}
```

This code generates 410 particles, with per-particle time of

```
$ ./maxw-mpw
Time per sample: 49.2463 ns
```

which is comparable to the $M = 3$ method of Birdsall. Yet despite this low particle count, the VDF is reproduced exactly. This is shown in Figure 4.6

4.4.2 Particle Merge

Unfortunately, this variable weight method suffers from two difficulties. First, performing DSMC collisions with variable weight particles is not trivial as discussed in section 4.7.2. Second, consider what happens when an energetic electron loses energy in an ionization collision. The particle, shown by the gray circle in Figure 4.6 joins the low-energy bulk population. However, the weight of this particle is now some 5 orders of magnitude smaller than that of the regular bulk electrons. This particle carries a negligible contribution to charge density, yet retains the overhead associated with every other particle. After a while, our simulation may become saturated by these low weight particles. Therefore, variable weight codes need to implement a *particle merge* algorithm, in which several low weight particles merge into a more massive one with a higher weight. In doing so, it is important to prevent introducing

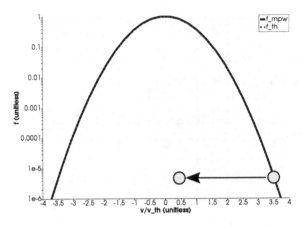

Figure 4.6: Maxwellian VDF sampled using function evaluation.

non-physical diffusion of the velocity distribution function. One approach is described by Fox [30]. We start by grouping particles into spatial cells. We then impose a velocity grid in cells with multiple particles. Particles are then sorted into the velocity bins. Each bin containing three or more particles has all particles removed and replaced by two new particles selected to retain velocity moments. Specifically, we have

$$n^0 = \sum i = 1^n w_i = w_{p1} + w_{p2} \tag{4.27}$$

$$p^0_{xyz} = \frac{\sum_{i=1}^n w_i v^i_{xyz}}{n^0} = \frac{w_{p1} v^{p1}_{xyz} + w_{p2} v^{p2}_{xyz}}{n^0} \tag{4.28}$$

$$t^0_{xyz} = \frac{\sum_{i=1}^n w_i (v^i_{xyz})^2}{n^0} - (p^0_{xyz})^2 = \frac{w_{p1}(v^{p1}_{xyz})^2 + w_{p2}(v^{p2}_{xyz})^2}{n^0} \tag{4.29}$$

We have seven equations for eight unknowns. We close the system by dividing the total weight evenly between the new particles,

$$w_{p1} = w_{p2} = \frac{1}{2} n^0 \tag{4.30}$$

The remaining equations then reduce to

$$v^{12}_{xyz} = pxyz^0 \pm \sqrt{t^0_{xyz}} \tag{4.31}$$

4.5 DIAGNOSTICS

4.5.1 Macroscopic Properties

Now that we have modified the source, we would like to make sure it is indeed sampling the correct velocity and temperature. Particle codes natively only contain positions and velocities of a multitude of particles. This by itself is not very useful. We are instead interested in the macroscopic flow properties such as density, stream velocity, and temperature. From Section 2.7.5, we already know how to compute number density by scattering particle positions to the grid. The other properties are obtained similarly. The bulk velocity comes from averaging individual particle velocities on each node,

$$\bar{v} = \frac{\sum_p w_{mp,p} v_p}{\sum_p w_{mp,p}} \tag{4.32}$$

Temperature is computed similarly by averaging $\vec{v}^2$. These three properties correspond to different *moments* of the velocity distribution function. Density is the zeroth moment, since we are effectively averaging $\vec{v}^0$. Velocity and temperature are the first and second moment, respectively.

To compute temperature, we need to make an assumption about the velocity distribution function. For a Maxwellian population, the following relationship

$$KE = \frac{1}{2} k_B T \tag{4.33}$$

holds for each degree of freedom. Hence

$$\frac{1}{2} m \left(\vec{u}^2 + \vec{v}^2 + \vec{w}^2 \right) = \frac{3}{2} k_B T \tag{4.34}$$

This equation is valid only for gas at rest. Otherwise, we need to first subtract the mean drift velocity component. We thus write

$$\frac{1}{2} m \sum_i^3 \overline{vv}_i^2 = \frac{3}{2} kT \tag{4.35}$$

where

$$\overline{vv}_i \equiv \frac{\sum_p w_{mp,p} \left(v_i^p - \bar{v}_i \right)^2}{\sum_p w_{mp,p}} \tag{4.36}$$

is the variance of the i-th velocity component.

We could use the above expression by first looping over all particles to compute the mean stream velocity $\bar{v}$. We would then loop through the particles one more time to compute the average of the $(\vec{v} - \bar{v})^2$ term. While this approach works, it is susceptible to noise. As we saw in Section 3.5, we use averaging to reduce numerical noise. Unfortunately, this is not viable for the above method. In a low density region, we may have cells that on average receive one, or fewer, particles per time step. We need, at a minimum, two particles to compute the variance in Equation 4.35. Temperature in cells with fewer than two particles evaluates to $T = 0$ K. Including these results in the running average leads to an artificial decrease in temperature.

Therefore, we need a more robust algorithm. Following [13], we start by expanding the variance term,

$$\overline{vv}_i \equiv \frac{\sum_p w_{mp}^p \left[(v_i^p)^2 - 2\bar{v}_i v_i^p + (\bar{v}_i)^2 \right]}{\sum_p w_{mp}^p} \tag{4.37}$$

This equation can be separated into three terms

$$\overline{vv}_i \equiv \frac{\sum_p w_{mp}^p \left(v_i^p\right)^2}{\sum_p w_p} - 2\bar{v}_i \frac{\sum_p w_{mp}^p v_i^p}{\sum_p w_{mp}^p} + (\bar{v}_i)^2 \frac{\sum_p w_{mp}^p}{\sum_p w_{mp}^p} \tag{4.38}$$

The middle fraction is simply $\bar{v}_i$. The nominator and denominator in the last fraction also cancel out. Hence

$$\overline{vv}_i \equiv \frac{\sum_p w_{mp}^p \left(v_i^p\right)^2}{\sum_p w_{mp}^p} - 2(\bar{v}_i)^2 + (\bar{v}_i)^2 \tag{4.39}$$

or

$$\overline{vv}_i \equiv \overline{(v_i)^2} - (\bar{v}_i)^2 \tag{4.40}$$

The first term on the right hand side is the average of squared velocity, while the second term is the average velocity, squared. Hence, instead of averaging temperature at each time step, we accumulate the *moments* w_{mp}, $w_{mp}v_i$, and $w_{mp}(v_i)^2$. When we need to compute temperature, we first average v_i and $(v_i)^2$. Then

$$T = \frac{m}{3k} \left(\overline{uu} + \overline{vv} + \overline{ww}\right) \tag{4.41}$$

These algorithms are implemented in the `sampleMoments` and `computeGas` `Properties` functions added to `Species`,

```
// aggregate particle velocity moments
void Species :: sampleMoments () {
  for (Particle &part : particles) {
    double3 lc = world.XtoL(part.pos);
    n_sum.scatter(lc, part.mpw);
    nv_sum.scatter(lc, part.mpw*part.vel);
    nuu_sum.scatter(lc, part.mpw*part.vel[0]*part.vel[0]);
    nvv_sum.scatter(lc, part.mpw*part.vel[1]*part.vel[1]);
    nww_sum.scatter(lc, part.mpw*part.vel[2]*part.vel[2]);
  }
}
```

and

```
// uses sampled data to compute velocity and temperature
void Species :: computeGasProperties () {
  vel = nv_sum/n_sum;    // stream velocity

  for (int i=0;i<world.ni;i++)
    for (int j=0;j<world.nj;j++)
      for (int k=0;k<world.nk;k++) {
        double count = n_sum(i,j,k);
        if (count<=0) {T[i][j][k] = 0; continue;}

        double u_ave = vel(i,j,k)[0];
        double v_ave = vel(i,j,k)[1];
        double w_ave = vel(i,j,k)[2];
        double u2_ave = nuu_sum(i,j,k)/count;
        double v2_ave = nvv_sum(i,j,k)/count;
```

```
    double w2_ave = nww_sum(i,j,k)/count;

    double uu = u2_ave - u_ave*u_ave;
    double vv = v2_ave - v_ave*v_ave;
    double ww = w2_ave - w_ave*w_ave;
    T[i][j][k] = mass/(2*Const::K)*(uu+vv+ww);
  }
}
```

The data is stored in new fields added to the `Species` class,

```
class Species {
public:
  Field den;            // number density
  Field T;              // temperature
  Field3 vel;           // stream velocity
  Field den_ave;        // averaged number density
protected:
  Field n_sum;          // number of particle
  Field3 nv_sum;        // number*velocity
  Field nuu_sum,nvv_sum,nww_sum; // number*vel. squared
  World &world;
};
```

We also add code to output these fields in `Output::field`,

```
void Output::fields(/*....*/) {
// species stream velocity
for (Species &sp:species) {
  out<<"<DataArray Name=\"vel."<<sp.name<<"\"
    NumberOfComponents=\"3\" format=\"ascii\" type=\"Float64\">\n";
  out<<sp.vel;
  out<<"</DataArray>\n";
}

// species temperature
for (Species &sp:species) {
  out<<"<DataArray Name=\"T."<<sp.name<<"\"
    NumberOfComponents=\"1\" format=\"ascii\" type=\"Float64\">\n";
  out<<sp.T;
  out<<"</DataArray>\n";
}
```

The resulting data computed after 300 time steps is shown in Figure 4.7. This plot shows the temperature and velocity for neutrals prior to the simulation reaching a steady state. As we can see, $T = 1000$ K is attained in the region away from the sphere. The high temperature in front of the sphere arises from the reflected particles moving in the opposite direction to the bulk beam. This region contains particles with large relative velocity, which the above algorithm interprets as temperature. The computed value is not the true temperature, since collisions are not yet included. The velocity plot in the bottom half indicates that the leading front of the gas population is formed by the higher velocity particles. In the bulk region, we obtain the expected 7 km/s velocity.

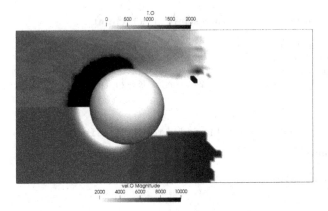

Figure 4.7: Temperature and stream velocity obtained from particle data.

4.5.2 Macroparticle Count

Usually of interest is also the number of particles per cell. This information is useful in diagnosing regions that may be susceptible to numerical noise. The rule of thumb I follow is to have at least 10 particles per cell, although plume expansion simulations often contain wake regions where this requirement is not satisfied. Macroparticle count is an example of *cell-centered* data. We add a new field to Species,

```
Field mpc;          // macroparticles per cell
```

We instantiate it in the **Species** constructor as follows,

```
Species (...) :
nd_ave(world.nn),mpc(world.ni-1,world.nj-1,world.nk-1),...
```

Note that the `mpc` field is allocated to contain $(ni - 1) \times (nj - 1) \times (nk - 1)$ items. These values correspond to the number of cells in the three dimensions. We next add a function to evaluate the counts. We simply loop through all particles, and for each, use the integer part of the logical coordinate to obtain the cell index. This cast is achieved using a `World::XtoIJK` function. We have

```
void Species::computeMPC() {
  mpc.clear();
  for (Particle &part:particles) {
    int3 ijk = world.XtoIJK(part.pos);
    int i = ijk[0], j = ijk[1], k=ijk[2];
    mpc[i][j][k] += 1;
  }
}
```

All previously exported mesh results were located within a `<PointData>` block. VTK also supports a `<CellData>` block for cell-centered data. The modification to `Output::fields` then reads

```
void Output::fields(World &world, vector<Species> &species) {
```

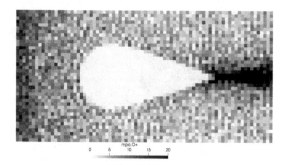

Figure 4.8: Number of macroparticles per cell.

```
/* ... */
// cell data
out<<"<CellData>\n";
// species temperature
for (Species &sp:species) {
  out<<"<DataArray Name=\"mpc."<<sp.name<<"\"
    NumberOfComponents=\"1\" format=\"ascii\"
    type=\"Float64\">\n";
  out<<sp.mpc;
  out<<"</DataArray>\n";
}
out<<"</CellData>\n";
/* ... */
}
```

The hook for computing the macroparticle count is added to the main loop,

```
// main loop in Main.cpp
for (Species &sp:species) {
  sp.advance(neutrals,spherium);
  sp.computeNumberDensity();
  sp.sampleMoments();
  sp.computeMPC();
}
```

These freshly computed counts are shown in Figure 4.8. This plot was obtained with ion macroparticle weight $w_{mp} = 50$ on a $41 \times 41 \times 81$ node grid. At steady state, the simulation contains over half a million ions. Yet, as we can see, most cells contain fewer than 10 particles. We can increase the number of particles per cell by running with a lower value of w_{mp}, but this approach may be computationally prohibitive. Alternatively, we can use a coarser mesh. The downside of that approach is the inability to resolve fine features.

4.5.3 Particle Plots

Another useful visualization technique involves outputting positions and velocities of individual particles. Using this data we can generate animations of particle motion as shown in Figure 4.9. The individual particles are colored by

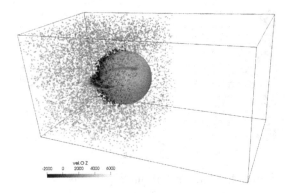

Figure 4.9: Scatter plot of neutral particles.

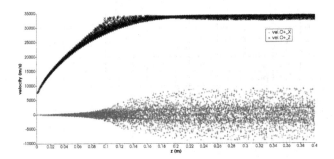

Figure 4.10: Phase plot of ion positions and velocities.

the z velocity. A slice is also used to visualize the mesh-based stream velocity in the region near the sphere. We can also generate two dimensional scatter plots of different velocity and position components plotted against each other. These types of plots are called *phase plots*. They are extremely useful for spotting trends in the data. For instance, Figure 4.10 shows a plot of the z component of position against the z component of velocity for individual particles. We can observe that ions accelerate from their initial 7 km/s velocity to about 34 km/s at the sphere. This value agrees with energy conservation,

$$\frac{1}{2}m_i(v_2^2 - v_1^2) = q\Delta\phi \qquad (4.42)$$

Evaluating with $\Delta\phi = 100$ V and $m_i = 16$ amu, we obtain $v_2 = 34,019$ m/s. The other set of points captures the $z - u$ phase plot. This plot compares the x component of velocity against z. We observe that while initially there is no u component of velocity due to $T_x = 0$, the focusing effect produced by the sphere introduces movement in the x (and similarly y) direction.

The particle output function is copied below,

```
void Output :: particles (World &world , vector<Species> &species , int
    num_parts) {
```

```
// loop over all species
for (Species &sp: species) {
  // open a phase_sp_it.vtp
  stringstream name;
  name<<"results/parts_"<<sp.name<<"_"<<setfill('0')<<setw(5)<<
    world.getTs()<<".vtp";
  ofstream out(name.str());
  if (!out.is_open()) {cerr<<"Could not open
    "<<name.str()<<endl;return;}

  // build a list of particles to output
  double dp = num_parts/(double)sp.getNp();
  double counter = 0;
  vector<Particle*> to_output;
  for (Particle &part : sp.particles)    {
    counter+=dp;
    if (counter>1) {to_output.emplace_back(&part);counter-=1;}
  }

  // header
  out<<"<?xml version=\"1.0\"?>\n";
  out<<"<VTKFile type=\"PolyData\" version=\"0.1\"
    byte_order=\"LittleEndian\">\n";
  out<<"<PolyData>\n";
  out<<"<Piece NumberOfPoints=\""<<to_output.size()<<"\"
    NumberOfVerts=\"0\" NumberOfLines=\"0\" ";
  out<<"NumberOfStrips=\"0\" NumberOfCells=\"0\">\n";

  // points
  out<<"<Points>\n";
  out<<"<DataArray type=\"Float64\" NumberOfComponents=\"3\"
    format=\"ascii\">\n";
  for (Particle *part: to_output)
  out<<part->pos<<"\n";
  out<<"</DataArray>\n";
  out<<"</Points>\n";

  // velocities
  out<<"<PointData>\n";
  out<<"<DataArray Name=\"vel."<<sp.name<<"\" type=\"Float64\"
    NumberOfComponents=\"3\" format=\"ascii\">\n";
  for (Particle *part: to_output) {
  out<<part->vel<<"\n";
  out<<"</DataArray>\n";
  out<<"</PointData>\n";

  out<<"</Piece>\n";
  out<<"</PolyData>\n";
  out<<"</VTKFile>\n";
  out.close();
}
}
```

The particles are output using VTK's PolyData structure. This file type is used for outputting general polygonal data such as those encountered in surface meshes. The file consists of arbitrarily located points (mesh nodes), and the associated lines, triangles, quadrangles, or generic polygons. Here we only use the `<points>` block to output the particle positions. We don't care about data connectivity, since the individual points are subsequently visualized using glyphs. Also, given that typical simulations contain millions of particles, it is impractical to output all of them. Therefore, we call this function with an approximate number of particles to save num_mp. We compute the average output probability, $dp = n_{p,output}/n_p$. We then loop through all particles, and for each, increment a `counter` variable by dp. We save a pointer to all particles for which the counter exceeds 1 to a `vector<Particle *>`. We also subtract 1 from the counter whenever adding a particle. This approach correctly handles cases in which n_p is not evenly divisible by $n_{p,output}$. We then loop through this vector and export the selected particles.

4.6 COLLISIONS

While our simulation now tracks a variety of species, the particles do not interact with each other except, in the case of ions, through the electric field. Ion-neutral and neutral-neutral interactions are handled through *collisions*. Collisions encountered in plasmas include

- **Momentum Exchange (MEX)**: These are the standard "billiard ball"-like interactions, in which a faster particle transfers some of its momentum to a slower molecule. They result in the thermalization of the velocity distribution function.

- **Charge exchange collisions (CEX)**: In these collisions, a valence electron "jumps" from an atom to a nearby ion without any momentum exchange between the two nucleii. These interactions are of concern to the electric propulsion community. The plume of a plasma thruster contains fast moving ions and slow moving neutrals arising from the un-ionized propellant. This interaction results in the formation of fast neutrals and slow ions, $A^{\circ}_{slow} + A^{+}_{fast} \rightarrow A^{+}_{slow} + A^{\circ}_{fast}$. The slow ions are susceptible to local electric fields and can backflow to regions with no direct line of sight to the thruster.

- **Ionization and Recombination**: These interactions primarily arise between high energy electrons and slow neutrals. Ionization, and hence its inverse, recombination, is a three body reaction. As such, recombination in the gas phase is very rare for low density discharges. Instead, most charged particles recombine on the surface, $e^{-1} + A^{+} + X \rightarrow A^{0} + X$. Here the surface X takes role of the third body needed for energy balance.

- **Excitation**: Electrons that are not energetic enough to knock off a valence electron from an atom may instead bring the valence electron to a higher shell. Subsequent impact by another low energy particle can then liberate the electron. Alternatively, the electron may decay back to the ground state by a spontaneous emission of a photon.

- **Coloumb collisions:** These are small angle perturbations between charged particles arising from the Coulomb force. They are often neglected in PIC simulations, since they act only on spatial scales smaller than the Debye length.

4.6.1 Collision Cross Section

Multiple collisional interaction may be present with a varying *collision frequency*. It is given by

$$\nu = n_a \overline{\sigma v} \tag{4.43}$$

Here n_a is the number density of the target species and σ is the *collision cross-section*. Imagine that a unit area slab contains n_a number density of atoms with a cross-sectional area σ. The area that is blocked over a thickness dx is $n_a \sigma dx$. Incoming flux Γ is reduced to Γ' after length dx according to

$$\Gamma' = \Gamma(1 - n_a \sigma dx) \tag{4.44}$$

We can then write

$$\frac{d\Gamma}{dx} = -\Gamma n_a \sigma \tag{4.45}$$

$$\frac{d\Gamma}{\Gamma} = -n_a \sigma dx \tag{4.46}$$

Next integrating from initial flux Γ_0 at $x = 0$, we obtain

$$\Gamma = \Gamma_0 \exp(-n_a \sigma x) \tag{4.47}$$

$$= \Gamma_0 \exp\left(-\frac{x}{\lambda_m}\right) \tag{4.48}$$

Here

$$\lambda_m = \frac{1}{n_a \sigma} \tag{4.49}$$

is the *mean free path*. As you remember from Chapter 1, it is used to define the Knudsen number, $Kn = \lambda_m / L$. If λ_m is greater than a characteristic length L, the flow is in collisionless (free molecular flow) regime.

The collision cross-section characterizes the likelihood of some collision type. It is generally a function of relative velocity or relative energy between the colliding particles, $\sigma = \sigma(v_r)$. Literature contains fits to experimental data for many common interaction pairs. For others, we may need to defer to analytical models derived from quantum mechanics. Lieberman [42] provides forms that can be used in the absence of better data. For momentum exchange collisions, we can use the Variable Hard Sphere (VHS) model of Bird [12]. The collision cross-section between two hard spheres is

$$\sigma = \pi d_{12}^2 \tag{4.50}$$

where $d_{12} = r_1 + r_2$, and r_1 and r_2 are the radii of the two molecules. For interactions within a single species, this is identical to πd^2, where d is the diameter. This hard sphere model is not realistic, since the likelihood of collision changes with the relative velocity. VHS improves the hard sphere model by letting the diameter be a function of relative velocity, $d = d(v_r)$ [12],

$$d = d_{ref} \sqrt{\frac{[2kT_{ref}/(m_r c_r^2)]^{\omega - 0.5}}{\Gamma(2.5 - \omega)}} \tag{4.51}$$

where d_{ref} is a reference diameter at the reference temperature T_{ref}, ω is a viscosity index, Γ is the Gamma function, and m_r is the reduced mass given by

$$m_r = \frac{m_1 m_2}{m_1 + m_2} \tag{4.52}$$

Values of d_{ref} and ω for selected materials can be found in Bird's Appendix A.

4.6.2 Collision Modeling

When it comes to actually modeling species interactions, we can choose one of three methods. We can model the interaction as a *chemical reaction*, or we can use the Monte Carlo Collisions (MCC) or Direct Simulation Monte Carlo (DSMC) algorithms. These approaches model the interactions as fluid-fluid, particle-fluid, and particle-particle, respectively. More specifically, only the gas macroscopic properties, such as density, bulk velocity, and temperature are needed to model chemical reactions. This is the most generic model. It is applicable to both kinetic and fluid species, since the macroscopic properties can always be computed from particle data. The MCC approach requires the source species to be kinetic, while the target species can be either fluid or kinetic. DSMC is applicable only when both reactants are kinetic.

4.6.3 Chemistry

A chemical reaction such as

$$A + e^- \rightarrow A^+ + 2e^- \tag{4.53}$$

can be represented as a series of rate equations,

$$\frac{dn_A}{dt} = -kn_An_e \tag{4.54}$$

$$\frac{dn_e}{dt} = kn_An_e \tag{4.55}$$

$$\frac{dn_i}{dt} = kn_An_e \tag{4.56}$$

In this reaction, we assume that a number density n_A of atoms interacts with a number density of electrons n_e with a rate coefficient k (m^3/s). The volumetric reaction rate is kn_An_e. In each reaction, the neutral is destroyed, and the neutral density decays according to $-kn_An_e$. On the other hand, each interaction results in the creation of one new electron and an ion. The number density of these two species increases. The reaction rate coefficient k is typically a function of energy, or, in more practial terms, temperature. For many interaction types, the rate coefficient can be found directly. Otherwise, we can estimate it from cross-section data by integrating over the velocity space

$$\bar{k}(T) = \int_0^\infty g(v,T)\sigma(v)dv \tag{4.57}$$

To demonstrate how this works in practice, let's assume that the "spherium" material is easily ionized. We use the Boltzmann electron model to obtain the local electron density. We use the cell-centered approach for modeling ionization. We loop through the cells, and evaluate the electron density and temperature (which is assumed to remain constant) at the cell center. We then compute $dn_i = (dn_i/dt)\Delta t$, from which the fractional number of particles to inject is obtained by multiplying by the cell volume. Note that the mathematical formulation does not prevent us from obtaining negative densities. Negative neutral density implies that we ionized more material than present. This is an indicator that the simulation time step is too large. Besides re-running with a smaller step, we may also need to implement a *limiter*. A simple limiter consists of setting dN_A to the smaller of the computed value from the chemical reaction and the actual neutral density.

This algorithm is implemented by class `ChemistryIonize` located in `Collisions.h`,

```
class ChemistryIonize: public Interaction {
public:
    ChemistryIonize(Species &neutral, Species &ions, World &world,
        double rate):
    neutrals{neutrals},ions{ions},world{world},rate{rate} {}
    void apply(double dt);
protected:
    Species &neutrals;
    Species &ions;
    World &world;
    double rate;
};
```

This class inherits from a base **Interaction** class

```
class Interaction {
public:
    virtual void apply(double dt) = 0;  // pure virtual function
    virtual ~Interaction() {}
};
```

This base class then allows us to store a generic list of material interactions, similar to what was done previously for sources in Section 4.4. In **main**, we add

```
// setup material interactions
vector<unique_ptr<Interaction>> interactions;
interactions.emplace_back(new
    ChemistryIonize(sph,sph_ions,world,1e-3));
```

We also add a hook to process all defined interactions in the main loop,

```
while(world.advanceTime()) {
    /* ... */
    for (auto &interaction:interactions)  // perform interactions
        interaction->apply(World.);
```

Updating the target population is not trivial for species modeled by particles. The approach we take is to define a field for all species to store the density loss in each cell arising from chemical interactions. After the density change is computed by applying all interactions of interest, we loop through the cells. In cells with a positive dn, we compute the integer number of particles to sample from

$$N = \text{int}(dn\Delta V + \mathcal{R}) \tag{4.58}$$

The field dn is then cleared. The alternate approach is to inject $\text{int}(dn)$ particles and retain the remaining fractional part. This approach eliminates the noise associated with stochastic sampling, but in reactions with a low rate, the ions are produced in "waves": time steps in which a particle is created are separated by long periods with no production. A similar stochastic approach can be used to destroy mass. We loop through the particles, and evaluate dN at the particle location. The probability of the particle surviving is

$$P = dN/w_{mp} \tag{4.59}$$

P will be greater than unity while there are particles to be killed. For fractional amounts, we again use a stochastic approach, $dN/w_{mp} + \mathcal{R}$. The particle is deleted as long as $P \geq 1$. Now of course, none of this is applicable if the source or the target species are modeled as a fluid. In that case, we simply increment or decrement the local species density.

4.6.4 Monte Carlo Collisions (MCC)

Monte Carlo Collisions (MCC) is an algorithm in which a source particle collides with a target "cloud". Only the source particle is affected by the collision, at least unless special corrections are implemented. The collisions thus do not conserve momentum and as such, this method is best suited for interactions involving trace species colliding with a much denser target. The MCC method is much faster than DSMC. It is often used by the electric propulsion community to model charge exchange (CEX) collisions in the near field of a plasma thruster. In this region, the neutral density exceeds ion density by several orders of magnitude and thus the momentum transfer from the ions to the neutrals is negligible.

MCC collision probability is computed per [14] as

$$P = 1 - \exp(n_a \sigma v_{rel} \Delta t) \tag{4.60}$$

where $\sigma(v_{rel})$ is the cross section and n_a is the target density. To implement MCC, we loop over all particles, and for each, compute collision probability per Equation 4.60. This calculation requires the density of the target species at the particle location. This value is obtained by gathering the appropriate field.

We also need to evaluate the relative velocity between the source particle and the target species. We have several options here. One common approach is to assume that the velocity of the target is insignificant compared to the source so that $v_{target} = 0$ and $v_{rel} = v_{source}$. This is generally valid for modeling electron or even ion collisions with neutrals. The other possibility is to assume that the target material is a fluid with some mean stream velocity $\vec{v}$ and a temperature T. We then sample a random virtual particle from the population. The collision happens if $P > \mathcal{R}$, where $\mathcal{R}$ is a random number. The strict nonequality assures that for $P = 0$, there will be no collisions. This algorithm is implemented as follows

```cpp
void MCC_CEX::apply(double dt) {
  // set pointers to target data
  Field &target_den = target.den;
  Field &target_temp = target.T;
  Field3 &target_vel = target.vel;

  // loop over all particles
  for (Particle &part:source.particles) {
    // get target velocity
    double3 lc = world.XtoL(part.pos);
    double3 vt = target_vel.gather(lc);

    // get target density
    double nn = target_den.gather(lc);

    // compute cross-section, function of relative velocity
    double3 v_rel = part.vel-vt;
    double v_rel_mag = mag(v_rel);

    // evaluate cross-section
    double sigma = evalSigma(v_rel_mag);

    // compute probability
```

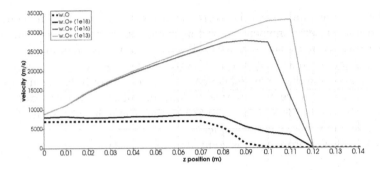

Figure 4.11: Ion velocity along a line centerline for three different values of neutral density.

```cpp
double P = 1 - exp(-nn*sigma*v_rel_mag*dt);

// compare to a random number to see if collision happened
if (P>=rnd()) {
    // sample a virtual target particle
    double T_target = target_temp.gather(lc);    // sample target
        temperature
    double3 v_target = target.sampleIsotropicVel(T_target);
    part.vel = v_target;
    }
  }
}
```

Here `evalSigma` is a placeholder for a function that returns σ based on some analytical or tabulated data. For this demo, we just return a constant value,

```cpp
double evalSigma(double rel_g) {
    return 1e-16;    // placeholder for a sigma expression
}
```

The MCC interaction is then included in the simulation by adding another entry to the `interactions` vector in `main.cpp`,

```cpp
interactions.emplace_back(new MCC_CEX(ions, neutrals, world));
```

Figure 4.11 shows an example result. Here we plot the mean velocity along the $(0, 0, z_0) : (0, 0, z_m)$ centerline after 300 time steps. The dashed line plots the velocity of the neutrals, while the three solid lines are ion velocities at different neutral densities. We can see that as the density of the background gas increases, the mean ion velocity becomes increasingly more coupled to the neutral flow.

4.7 DSMC

The Direct Simulation Monte Carlo (DSMC) method is the most rigorous, and also the most computationally expensive, of the three interaction methods.

Unlike MCC, DSMC actually collides two simulation particles. Only particles "near by" can collide with each other. Typically, this implies that the particles are located in the same cell. The simple example here utilizes the PIC mesh for collisions, however, nowadays it is customary to utilize adaptive collision cells that respond to local flow properties [13]. Our codes so far relied on the function XtoL to translate a physical particle position to a cell index. This function allows us to determine which cell a particle resides in. There is however no function that gives us a list of all particles located within a single cell.

4.7.1 Particle Sorting

Therefore, we need to implement an algorithm for sorting particles to cells. Since our particles are stored in an array, we can simply reference an individual particle using the array index. Alternatively, a Particle* pointer can be used. We take the first approach, and define a vector<int> of particle indexes. Since we need this data in each cell, we actually instantiate an array of these vectors,

```
vector<Particle*> *parts_in_cell;        // array of vectors
int n_cells = (world.ni-1)*(world.nj-1)*(world.nk-1);
parts_in_cell = new vector<Particle*> [n_cells];
/* ... */
delete[] parts_in_cell;    // memory cleanup
```

Alternatively, we could have defined a vector of vectors, but I found the above approach slightly simpler. The index into the parts_in_cell array is the linear cell index,

$$c = k(n_{ci}n_{cj}) + jn_{ci} + i \qquad (4.61)$$

where n_c is the number of cells in the appropriate direction. This equation is implemented in World::XtoC. We then loop through all particles, and compute the cell the particle is located in. We store the particle identification number in that cell's lookup table.

4.7.2 Collision Groups

Next we loop through all cells, and in each, compute the number of collision groups to *check* from

$$N_g = \frac{1}{2}N_1 N_2 w_{mp} \left(\sigma c_r\right)_{\max} \frac{\Delta t}{V} \qquad (4.62)$$

This method is known as No Time Counter, or NTC [12]. Here $(\sigma c_r)_{\max}$ is a variable that keeps track of the largest product of cross section and relative velocity. It is initialized to some best guess and then updated at each time step. The N_g term is generally not an integer. We again use a stochastic approach in which we add a random number and truncate. Alternatively, we can keep track of the fractional reminder for each cell,

```
Ng_int = (int)(Ng+Ng_rem);    // add prior reminder
Ng_rem += Ng-Ng_int;          // update
```

In Equation 4.62, N_1 and N_2 are the numbers of simulation macroparticles of the two interacting species. The two species need to share the same macroparticle weight w_{mp}. This is a major limitation of this scheme. Other authors have devised methods for simulating interactions between species of different weight but these schemes involve creating fractional particles or are energy conserving only in average [18, 43], and are generally very inefficient for large differences in species densities.

4.7.3 Collision Probability

Now that we have the number of possible collision pairs, we select that many random pairs. The pairs are selected by picking two random particle ids p_1 and p_2 from the `parts_in_cell` vector, making sure that $p_2 \neq p_1$. The relative velocity is computed from $\vec{v}_r = \vec{v}_1 - \vec{v}_2$. Collision cross-section is obtained per Equation 4.51. The collision occurs if $P > \mathcal{R}$, where

$$P = \frac{\sigma v_r}{(\sigma v_r)_{max}} \tag{4.63}$$

The denominator is updated at the end of the algorithm if we encounter a larger value during the loop. The resulting code is listed below

```
void DSMC_MEX::apply(double dt) {
    vector<Particle*> *parts_in_cell;
    int n_cells = (world.ni-1)*(world.nj-1)*(world.nk-1);
    parts_in_cell = new vector<Particle*> [n_cells];

    // sort particles to cells
    for (Particle &part:species.particles) {
        int c = world.XtoC(part.pos);
        parts_in_cell[c].push_back(&part);
    }

    double sigma_vr_max_temp = 0;    // reset for max computation
    double dV = world.getCellVolume();    // internal cell volume
    double Fn = species.mpw0;    // specific weight using Bird's
        notation
    int num_cols=0;    // reset collision counter

    // loop over cells
    for (int c=0;c<n_cells;c++) {
        vector<Particle*> &parts = parts_in_cell[c];
        int np = parts.size();
        if (np<2) continue;    // need at least two particles per cell

        // compute number of groups according to NTC
        double ng_f = 0.5*np*np*Fn*sigma_cr_max*dt/dV;
        int ng = (int)(ng_f+0.5);    // number of groups, round

        // assumes at least two particles per cell
```

```
for (int g=0;g<ng;g++) {
    int p1, p2;
    p1 = (int)(rnd()*np);    // random number in [0,np-1]
    do { p2 = (int)(rnd()*np);
    } while (p2==p1);    // select p2 that is different from p1

    // compute relative velocity
    double3 vr_vec = parts[p1]->vel - parts[p2]->vel;
    double vr = mag(cr_vec);

    // evaluate cross section
    double sigma = evalSigma(cr);

    // eval sigma_vr
    double sigma_vr=sigma*vr;

    // update sigma_vr_max
    if (sigma_vr>sigma_vr_max_temp)
        sigma_vr_max_temp=sigma_vr;

    // eval prob
    double P=sigma_vr/sigma_vr_max;

    // did the collision occur?
    if (P>rnd()) {
        num_cols++;
        collide(parts[p1]->vel, parts[p2]->vel, species.mass,
            species.mass);
    }
    }
}
delete[] parts_in_cell;
if (num_cols)
    sigma_vr_max = sigma_vr_max_temp;
}
```

DSMC collisions are enabled in **main** with

```
interactions.emplace_back(new DSMC_MEX(neutrals, world));
```

4.7.4 Momentum Exchange Collision Handler

The above code calls a `collide` function to perform the actual collision. DSMC is most often used to model momentum transfer (MEX) collisions. These are the standard collisions in which two molecules bump into each other, and the faster molecule transfers some of its momentum to the slower one. Consider two molecules colliding with each other. Per [12], momentum conservation dictates that

$$m_1\vec{v}_1 + m_2\vec{v}_2 = m_1\vec{v}_1^* + m_2\vec{v}_2^* = (m_1 + m_2)\vec{v}_m \qquad (4.64)$$

Here $\vec{v}_m$ is known as the center of mass velocity. The relative velocity is $\vec{v}_r = \vec{v}_1 - \vec{v}_2$. To conserve energy, the magnitude of relative velocity needs to remain constant during the collision, so $|\vec{v}_r^*| = |\vec{v}_r|$. Then

$$\vec{v}_1^* = \vec{v}_m + \frac{m_2}{m_1 + m_2}\vec{v}_r^* \tag{4.65}$$

$$\vec{v}_2^* = \vec{v}_m - \frac{m_1}{m_1 + m_2}\vec{v}_r^* \tag{4.66}$$

While we have the magnitude, we do not yet have an expression for the direction of $\vec{v}_r^*$. One option is to assume that molecules interact as hard spheres. The scattering is then isotropic. We start by sampling the cosine of a random elevation angle,

$$\cos(\xi) = 2\mathcal{R}_1 - 1 \tag{4.67}$$

Next we select a random azimuth angle,

$$\epsilon = 2\pi\mathcal{R}_2 \tag{4.68}$$

Velocity components are given by

$$\vec{v}_r^* = < v_r \cos\xi, v_r \sin\xi \cos\epsilon, v_r \sin\xi \sin\epsilon > \tag{4.69}$$

This code is implemented as follows,

```
void DSMC_MEX:: collide (double3 &vel1 , double3 &vel2 , double mass1,
    double mass2) {
    double3 cm = (mass1*vel1 + mass2*vel2)/(mass1+mass2);

    // relative velocity , cr_mag remains constant in the collision
    double3 cr = vel1 - vel2;
    double cr_mag = mag(cr);

    // pick two random angles , per Bird's VHS method
    double cos_chi = 2*rnd()-1;
    double sin_chi = sqrt(1-cos_chi*cos_chi);
    double eps = 2*Const::PI*rnd();

    // perform rotation
    cr[0] = cr_mag*cos_chi;
    cr[1] = cr_mag*sin_chi*cos(eps);
    cr[2] = cr_mag*sin_chi*sin(eps);

    // post collision velocities
    vel1 = cm + mass2/(mass1+mass2)*cr;
    vel2 = cm - mass1/(mass1+mass2)*cr;
}
```

Note that the two velocities are passed in as reference so that they can be modified by the handler.

This section really just barely scratched the surface on DSMC. There are many improvements that could be made to the code. These include the use of subcells or automatic gridding, support for variable weight, and the use of the Variable Soft Sphere (VSS) model to capture non-isotropic scattering

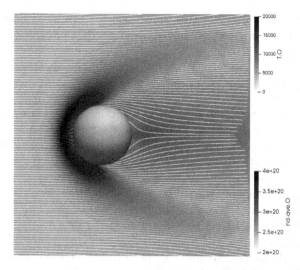

Figure 4.12: DSMC simulation of neutral flow past a sphere.

angle [13]. Still, even with our simple algorithm, we obtain the improvement to the neutral population shown in Figure 4.12. Here the neutral density is increased to $n_a = 2 \times 10^{20}$ m^{-3}. Unlike in the collisionless, low-density case in Figure 4.3, neutrals scattering off the sphere now become entrained in the free stream, leading to a formation of a bow shock. This figure visualizes the neutral velocity streamlines colored by number density. The background contour plots temperature.

4.8 SUBCYCLING

As we start introducing additional species and interactions, it quickly becomes obvious that different processes take place at different temporal scales. Electrons move several orders of magnitude faster than the ions, which in turn move several orders of magnitude faster than the neutrals. The collision rate for some collisions may be too low to be modeled at every time step. PIC codes thus often implement *subcycling*. The common approach is to run the simulation at the fastest time scale, and then use the modulus (%) to perform some operation only some specified number of time steps apart. For example, a simulation containing electrons, ions, and neutrals may contain the following logic

```
// main loop
if (ts%100==0) neutrals.advance(dt*100); // neutrals move every
    100 time steps
if (ts%10==0) ions.advance(dt*10);   // ions move every 10 time
    steps
electrons.advance(dt); // electrons move every time step
```

It is important to remember that since ions are advanced only once every 10 electron time steps, the ion velocity and particle integration needs to be integrated through $10\Delta t$.

4.9 DIELECTRIC BOUNDARIES

In our example, the sphere is modeled as a conductor held at a fixed potential by some virtual power supply. While such a setup may be valid in a laboratory setting, for space applications we may be interested in starting with $\phi = 0$ and self-consistently determining the potential arising from the impinging ion and electron current. Resolving *spacecraft charging* is imperative for space weather instruments that may produce erroneous reading if the potential gradient in respect to the ambient plasma is not taken into account. Large potential differences can also lead to destructive arcing.

Self-consistent charging calculation involves solving potential both "inside" and "outside" the object. To do so, we need to treat the object as a dielectric. In a dielectric material, electric field $\vec{E}$ leads to a separation of bound charges, giving a rise to a local electric dipole moment. The macroscopic summation of these individual moments is given by the displacement field

$$\vec{D} = \epsilon_0 \vec{E} + \vec{P} \tag{4.70}$$

Here $\vec{P}$ is the polarization density field. Taking divergence of both sides, we obtain

$$\nabla \cdot \vec{D} = \epsilon_0 \nabla \cdot \vec{E} + \nabla \cdot \vec{P} \tag{4.71}$$

or

$$\epsilon_0 \nabla \cdot \vec{E} = \nabla \cdot \vec{D} - \nabla \cdot \vec{P} \tag{4.72}$$

Comparing to the Gauss' Law, $\epsilon_0 \nabla \cdot \vec{E} = \rho$, we see

$$\rho = \nabla \cdot \vec{D} - \nabla \cdot \vec{P} \equiv \rho_f + \rho_b \tag{4.73}$$

Charge density can be decomposed into "free" and "bound" charges. Previously, we considered only the "free" contribution. From this definition, we also see that

$$\nabla \cdot \vec{D} = \rho_f \tag{4.74}$$

Next, if the object is composed of a linear (polarization directly proportional to the electric field), homogeneous (no spatial variation), and isotropic (no directional variation) material, then polarization density can be written in terms of electric field,

$$\vec{P} = \epsilon_0 \chi \vec{E} \tag{4.75}$$

where χ is the *electron susceptibility*.

Substituting this definition into the displacement field equation, we obtain

$$\vec{D} \equiv \epsilon_0 (1 + \xi) \vec{E} = \epsilon_0 \epsilon_r \vec{E} \tag{4.76}$$

where e_r is the relative permittivity. With $\epsilon = \epsilon_0 \epsilon_r$, we have

$$\vec{D} = \epsilon \vec{E} \tag{4.77}$$

Taking divergence,

$$\nabla \cdot \vec{D} \equiv \nabla \cdot (\epsilon \vec{E}) \tag{4.78}$$

we obtain

$$\epsilon \nabla \cdot \vec{E} + \vec{E} \cdot \nabla \epsilon = \rho_f \tag{4.79}$$

For cases with no variation in permittivity, and $\epsilon_r = 1$ (vacuum), the above equation reduces to the familiar $\nabla \cdot \vec{E} = \rho_f / \epsilon_0$.

Next, applying the electrostatic assumption, $\vec{E} = -\nabla \phi$, we arrive at

$$\epsilon \nabla^2 \phi + \nabla \phi \cdot \nabla \epsilon = -\rho_f \tag{4.80}$$

We next assume that the domain consists of multiple pieces, with each piece having uniform permittivity. Equation 4.80 reduces to the standard $\epsilon \nabla^2 \phi = -\rho_f$ Poisson's equation within each piece.

At each piece boundary, there is a discontinuity in ϵ. This discontinuity provides a piece-wise boundary condition. To incorporate it, we define a control volume centered on the interface, and use the Divergence Theorem to rewrite $\nabla \cdot (\epsilon \nabla \phi) = -\rho_f$ as

$$\oint_S \epsilon \nabla \phi \cdot n dA = -\rho_f \Delta V \tag{4.81}$$

We then evaluate the integral on all faces of this control volume "pill box". Here we assume that the variation is negligible in the tangential directions. Then

$$(\epsilon \nabla_x \phi)^+ - (\epsilon \nabla_x \phi)^- = -\rho_f \Delta x = -\sigma_f \tag{4.82}$$

where σ_f is the surface charge. Allowing for a variable grid spacing, we have

$$\frac{\epsilon^- (\phi_i - \phi_{i-1})}{\Delta x^-} - \frac{\epsilon^+ (\phi_{i+1} - \phi_i)}{\Delta x^+} = \rho_f \left(\frac{\Delta x^+ + \Delta x^-}{2} \right) \tag{4.83}$$

This expression can be solved for ϕ_i,

$$\phi_i = \left[\rho_f \left(\frac{\Delta x^+ + \Delta x^-}{2} \right) (\Delta x^+ \Delta x^-) + \epsilon^- \Delta x^+ \phi_{i-1} + \epsilon^+ \Delta x^- \phi_{i+1} \right]$$
$$\left(\frac{1}{\epsilon^- \Delta x^+ + \epsilon^+ \Delta x^-} \right) \tag{4.84}$$

This expression is used to update potential on each interface node at each solver iteration. Conductors can be included in the simulation using arbitrarily high permittivity. This is the approach utilized by Kafafy in [40].

4.10   DISCRETIZED SURFACES

Finally, simulations involving analytical objects such as the sphere can be quite limiting. It is often desired to run simulations using arbitrary surfaces. Just as we use a volume mesh to discretize the spatial domain, a surface mesh can be used to approximate an arbitrary surface. This mesh is composed of basic elements, such as triangles or quadrangles. Surface meshes are generated by first drawing the geometry in a CAD program and then running the solid model through a meshing package. Chapter 6 discusses steps for loading meshes from a file. Let's for now assume that we have generated and imported a surface mesh for some arbitrary geometry. This mesh consists of various triangular elements representing the "skin" of the object. Let's also assume that surface normals point out of the object into the plasma domain. This orientation of normals needs to be set (or at least verified) in the meshing program. The next task is using this collection of triangles to replace the analytical geometry.

Just as with the analytical sphere, we need to account for the meshed surface in the field solver and in the particle pushing code. We account for both by implementing a discretized version of the inSphere function. Consider any arbitrary point representing either a node location or a particle. A point can be classified as internal or external to a surface by casting a ray from the point to any visible surface triangle, and then computing the dot product with the surface normal,

$$\vec{r} = \vec{x}_c - \vec{x}_p \tag{4.85}$$

$$\cos\theta = \frac{\vec{r} \cdot \hat{n}}{|\vec{r}|} \tag{4.86}$$

Here $\vec{x}_c$ is the centroid of a visible triangle. Since the normals $\hat{n}$ point outward, a negative dot product indicates the node is external. The challenge is to find the visible triangle. A triangle can be deemed visible if a ray to its centroid is not blocked by any other triangle. This is visualized in Figure 4.13. In a system with N triangles, we may need to compute up to $N(N-1)$ *line triangle intersection* checks to determine visibility. Of course, the actual number of triangles to check is often smaller since we terminate the search once a blocking triangle is found. Still, due to the N^2 dependence, this search introduces a significant overhead for large meshes. Production codes need to utilize additional approaches to speed up the algorithm. For instance, we can use an octree to store the surface triangles according to their spatial location. We then only check the elements in the vicinity of the node. We can also utilize different algorithms such as *z-buffering* initially developed for computer graphics. These methods may further benefit from performing the computation on a graphics card (GPU).

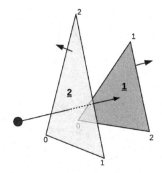

Figure 4.13: Triangle is visible if a ray to it is not intersected by any other triangle.

4.10.1 Line Triangle Intersection

The *line-triangle intersection* algorithm consists of three steps:

1. Check for a line-plane intersection

2. Compute the intersection position on the plane

3. Test whether the point lies within the bounds of the triangle

Equation for a line between two points $\vec{x}_a$ and $\vec{x}_b$ can be cast into a parametric form as

$$\vec{x} = \vec{x}_a + t(\vec{x}_b - \vec{x}_a) \tag{4.87}$$

This relationship can be then substituted into the equation for a plane,

$$(\vec{x} - \vec{x}_0) \cdot \hat{n} = 0 \tag{4.88}$$

to obtain

$$t = \frac{(\vec{x}_0 - \vec{x}_a) \cdot \hat{n}}{(\vec{x}_b - \vec{x}_a) \cdot \hat{n}} \tag{4.89}$$

Here $\vec{x}_0$ is an arbitrary point that we know lies on the plane. Some obvious choices are any of the three triangle vertexes or the triangle centroid. $\hat{n}$ is the normal vector. We compute the normals for a triangle composed of nodes $\vec{x}_0$, $\vec{x}_1$ and $\vec{x}_2$ using

$$\hat{n} = \text{unit}\left[(\vec{x}_1 - \vec{x}_0) \times (\vec{x}_2 - \vec{x}_0)\right] \tag{4.90}$$

The node ordering follows the *right-hand rule*. This same convention is used by meshing programs to set the node orientation. Flipping surface normals implies permuting the node ordering in the element connectivity entry.

The plane intersects the line if $t \in [0, 1]$. We next need to determine if the intersection point lies inside the triangle or not. There are several approaches that can be considered here. One option is to compute the logical (triangle) coordinate of the point by subdividing the triangle using the point as one of

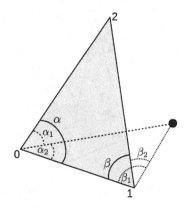

Figure 4.14: Illustration of possible approaches for checking if a point is located inside a triangle.

the vertices. These fractional areas add up to 1 for an internal point. I however prefer to use an approach based on vertex angles. This method is shown in Figure 4.14. We loop through the three vertexes, and at each form a ray to the point. We then compare the two angles formed using the ray and the vertex edges α_1 and α_2 to the vertex angle α. The benefit of this method is that we terminate the check once an angle that is larger than the vertex angle, $\beta_1 > \beta$, is found. The point is inside only if $\alpha_1 \leq \alpha$ and $\alpha_2 \leq \alpha$ holds for all three vertexes. Instead of comparing angles, we compare the cosines,

$$\cos(\alpha_1) = \text{unit}\left[(\vec{x}_p - \vec{x}_0) \cdot (\vec{x}_2 - \vec{x}_0)\right] \tag{4.91}$$

Other schemes exist. Another popular approach involves computing the orientation of the cross products (normal vectors) formed between the ray and the edges. On node 0 we have

$$\vec{n}_1 = (\vec{x}_1 - \vec{x}_0) \times (\vec{x}_p - \vec{x}_0) \tag{4.92}$$

$$\vec{n}_2 = (\vec{x}_p - \vec{x}_0) \times (\vec{x}_2 - \vec{x}_0) \tag{4.93}$$

The point is outside the triangle if the two normals point in the opposite direction, $\hat{n}_1 \cdot \hat{n}_2 < 0$. This method also allows early termination, however, computing the cross-product involves more operations than computing the dot product. In the end, the choice should be made by performing timing studies for your particular setup.

4.10.2 Particle Impacts

This algorithm is also used to check for particle impacts. The objective is to determine if particle impacted a surface during the push. This involves testing the line segment for a particle's pre- to post-push position for surface triangle intersections. Many of my simulations use surface meshes with million or more triangles, and you can surely imagine that a brute force approach in which the check is made against the entire surface mesh would be prohibitively slow. Therefore, it is crucial to reduce the search space. The approach used in my thesis [21] was to store an array of references to all surface elements crossing *interface* mesh cells. The surface impact algorithm then considers only the triangles located in the cells within the bounding box defined by the particle's starting and ending position, $[\vec{x}_1, \vec{x}_2]$. Another approach used in my contamination transport code [22] involves storing the elements in an octree.

We also need to remember that the line segment to the final position can cross multiple surface elements. Consider a particle interacting with a thin beam. It is possible for the particle to both enter and exit during a single push. This is clearly not physical. Therefore, simply finding the first intersection is not sufficient. We need to consider all elements the particle could have crossed and find the element that was hit first. This can be done by initializing a t_{min} variable to > 1. We then loop through the elements, and whenever we find an intersection with $t \in [0, t_{min}]$, we update the value of t_{min} and also store a reference to the impacted triangle. This can be a pointer to the triangle object, or simply an integer index if surface elements are stored within a vector.

4.10.3 Caveats

With this in mind, it is important to point out that utilizing discretized surfaces is not fool proof. Due to the finite precision of computer math, we need to include some small tolerance ϵ_{tol} when performing line-triangle checks. For instance, we should use

$$\alpha_1 > \alpha + \epsilon_{tol} \tag{4.94}$$

to flag an out of triangle point. Without this tolerance, we may encounter *particle leaks* by missing intersections located on or just inside the triangle edges. However, this tolerance also effectively increases the size of the element, which brings with it a new set of difficulties as triangles now technically overlap. Also, consider a particle impacting a corner between two elements. Diffusely reflecting the particle exactly from the surface location can lead to a leak, since the randomly sampled post-collision vector can take the particle to the inside part of the other element forming the corner. Therefore, instead of pushing the particle completely to the surface, we prefer to push it only most of the way,

$$\vec{x_p} = \vec{x}_1 + (1 - \epsilon)t_p(\vec{x}_2 - \vec{x}_1) \tag{4.95}$$

A good starting point may be to use $\epsilon \approx 10^{-4}$, however, deciding on a proper value may not be trivial. If the value is too small, particles may leak through the surface. For larger values, the particles may be bouncing off too far away from the surface. This behavior preferentially affects faster particles, which may lead to non-physical results.

4.11   SUMMARY

In this chapter we saw how to introduce multiple species that can interact with each other. We also learned how to model surface interactions and how to output macroscopic flow velocities and temperatures. We also learned how to sample the Maxwellian velocity distribution function. We then saw how to include material interactions using chemical reactions, Monte Carlo Collisions (MCC), and Direct Simulation Monte Carlo (DSMC). We then reviewed sub-cycling and the use of dielectric surfaces. We concluded with a brief review of steps needed to include discretized surface geometry.

EXERCISES

4.1   *Collision Comparison.* Write a program to compare the MCC and DSMC handling of collisions. Modify the sphere program to model a 0.2 m long domain in z and only 1 cell in the x and y dimension. This simulation considers neutral gas and hence the field solver can be turned off. Initialize two types of molecules. Molecules of type A are injected at $z = 0$ with $v = 1000$ m/s, $n_A = 10^{16}$ m^{-3} and $T_A = 0$ K. Molecules of type B generate uniform density n_B between $z = 0.1$ and $z = 0.2$ m. These molecules are cold, $T_B = 0$ K. Run the simulation for 1000 time steps with $\Delta t = 10^{-6}$ s with MCC and DSMC collisions. Use constant $\sigma = 10^{-16}$ m^2 cross-section. For the MCC run, type B molecules can be replaced with a density field. Compare the results for three values of B density, $n_B \in \{10^{16}, 10^{17}, 10^{18}\}$ m^{-3}. For each case, compute the average momentum of each species, as well as the total momentum. Does it stay conserved? Also visualize the $z - w$ and $z - \sqrt{u^2 + v^2}$ phase plots, and density of both species versus z.

4.2   *Impinging Beams.* Write code to simulate two impinging beams. Modify the above code to inject the B species from the z^+ face with a $-w$ velocity. Compare MCC and DSMC results.

4.3   *Velocity Histogram.* Implement a virtual probe sampling particle VDF within a small region. Start by defining boundaries of a box to be sampled. Initialize a histogram (a 1D vector of floats) to store the number of particles in each velocity bin. Compute bin spacing from $dv = (v_{max} - v_{min})/n_{bins}$. Loop through all particles of a given species, and for each, determine if it is located within the probe's sampling box. If so, determine which bin the velocity falls in. Capture a histogram for the three velocity directions, as well as for velocity magnitude (or energy). To reduce noise, you may want to start sampling only at steady state and collect samples for large number of time step.

4.4   *Particle Merge.* Write an algorithm to merge particles in a cell. Sample a random $u - v$ Maxwellian population using large number of variable weight particles. Then perform the merge using the algorithm from Section 4.4.2. Compare a histogram of the pre- and post-merge velocity distributions.

4.5 *Dielectric Boundaries.* Write a one-dimensional potential solver for a 0.1 m long domain with $\phi(0) = 0$ and $\phi(0.1) = 10$ V. Let the region $0.05 \leq x \leq 0.07$ be composed of a dielectric wall with a user defined value of ϵ_r. The regions to the left and right are composed of vacuum. Obtain potential for $\epsilon_r = 3$ and $\epsilon_r = 30$. You should observe that the second case results in constant potential on the wall.

4.6 *Discretized surface.* Use a meshing program such as Salome (`https://www.salome-platform.org/`) to generate a discretized representation of a sphere. Export the mesh in the `.dat` or `.unv` (see Chapter 6 for more information) and write a mesh loader. Then modify the code to use the discretized surface to flag the internal Dirichlet nodes and to perform particle surface hit checks. Compare the results to the case with the analytical sphere.

CHAPTER 5

Symmetry

5.1 INTRODUCTION

I N THIS chapter we learn how to utilize symmetry that is inherent in many simulation problems to reduce the computational time. We start by implementing symmetric and periodic boundaries in the three-dimensional code from the prior chapters. We then develop two two-dimensional versions. The first version considers a planar domain in which we simulate an infinitely long cylinder. In the second version, we model the flow past a sphere using the axisymmetric formulation.

5.2 SYMMETRY

Many engineering problems exhibit spatial symmetry. The flow around the sphere is one such example. The injected plasma is spatially uniform. The sphere geometry can also be cut along the $x-z$ and $y-z$ plane without any loss of detail and there are no external forces applied. Therefore, the results should be identical for any of the four quadrants obtained by dividing the $x-y$ plane, as shown in Figure 5.1. We can expect the symmetric simulation to run faster due to the reduced number of mesh nodes and particles. Alternatively, if we are satisfied with the time required for the full run, we can redo the simulation with a finer mesh or with more particles without increasing the wall time.

5.2.1 Domain Boundary Condition

Implementing symmetry in three-dimensional PIC codes is quite trivial. We simply need new boundaries for the field solver and the particles. Consider a problem that is symmetric about the $x = 0$ plane. For any spatially varying scalar such as potential or charge density, the following relationship holds

$$f(x, y, z) = f(-x, y, z) \tag{5.1}$$

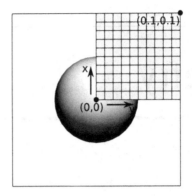

Figure 5.1: Only one quarter domain needs to be simulated due to symmetry.

The first order derivative across the plane of symmetry can be evaluated by considering two points with $x = \delta$ being some small distance,

$$\frac{\partial f}{\partial x}\bigg|_{x=0} \approx \frac{f(\delta, y, z) - f(-\delta, y, z)}{2\delta x} = 0 \tag{5.2}$$

Gradient of any scalar field needs to vanish. In other words, the symmetry boundary condition for field quantities is identical to the zero Neumann condition. Since our code already sets this condition by default, we don't actually have to make any changes besides adjusting the domain dimensions in `main`,

```
World world(11,11,41);
world.setExtents({0,0,0},{0.1,0.1,0.4});  // origin is (0,0)
```

5.2.2 Particles

Consider the illustration in Figure 5.2. Symmetry implies that any event on part of the domain has an identical counterpart across the plane of symmetry. For any particle moving across the boundary, there is an identical particle moving in the opposite direction. Instead of destroying one particle and injecting another one, we reproduce this behavior by *specularly* reflecting the incident ion. We already saw how to implement specular reflection in Chapter 2. We now simply hardwire a bit of new logic to the particle boundary check to reflect particles crossing the $i = 0$ and $j = 0$ ($x = x_0$ and $y = y_0$) planes,

```
void Species::advance() {
  // loop over all particles*/
  for (Particle &part: particles) {
    /* ... */
    // check for symmetric boundary
    if (part.pos[0]<x0[0]) {
      part.pos[0] = 2*x0[0] - part.pos[0];
      part.vel[0] *= -1;  // flip velocity
    }
```

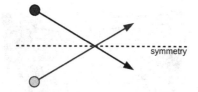

Figure 5.2: Visualization of a specular reflection as a model for particles crossing a symmetric boundary.

```
if (part.pos[1]<x0[1]) {
    part.pos[1] = 2*x0[1] - part.pos[1];
    part.vel[1] *= -1;  // flip velocity
}

// did this particle leave the domain or hit the sphere?
if (!world.inBounds(part.pos) ||
    world.inSphere(part.pos))  part.mpw = 0;  // kill the
        particle
part.dt = 0;  // this particle finished the entire step
}
```

A generalized version of this code would utilize some flags to allow setting reflecting boundaries from **main**. This code also uses the simple integrator from Chapter 3 in which particles are pushed through an entire time step. The check for the symmetric boundary is applied prior to the check for particles hitting the sphere or leaving the domain. This ordering is necessary since it is possible that a reflected particle may also leave the computational domain or impact the sphere.

5.2.3 Results

This new version of the code runs in 99 seconds versus 200 needed for the full domain, and contains 165,000 particles instead of 661,000 with the full domain. Of course, we need to verify the results agree with each other. This is indeed the case, as shown in Figure 5.3. The small differences arise from the domain not being completely symmetric due to the inability fo the sugar-cubed representation to resolve the smooth surface geometry.

5.3 PERIODIC SYSTEMS

Now let's assume that you want to model a system containing many identical and uniformly spaced spheres along the y direction. Let's also assume that there is an electric field $E_y = -500$ V/m produced by some external electrodes. While the geometry is symmetric, the problem is no longer symmetric due to the presence of the applied field. This field can be expected to accelerate ions in the $-y$ direction. Symmetry would imply that any ion crossing the $-y$ domain boundary is elastically reflected back into the domain, which

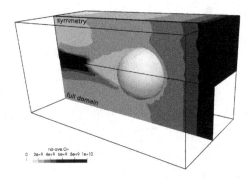

Figure 5.3: Comparison of full domain and symmetric results.

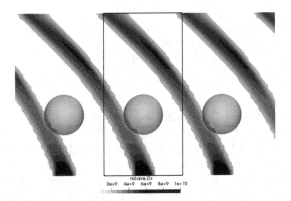

Figure 5.4: Example of a periodic system. Only the region inside the black rectangle is simulated.

would be non-physical. Instead, this system is *periodic*. Periodic simulations generally model annular configurations that wrap on themselves, but they are also applicable to linear cases such as the one here. Again, as was the case with the symmetric simulation above, implementing periodicity involves modifying boundary conditions for the field solver and for the particles.

5.3.1 Domain Boundary

The important distinction with periodic systems is that the "left-most" and the "right-most" nodes are identical. For a system periodic in y, we have

$$\phi_{i,0,k} = \phi_{i,nj-1,k} \tag{5.3}$$

This relationship then implies that the standard stencil can be used across the boundary. The y component of the Laplace operator is discretized as

$$\frac{\partial^2 \phi}{\partial y^2}\bigg|_{i,j,k} = \frac{\phi_{i,j-1,k} - 2\phi_{i,j,k} + \phi_{i,j+1,k}}{\Delta^2 y} \tag{5.4}$$

For any node on the $j = 0$ face we have

$$\frac{\partial^2 \phi}{\partial y^2}\bigg|_{i,0,k} = \frac{\phi_{i,n_j-2,k} - 2\phi_{i,0,k} + \phi_{i,1,k}}{\Delta^2 y} \tag{5.5}$$

Similarly, on the $j = n_j - 1$ side,

$$\frac{\partial^2 \phi}{\partial y^2}\bigg|_{i,n_j-1,k} = \frac{\phi_{i,n_j-2,k} - 2\phi_{i,n_j-1,k} + \phi_{i,1,k}}{\Delta^2 y} \tag{5.6}$$

These expressions are identical, and hence we expect to recover the relationship from Equation 5.3. We implement the boundary by modifying code in PotentialSolver::buildMatrix,

```
void PotentialSolver :: buildMatrix () {
  /* ... */
  // set matrix coefficients
  for (int k=0;k<nk;k++)
    for (int j=0;j<nj;j++)
      for (int i=0;i<ni;i++) {
        /* ... */
        node_type[u] = NEUMANN;                            // default
        if (i==0) {A(u,u)=idx;A(u,u+1)=-idx;}              // x-min
        else if (i==ni-1) {A(u,u)=idx;A(u,u-1)=-idx;}      // x-max
        else if (k==0) {A(u,u)=idz;A(u,u+ni*nj)=-idz;}     // z-min
        else if (k==nk-1) {A(u,u)=idz;A(u,u-ni*nj)=-idz;}  // z-max
        else {                          // standard internal stencil
          int uy_minus = (j==0)?world.U(i,nj-2,k):u-ni;
          int uy_plus = (j==nj-1)?world.U(i,1,k):u+ni;
          A(u,u-ni*nj) = idz2;
          A(u,uy_minus) = idy2;
          A(u,u-1) = idx2;
          A(u,u) = -2.0*(idx2+idy2+idz2);
          A(u,u+1) = idx2;
          A(u,uy_plus) = idy2;
          A(u,u+ni*nj) = idz2;
          node_type[u] = REG;          // set regular node type
        }
      }
}
```

We start by removing the application of Neumann boundaries to the y faces. We then modify the code setting the regular internal stencil to use variables uy_minus and uy_plus in place of the previously hard coded $u - n_i$ and $u + n_i$. On boundaries, where $i = 0$ or $i = n_j - 1$, we set these variables to $n_j - 2$ and 1, respectively. On these nodes, the row, while still consisting of seven non-zero values, contains coefficients located off the primary bands. A sparse matrix container based on predefined banded vectors would not be able to support this periodic system.

The fact that the nodes wrap around also extends to the electric field calculation. Instead of applying a one-dimension form, we use the standard central-difference on the periodic boundaries,

$$E_y|_{i,0,k} \equiv -\frac{\partial \phi}{\partial x} = \frac{\phi_{i,1,k} - \phi_{i,ni-2,k}}{2\Delta x} \tag{5.7}$$

Similarly, on the $y+$ face,

$$E_y|_{i,n_j-1,k} = \frac{\phi_{i,1,k} - \phi_{i,ni-2,k}}{2\Delta x} \tag{5.8}$$

These two expressions are identical. We thus write

```
// void PotentialSolver::computeEF()
// y component, periodic boundary
if (j==0 || j==world.nj−1)
    ef[i][j][k][1] = −(phi[i][1][k] − phi[i][world.nj−2][k])/(2*dy);
else
    ef[i][j][k][1] = −(phi[i][j+1][k] − phi[i][j−1][k])/(2*dy);
```

5.3.2 Particles

Particles leaving through a periodic boundary are re-injected from the opposite end with no change in their velocity. We add the following code to Species::advance boundary check:

```
// did the particle leave through the periodic boundary?
if (part.pos[1]<x0[1]) part.pos[1] += Ly;
else if (part.pos[1]>=xm[1]) part.pos[1] −= Ly;

// did this particle leave the domain or hit the sphere?
if (!world.inBounds(part.pos) ||
world.inSphere(part.pos))  part.mpw = 0;   // kill the particle
```

The teleportation to the opposite end is accomplished by adding, or subtracting, the y dimension domain length, $L_y = (n_j - 1)\Delta y$ to the particle y position.

5.3.3 Density Synchronization

There is one more piece missing, and that is the need to synchronize the macroscopic field properties, such as density. Prior to this change, the density on the $y-$ nodes is computed using only particles located in the $j = 0$ cells. But due to the periodicity, particles in the right-most cell $j = n_j-2$ also contribute to this node, since $n_{i,0,k}$ and $n_{i,n_j-1,k}$ are effectively the same point. Therefore we simply add the two contributions to set the $j = 0$ node. We then copy these values to nodes on the $n_j - 1$ boundary. This is accomplished by adding the following code to Species::computeNumberDensity.

```
// add up contributions along the periodic boundary
for (int i=0;i<world.ni;i++)
  for (int k=0;k<world.nk;k++) {
    den[i][0][k] += den[i][world.nj-1][k];
    den[i][world.nj-1][k] = den[i][0][k];
}
```

This change required that the code computing node volumes in World::
computeNodeVolumes is modified by removing the scaling by 0.5 in the y
direction. Alternatively, had we kept the half volumes on the y boundaries,
we would need to set the density to the average,

```
den[i][0][k] = 0.5*(den[i][0][k]+den[i][world.nj-1][k]);
```

The approach taken here is more general, as it applies to cases with non-
uniform mesh spacing.

5.3.4 Results

Finally, to make the visualization more interesting, we reduce the particle
injection source area to a smaller rectangle with bounds $[x_0 + 0.4L_x, x_0 + 0.6L_x] \times [y_0 + 0.4L_y, y_0 + 0.6L_y]$ by setting

double A = 0.2*Lx*0.2*Ly;

and

double3 pos {x0[0]+(0.4+0.2*rnd())*Lx, x0[1]+(0.4+0.2*rnd())*Ly,
    x0[2]};

in ColdBeamSource::sample(). With these changes, we obtain the result in
Figure 5.4. The code only simulates the region highlighted by the dark rectangle. The surrounding solutions were added during post processing by utilizing
Paraviews transform filter. Note that there is no discontinuity along the
boundaries.

5.4 TWO-DIMENSIONAL CODES

Yet an even greater speed up is possible by switching to two dimensions.
Often, even if the problem exhibits localized three-dimensional effects, we can
obtain an approximate two-dimensional solution in a fraction of time needed
for the full 3D run. Two-dimensional codes can be used as proof of concept
studies prior to committing computational resources needed for the full run.
The speed up arises from fewer mesh cells, since effectively the n_k dimension
reduces to 1. The number of unknowns scales with $1/\Delta_h^2$ instead of the $1/\Delta_h^3$
dependence in 3D codes. Refining the mesh to use half-sized cells increases
the number of nodes by a factor of four. Similar refinement applied to a 3D
problem increases the node count eight fold. Not only is the memory footprint
reduced, the solver convergence time also decreases. The number of particles is
also greatly reduced due the lower cell count. Of course, an even greater speed
up is possible by switching to one dimension, but there are simply not as many

cases for which the 1D approximation is appropriate. They generally involve studies of plasma sheath near a large planar wall, or discharge formation between two large electrodes.

Two- and one-dimensional codes can be characterized by the number of spatial and velocity dimensions they retain. We focus on *2D-3V* codes in which all three velocity components are retained, but spatial properties are a function of only two coordinates. For generality, I like to retain all three position coordinates within the `Particle` data structure. The "extra" coordinate is simply ignored when performing particle to mesh, and mesh to particle interpolations.

Two-dimensional codes can be classified according to which spatial coordinates are modeled. The most common types are:

- **Planar (XY)**: The simulation domain is a plane corresponding to a slice of an infinitely deep 3D Cartesian "brick". This may be the case when simulating the flow of plasma past a very long cylinder.

- **Axisymmetric (RZ)**: This common scenario simulates cylindrical devices in which there is no variation with the azimuthal rotation about the z axis, $\partial()/\partial\theta = 0$.

- **Azimuthal (Rθ)**: This is another possible configuration for cylindrical devices. We now model an azimuthal plane, and assume there is no variation in the axial direction, $\partial()/\partial z = 0$.

It is important to keep in mind that even if the geometry is symmetric, the problem may involve three-dimensional effects. One common example from the field of electric propulsion involves electron transport in Hall effect thrusters. These devices consist of an annular channel in which electrons are trapped in a closed azimuthal drift due to a crossed $\vec{E} \times \vec{B}$ configuration. The geometry is axisymmetric and historically, these devices have been modeled with RZ codes. The codes however fail to recover the experimentally observed electron diffusion across the magnetic field lines. Recent work indicates this "anomalous diffusion" may be arising from ionization waves that develop in the θ direction [6]. Therefore, despite the device being completely axisymmetric, the resulting plasma state contains spatial features that vary with θ.

5.5 PLANAR (XY) CODES

Assume that instead of modeling the flow past a sphere, we are interested in studying the flow past a very long cylinder, as shown in Figure 5.5. The cylinder is oriented with the long axis in the y direction. Since the geometry no longer varies in y, and since the injection source is also spatially invariant, we can expect that $\partial()/\partial y$ vanishes for any property of interest. We thus reduce the simulation to a single $x - z$ slice. This is an example of a planar, "XY" code, despite the simulation actually modeling an $x - z$ slice due to our choice of coordinates.

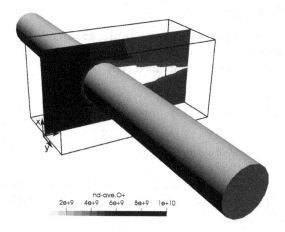

Figure 5.5: Example of a problem suitable to 2D planar modeling.

5.5.1 Node Volumes

Converting a 3D Cartesian code a 2D planar code is very simple. We disregard the third spatial coordinate when performing the interpolation to and from the grid. In doing so, we need to decide on the meaning behind the coordinates. Here we let $x = \vec{x}_0$ and $z = \vec{x}_1$. Parts of the code that depend on the now discarded y dimension are modified to assume unit depth. For instance, node volumes are now computed as

$$V_c = \Delta x \Delta z (1\text{m}) \tag{5.9}$$

Similarly, the source area for particle injection is

$$A = L_x \cdot (1\text{m}) \tag{5.10}$$

Since in our 3D simulation $\Delta y = 0.01$ m, this change results in cell volumes growing by two orders of magnitude. While this change has no impact on the math behind the simulation, it impacts the number of created macroparticles. In our approach, the macroparticle weight is fixed, and the number of simulation particles to inject is obtained from Equation 3.19. We thus need to remember to scale the macroparticle weight when switching from 3D to 2D. Alternatively we could fix the number of macroparticles to generate per time step, and let the code compute w_{mp} accordingly. This change may be especially warranted for generalized codes that support both planar and axisymmetric configuration. The depth dimension can be vastly different in these two approaches.

5.5.2   Data Containers

To support the new operations, the `World::XtoL` code is modified to return a two-dimensional coordinate of type `double2`, which is defined as `vec2<double>`. This type is a simplified copy of `vec3`,

```
template <typename T>
struct vec2 {
  vec2 (const T u, const T v) : d{u,v} {}
  /* ... */
};
using double2 = vec2<double>;
using int2 = vec2<int>;
```

We also update the overloaded `<<` operator in case we need to write out `vec2` data to a file. The extra zero is written out since VTK does not support two-dimensional vectors.

```
template<typename T>                      // ostream output
std::ostream& operator<<(std::ostream &out, vec2<T>& v) {
  out<<v[0]<<" "<<v[1]<<" 0";       // write out 3D vector
  return out;
}
```

The `XtoL` function in `World.h` now reads

```
double2 XtoL(double3 x) const {
  double2 lc;
  lc[0] = (x[0]-x0(0))/dh(0);
  lc[1] = (x[1]-x0(1))/dh(1);
  return lc;
}
```

Note that the input is a `double3` position to allow passing in particle data. The third coordinate is simply ignored. We also modify the `Field_` data object to store only two-dimensional data,

```
template <typename T>
class Field_ {
public:
  Field_(int ni, int nj) : ni{ni}, nj{nj} {
    // allocate memory for a 2D array
    data = new T*[ni];
    for (int i=0;i<ni;i++)
      data[i] = new T[nj];
  }
  /* ... */
protected:
  T **data;   // data held by this field
};
```

The `Field3` object used to store electric fields and particle velocities retains `double3` as its contained type. This allows us to set a fixed E_y electric field, if needed, but with data varying only in two dimensions,

$$\vec{E}_{i,j} = (E_x)_{i,j}\hat{i} + (E_y)_{i,j}\hat{j} + (E_z)_{i,j}\hat{k} \tag{5.11}$$

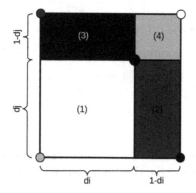

Figure 5.6: First order mesh to particle interpolation in two dimensions.

5.5.3 Scatter and Gather

We also make changes to the scatter and gather routines. Since the mesh is now two-dimensional, the mesh-to-particle interpolation is based on *area* ratios, as depicted in Figure 5.6. We have the following

```
// World.h
void scatter(double2 lc, T value) {
    int i = (int)lc[0];
    double di = lc[0]-i;

    int j = (int)lc[1];
    double dj = lc[1]-j;

    // use area ratios to scatter data
    data[i][j]     += (T)value*(1-di)*(1-dj);
    data[i+1][j]   += (T)value*(di)*(1-dj);
    data[i+1][j+1] += (T)value*(di)*(dj);
    data[i][j+1]   += (T)value*(1-di)*(dj);
}

T gather(double2 lc) {
    int i = (int)lc[0];
    double di = lc[0]-i;

    int j = (int)lc[1];
    double dj = lc[1]-j;

    // gather data using area ratios
    T val = data[i][j]*(1-di)*(1-dj)+
        data[i+1][j]*(di)*(1-dj)+
        data[i+1][j+1]*(di)*(dj)+
        data[i][j+1]*(1-di)*(dj);

    return val;
}
```

The electric field at the particle position is computed per

```
double2  lc = world.XtoL(part.pos);
double3  ef_part = world.ef.gather(lc);
```

The change to the code computing number density is also minimal,

```
double2  lc = world.XtoL(part.pos);
den.scatter(lc, part.mpw);
```

5.5.4 Output

As the final change, we modify the Output::fields output routine to represent a 2D mesh by setting the third dimension of Extent to 0 0,

```
out<<"WholeExtent=\"0 "<<world.ni-1<<" 0 "<<world.nj-1<<" 0
    "<<0<<"\">\n";
```

With these changes, we obtain the result shown in Figure 5.5. Comparison to the 3D simulation is left as one of the suggested exercises.

5.6 AXISYMMETRIC (RZ) CODES

I am a big fan of axisymmetric codes. There are surprisingly many cases for which this approximation is appropriate. These codes represent a three-dimensional domain generated by revolving a plane about the z axis. The other coordinate is the radial distance r. Because of this rotation, cell sizes are no longer uniform. This is visualized in Figure 5.7. This radial variation in cell volumes requires us to make several corrections. These include the need to update coefficients in the potential solver. We also need to modify the particle injection and the particle-to-mesh interpolation algorithms. Particle motion is also affected.

5.6.1 Node Volumes

We start by modifying the code for computing node volumes. It is important to realize that in the axisymmetric formulation, each node control volume corresponds to an *annulus* with some finite width Δz, and inner and outer radius r_1 and r_2. We can visualize this volume by imagining a single two-dimensional cell is rotated 360° about the axis. The volume is given by

$$V_{i,j} = \pi(r_{i,j+0.5}^2 - r_{i,j-0.5}^2)\Delta z \tag{5.12}$$

I am using j to correspond to the radial coordinate, and hence we are actually developing a "ZR" code. The ordering of coordinates is not important, as long as we are consistent. The direction of the θ coordinate follows the right hand rule. Assuming uniform spacing, $r_2 = r_{i,j+0.5} = r_{i,j} + 0.5\Delta r$ and similarly, $r_1 = r_{i,j} - 0.5\Delta r$. The above equation can then be rewritten as

$$V_{i,j} = \pi(2r_j\Delta r + \Delta^2 r)\Delta z \tag{5.13}$$

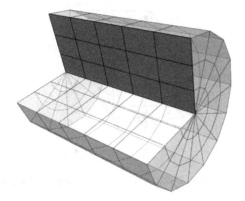

Figure 5.7: Cell volumes in axisymmetric codes vary with radial distance from the axis of rotation.

which is identical to

$$V_{i,j} = 2\pi r_{i,j+0.5}\Delta r\Delta z \tag{5.14}$$

In other words, the cell volume is also the circumference of the circle at the control volume centroid multiplied by the axial cell length, Δz. These two schemes are identical, but the first approach based on subtracting surface areas of two circles makes setting the volume on domain boundaries more natural. We simply make sure that $j - 0.5 \geq 0$ and $j + 0.5 \leq n_j - 1$. We perform this calculation using a `getR` function that evaluates the r coordinate per $r = r_0 + j\Delta r$. The resulting code is

```
void World::computeNodeVolumes() {
  for (int i=0;i<ni;i++)
    for (int j=0;j<nj;j++) {
      double r2 = getR(min(j+0.5,nj-1.0));
      double r1 = getR(max(j-0.5,0.0));
      double dz = dh[0];
      if (i==0 || i==ni-1) dz*=0.5;
      double V = dz*Const::PI*(r2*r2-r1*r1);
      node_vol[i][j] = V;
    }
}
```

The inclusion of the π term is not required as long as it is also omitted in other parts of the program.

5.6.2 Finite Volume Method

Next, we apply a correction to the potential solver. The Laplacian operator in cylindrical coordinates is given by

$$\nabla_r^2\phi = \frac{1}{r}\frac{\partial}{\partial r}\left(r\frac{\partial\phi}{\partial r}\right) + \frac{1}{r^2}\frac{\partial^2\phi}{\partial\theta^2} + \frac{\partial^2\phi}{\partial z^2} \tag{5.15}$$

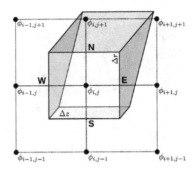

Figure 5.8: Visualization of the node control volume used by the Finite Volume Method.

For an axisymmetric domain with $\partial()/\partial\theta = 0$, we can also write an alternate form

$$\nabla_r^2\phi = \frac{1}{r}\frac{\partial\phi}{\partial r} + \frac{\partial^2\phi}{\partial r^2} + \frac{\partial^2\phi}{\partial z^2} \tag{5.16}$$

While we could rewrite this formulation directly with the help of the Finite Difference Method (FDM), we now introduce a new scheme for discretizing differential equations known as the *Finite Volume Method* or FVM. The great benefit of FVM is that, unlike FDM, it can be used on non-orthogonal meshes in which cell edges are not aligned with the coordinate axes. In the case of an orthogonal mesh, the FVM formulation reduces to the same set of equations obtained with FDM.

We start by dividing the domain into non-overlapping control volumes. In our node-centered formulation, these are the same node volumes used previously when computing number densities. Each control volume is centered on node (i, j) and spans the region from $(i - 0.5, j - 0.5)$ to $(i + 0.5, j + 0.5)$. We integrate the governing equation over each volume,

$$\int_V \nabla \cdot \nabla\phi dV = \int_V -\frac{\rho}{\epsilon_0}dV \tag{5.17}$$

We assume that the control volume is small enough so that ρ is uniform. The right hand side can be integrated as $(-\rho/\epsilon_0)\Delta V$. Next, the Divergence Theorem,

$$\int_V \nabla \cdot \vec{f}dV = \oint_S \vec{f} \cdot \hat{n}dA \tag{5.18}$$

is applied to the left hand side. This theorem relates the divergence of a vector field to the normal flux through the enclosing surface. We thus have

$$\oint_S \nabla\phi \cdot \hat{n}dA = -\frac{\rho}{\epsilon_0}\Delta V \tag{5.19}$$

The surface integral is evaluated along the volume boundaries. Again, we assume that the control volume is small enough so that $\nabla\phi$ is constant along each face. To simplify the subsequent equations, we label the four faces (there are no faces in the θ dimension since the domain wraps on itself) as East, North, West, and South. This is depicted in Figure 5.8. Note that the control volume is an annulus, despite for simplicity being shown as a wedge. The left hand side is written as

$$\oint_S \nabla\phi \cdot \hat{n} dA = (\nabla\phi)_E \cdot (\hat{n}\Delta A)_E + (\nabla\phi)_N \cdot (\hat{n}\Delta A)_N + \tag{5.20}$$
$$(\nabla\phi)_W \cdot (\hat{n}\Delta A)_W + (\nabla\phi)_S \cdot (\hat{n}\Delta A)_S$$

In our setup, the grid cells are aligned with the $\hat{e}_z$ and $\hat{e}_r$ directions. Therefore,

$$(\hat{n}\Delta A)_E = \hat{e}_z 2\pi r_{i,j} \Delta r \tag{5.21}$$
$$(\hat{n}\Delta A)_N = \hat{e}_r 2\pi r_{i,j+0.5} \Delta z \tag{5.22}$$
$$(\hat{n}\Delta A)_W = -\hat{e}_z 2\pi r_{i,j} \Delta r \tag{5.23}$$
$$(\hat{n}\Delta A)_S = -\hat{e}_r 2\pi r_{i,j-0.5} \Delta z \tag{5.24}$$

With the definition $\Delta V = 2\pi r_{i,j} \Delta r \Delta z$, Equation 5.19 reduces to

$$(\nabla\phi_N \cdot \hat{e}_r) \frac{r_{i,j+0.5}}{r_{i,j}\Delta r} - (\nabla\phi_S \cdot \hat{e}_r) \frac{r_{i,j-0.5}}{r_{i,j}\Delta r} + \tag{5.25}$$
$$(\nabla\phi_E - \nabla\phi_W) \cdot \hat{e}_z \frac{1}{\Delta z} = -\frac{\rho}{\epsilon_0}$$

The gradient in cylindrical coordinates has the same form as the Cartesian counterpart, and thus we directly substitute

$$\nabla\phi \cdot \hat{e}_z = \frac{\partial\phi}{\partial z} \tag{5.26}$$

and

$$\nabla\phi \cdot \hat{e}_r = \frac{\partial\phi}{\partial r} \tag{5.27}$$

For non-orthogonal meshes or meshes not aligned with the coordinate axes, we instead use the Stokes Theorem

$$\oint_S \nabla\phi dA = \oint_C \phi\hat{n}ds \tag{5.28}$$

Here the integration is over a two-dimensional control volume centered on the point where the gradient needs to be evaluated and the $\hat{n}$ normal vector points outward for a counterclockwise rotation. For example, for $\nabla\phi_E = (\nabla\phi)_{i+0.5,j}$ we have

$$\nabla\phi_E \Delta z \Delta r = \phi_{i+1,j}\Delta r\hat{e}_z + \phi_{i+0.5,j+0.5}\Delta z\hat{e}_r - \phi_{i,j}\Delta r\hat{e}_z - \phi_{i+0.5,j-0.5}\Delta z\hat{e}_r \tag{5.29}$$

or

$$\nabla\phi_{i+0.5,j} = \frac{\phi_{i+1,j} - \phi_{i,j}}{\Delta z}\hat{e}_z + \frac{\phi_{i+0.5,j+0.5} - \phi_{i+0.5,j-0.5}}{\Delta r}\hat{e}_r \qquad (5.30)$$

and

$$\nabla\phi_E \cdot \hat{e}_z = \frac{\phi_{i+1,j} - \phi_{i,j}}{\Delta z} \qquad (5.31)$$

which is the standard central difference form of the first derivative. By making similar substitutions for $\phi_N \cdot \hat{e}_r$, $\phi_W \cdot \hat{e}_z$, and $\phi_S \cdot \hat{e}_r$, we obtain

$$\left(\frac{\phi_{i,j+1} - \phi_{i,j}}{\Delta^2 r}\right)\frac{r_{i,j+0.5}}{r_{i,j}} - \left(\frac{\phi_{i,j} - \phi_{i,j-1}}{\Delta^2 r}\right)\frac{r_{i,j-0.5}}{r_{i,j}} \qquad (5.32)$$

$$\left(\frac{\phi_{i+1,j} - \phi_{i,j}}{\Delta^2 z}\right) - \left(\frac{\phi_{i,j} - \phi_{i-1,j}}{\Delta^2 z}\right) - = -\frac{\rho}{\epsilon_0}$$

Finally, by letting $r_{i,j\pm0.5} = r_{i,j} \pm \Delta r/2$, we arrive at

$$\left(\frac{\phi_{i,j-1} - 2\phi_{i,j} + \phi_{i,j+1}}{\Delta^2 r}\right) + \frac{\phi_{i,j+1} - \phi_{i,j-1}}{2\Delta r} + \qquad (5.33)$$

$$\left(\frac{\phi_{i-1,j} - 2\phi_{i,j} + \phi_{i+1,j}}{\Delta^2 z}\right) = -\frac{\rho}{\epsilon_0}$$

which is identical to the Finite Difference discretization of Equation 5.16. In a generalized simulation code, we would let the computer calculate the areas and gradients based on the local cell geometry. The algorithm would then be applicable to arbitrarily shaped mesh cells. In our code, for simplicity, we directly implement the above correction in **PotentialSolver::buildMatrix**,

```
else {
    // standard internal stencil
    A(u,u-ni) = idr2 - 0.5*idr;  // 1/dr^2 - 1/(2*dr)
    A(u,u-1) = idz2;
    A(u,u) = -2.0*(idz2+idr2);
    A(u,u+1) = idz2;
    A(u,u+ni) = idr2 + 0.5*idr;  // 1/dr^2 + 1/(2*dr)
    node_type[u] = REG;          // set regular node type
}
```

5.6.3 Particle Source

Due to the radial variation in cell volumes, we need to sample more particles in cells farther away from the axis of rotation in order to obtain uniform number density, $n = N/V$. Consider a small strip of width dr centered on r with area $V = 2\pi r dr$. The relative number of particles to create in this strip is proportional to the ratio between the strip and the total injection surface area. The cumulative probability of finding a particle in $[r_i, r]$ is given by

$$P = \frac{\int_{r_1}^{r} 2\pi r dr}{\int_{r_1}^{r_2} 2\pi r dr} = \frac{r^2 - r_1^2}{r_2^2 - r_1^2} \qquad (5.34)$$

Here r_1 and r_2 is the inner and outer radius of the injection surface. As we saw in Chapter 4, we can sample from an arbitrary distribution using uniformly distributed random numbers to correspond to a cumulative probability value. The particle radial positions are sampled per

$$r = \sqrt{A\mathcal{R} + r_1^2} \tag{5.35}$$

This change is implemented in the `ColdBeamSource::sample` function as

```
double3 pos {x0[0], sqrt(A*rnd()+x0[1]*x0[1]), 0};
```

For our domain with $r_1 = 0$, the above expression simplifies to $r_m\sqrt{\mathcal{R}}$.

5.6.4 Scatter and Gather

The variation in cell volumes also affects the scatter and gather operations. We start with the diagram in Figure 5.6, but take into account variation in "depth" with an increasing j coordinate. The normalized volume for section (1) is

$$V_1 = d_i d_j \left(\frac{r_j + 0.5 d_j \Delta r}{r_j + 0.5 \Delta r} \right) \tag{5.36}$$

where the correction in the parentheses is the ratio between the circumference evaluated the midpoint of section (1) and the entire cell. The term r_j is the radius along the cell bottom edge. We can similarly derive the correction for section (3)

$$V_3 = d_i(1 - d_j) \left(\frac{r_j + 0.5(d_j + 1)\Delta r}{r_j + 0.5 \Delta r} \right) \tag{5.37}$$

The correction factor for section (1) also applies to section (2), and similarly the factor from (3) applies to (4). The modified scatter algorithm in `Field.h` reads

```
void scatter(double2 lc, T value, double r0, double dr) {
    int i = (int)lc[0];
    double di = lc[0]-i;

    int j = (int)lc[1];
    double dj = lc[1]-j;

    // compute correction factors
    double rj = r0+j*dr;
    double f1 = (rj+0.5*dj*dr)/(rj+0.5*dr);
    double f2 = (rj+0.5*(dj+1)*dr)/(rj+0.5*dr);
    data[i][j]    += (T)value*(1-di)*(1-dj)*f2;
    data[i+1][j]  += (T)value*(di)*(1-dj)*f2;
    data[i+1][j+1] += (T)value*(di)*(dj)*f1;
    data[i][j+1]  += (T)value*(1-di)*(dj)*f1;
}
```

Note that we now need to provide the radial component of the mesh origin, r_0, as well as the cell spacing Δr. These parameters are provided directly, instead of a reference to the `World` object. Since `Field` is defined as a templated function with all code listed directly in the header, we cannot easily include the `World` object. `World` includes `Field.h` and thus including `World.h` from `Field` would result in a circular reference. We could get around this limitation by, for instance, moving the `Scatter` function out of `Field`, but the above approach seemed like a good work around.

The gather function is adjusted similarly,

```
T gather(double2 lc ,double r0, double dr) {
    int  i = (int)lc[0];
    double di = lc[0]-i;

    int j = (int)lc[1];
    double dj = lc[1]-j;

    // compute correction factors
    double rj = r0+j*dr;
    double f1 = (rj+0.5*dj*dr)/(rj+0.5*dr);
    double f2 = (rj+0.5*(dj+1)*dr)/(rj+0.5*dr);

    // gather electric field onto particle position
    T val = data[i][j]*(1-di)*(1-dj)*f2+
            data[i+1][j]*(di)*(1-dj)*f2+
            data[i+1][j+1]*(di)*(dj)*f1+
            data[i][j+1]*(1-di)*(dj)*f1;
    return val;
}
```

Treatment of particle weighing schemes for axisymmetric codes for more complex configurations is discussed in detail in [9].

5.6.5 Particle Rotation

Finally, we need to modify the particle push algorithm to simulate an axisymmetric domain. There are two routes we can consider here. One option is to transform the equations of motion into cylindrical coordinates. However, a simpler option for unmagnetized plasmas is to push the particles in 3D Cartesian (x, y, z) coordinates, but then perform *rotation* back to the computational plane after each push [15].

Consider a particle initially positioned on the computational $z - r$ plane. If the velocity contains an off-plane component, $v'_a \neq 0$, the particle will find itself somewhere off-plane at the completion of the push. This is visualized in Figure 5.9. Here A is the distance traveled off plane, while B is the displacement in the $\hat{e}_r$ direction. Particle's new radial distance from that axis of rotation r is

$$r = \sqrt{A^2 + B^2} \tag{5.38}$$

Instead of setting the distance directly, we can rotate the position through $\theta_r = -\theta$. The angle θ is computed from

$$\theta = \operatorname{atan}\left(\frac{A}{B}\right) = \operatorname{asin}\left(\frac{A}{R}\right) \tag{5.39}$$

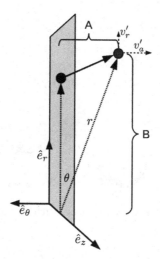

Figure 5.9: Visualization of post-push particle rotation.

Velocities need to be rotated similarly,

$$\left[\begin{array}{c} u_r \\ u_y \end{array}\right] = \left[\begin{array}{cc} \cos\theta_r & -\sin\theta_r \\ \sin\theta_r & \cos\theta_r \end{array}\right] \left[\begin{array}{c} u'_r \\ u'_y \end{array}\right] \tag{5.40}$$

Here u_r and u_y are the new velocity components. This algorithm is implemented in a new **rotateToZR** function added to **Species**,

```
void Species::rotateToZR(Particle &part) {
    double A = part.vel[2]*part.dt;    // movement off plane
    double B = part.pos[1];            // initial radius
    double r = sqrt(A*A + B*B);        // new radius

    double cos = B/r;
    double sin = A/r;

    part.pos[2] += acos(cos);      // update theta for plotting

    // rotate velocity through theta
    double v1 = part.vel[1];
    double v2 = part.vel[2];
    part.pos[1] = r;
    part.vel[1] = cos*v1 + sin*v2;
    part.vel[2] = -sin*v1 + cos*v2;
}
```

We add a call to this function to **Species::advance**,

```
// update position from v=dx/dt
part.pos += part.vel*part.dt;

// rotate particle back to ZR plane
rotateToZR(part);
```

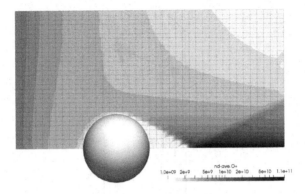

Figure 5.10: Ion density computed with the axisymmetric code.

```
// did this particle leave the domain?*/
if (!world.inBounds(part.pos)) { /*...*/ }
```

5.6.6 Results

Finally, with the above changes, we obtain the result visualized in Figure 5.10. Note that ion density is uniform in the upstream region near the injection boundary, as expected, despite the radial variation in cell volumes. This simulation was run with the the computational domain extended in the r dimension,

```
World world(41,31);
world.setExtents({0,0.0},{0.4,0.2});
```

For a smaller domain, you may observe some discrepancy between this and the full solution. In both simulations, we are assuming that the outer boundary is sufficiently removed from the sphere not to affect the solution. But that is not quite the case, especially for the domain spanning only ± 0.1 m in the x and y (or r) direction. The 3D simulation is then effectively modeling the flow of plasma in a cube, while the axisymmetric case is modeling the flow in a cylinder.

5.7 SUMMARY

In this chapter we saw how to speed up simulations by taking advantage of inherent symmetry. We started by modifying the 3D code to simulate only one quarter of the domain. We then discussed adding periodic boundaries prior to moving to two-dimensional codes. We developed a planar and an axisymmetric variant. We discussed the appropriate changes that need to be implemented in axisymmetric codes to account for the radial variation in cell volumes.

EXERCISES

5.1 *Flow Past a Cylinder.* Verify that the planar simulation produces results identical to those obtained from the full 3D run. Modify the three-dimensional code to model a cylinder instead of a sphere. This change mainly involves changing the `World::inSphere` algorithm.

5.2 *Effect of Boundaries.* Rerun the three-dimensional simulation with the domain increased by a factor of two in the x and y dimension and compare the result to the original $(-0.1, -0.1)$ to $(0.1, 0.1)$ span. Can you observe any difference in the overlapping region? If so, how large does the domain need to grow before the differences become negligible? Next, compare these results to those obtained with the axisymmetric simulation. How large does the axisymmetric domain need to grow before the results are identical to the 3D counterpart?

5.3 *One-Dimensional Sheath Simulation.* Use the 2D planar code as a starting point to develop a fully-kinetic one-dimensional simulation of the plasma sheath. You may want to use the direct solver from Chapter 1 to decrease run time. Let the simulation model the region between two large flat plates separated by a distance L. Potential on both walls is set to 0 V. Initialize the simulation by preloading uniform ion and electron density, with $kT_i = 0.1$ eV and $kT_e = 1$ eV. Ions and electrons impacting the walls are removed from the simulation, and hence the bulk density will decay. Set the time step such that electrons do not move more than one cell per time step. Run the simulation long enough until net current disappears, $I_e + I_i = 0$ (within numerical limits). Overlay the plot of electron and ion densities with the plot of plasma potential. You should observe formation of three regions: a neutral bulk plasma with $n_e = n_i$ and $E_x = 0$, a neutral pre-sheath with $E_x \neq 0$, and a non-neutral sheath with $n_e \neq n_i$ and $E_c \neq 0$. Feel free to experiment with different plasma densities, initial temperatures, applied voltage, and domain dimensions.

5.4 *Secondary Electron Emission.* Next modify the prior code to include secondary electron emission. Let every incident electron, on average, generate $\gamma = \in [0, 1]$ new electrons. Inject these electrons by sampling velocity from a Maxwellian at the wall temperature $T_{wall} = 1000$ K. Observe the electron velocity distribution function. You should observe some cooling of the primary population by the cold wall electrons. Furthermore, study the impact of different γ value on the sheath profile. With a sufficiently high secondary electron emission coefficient, you should observe a sheath inversion, which ends up trapping the low temperature population in a near wall region.

CHAPTER 6

Unstructured Meshes

6.1 INTRODUCTION

I N THIS chapter, we learn how to develop PIC simulations that run on unstructured tetrahedral meshes. These meshes allow us to resolve detailed surface geometries. But they also add a lot of complexity to both the particle pusher and the potential solver. We start by covering mesh import and export. We then explore how to move particles and how to use their positions to compute number density. Finally, we cover the basics of the Finite Element Method that can be used to solve potential and calculate electric field.

6.2 MESH GEOMETRY

The structured mesh that has been used so far simplifies the particle advance algorithm as it allows us to easily determine the cell in which a particle is located based on some physical-to-logical *mapping function*. Since this information is needed to gather the electric field and to scatter particle positions to compute densities, it is imperative that this mapping is fast. For the uniform Cartesian mesh, this map is the trivial $i = (x - x_0)/\Delta x$ relationship we have been using so far. But as we have also seen, uniform Cartesian meshes are not able to resolve complex geometries with curved surfaces. All our simulations of flow past a sphere truly modeled the flow past a collection of sugarcubes. Furthermore, even if the geometrical considerations are put aside, we may want to vary cell spacing based on local variation in plasma density. Some approaches that could be used include the use of *mesh stretching, multiple-domains*, or *adaptive mesh refinement*. For simple geometries, we may be able to use a *body-fitted* mesh with an analytical physical-to-logical mapping. Yet, for more complex configurations, we may need to move to an *unstructured mesh*.

As the name indicates, these meshes lack any inherent structure that would relate the physical and logical coordinates. Figure 6.1 plots an example. This picture actually shows a slice through a tetrahedral volume mesh, and hence the slice contains both triangular and quadrilateral elements. As can be seen,

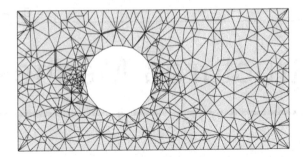

Figure 6.1: Slice through a tetrahedral unstructured mesh.

the mesh contains elements of varying sizes that are arranged without any particular ordering. Simply knowing a physical coordinate (x, y, z) does not allow us to easily determine which cell the point is located in. Of course, we could loop through every single element, and for each, test the location with some `isPointInTriangle` or `isPointInTetrahedron` function. This approach is very inefficient, given the millions of particles and cells in a typical PIC simulation. The algorithm developed in this chapter is faster than this brute force approach, but there is simply no way to avoid the added computational overhead of unstructured meshes. The potential solver also needs to be modified. We could utilize the Finite Volume Method from Chapter 5, but it is more common to deploy the *Finite Element Method* (FEM). The solver ends up being much more complicated than what we have seen so far, despite utilizing only first-order elements. This formulation results in a constant (zeroth order) electric field per cell. Therefore, before moving all our simulations to unstructured meshes, it is important to weigh the pros and cons. Does the ability to resolve curved surfaces outweigh the slower run time, added computational complexity, and the lack of smoothly varying electric field? Sometimes the optimal solution may be to utilize a hybrid approach, such as one using *cut cells* to resolve the interface cells [57].

6.2.1 Mesh Generation

Before we can start discussing moving particles or solving the field, we need to load the mesh geometry. Unstructured meshes need to be generated in external meshing programs. Some tools I have used in the past include Salome (`https://www.salome-platform.org`) and Pointwise (`https://www.pointwise.com`), but there are many more. We generally start by drawing the geometry in a Computer Aided Design (CAD) program and exporting it in some mutually understood format, such as IGES or STEP. This geometry is then loaded into the meshing application. Some CAD programs include their own meshers, and some meshers have their own rudimentary CAD capabilities. We may then need to perform a geometry cleanup. For instance, solid

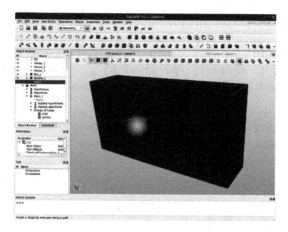

Figure 6.2: Screenshot of the geometry used to create the mesh in the Salome.

bodies may need to be broken down to individual faces so that different mesh-ing parameters can be applied on each. Sometimes we also may want to delete small features. The geometry is then meshed. In Section 4.10 we discussed the use of *surface meshes* that only capture the "skin" of a solid body. Here we are interested in creating a *volume mesh* that also resolves the internal region using elements such as tetrahedrons, hexahedrons, or wedges. For the example in this chapter, I used Salome's Geometry module to first draw a solid brick extending from $(-0.1, -0.1, 0)$ to $(0.1, 0.1, 0.4)$. I then drew a sphere, and per-formed a *subtraction* operation to make a spherical gap inside the solid brick, as shown by the clip in Figure 6.2. This is the part that could be done in an external CAD program, but Salome's Geometry module is plenty capable for simple tasks such as this. This geometry is then meshed using tetrahedral elements. I specified discretization parameters that generated what seemed like an acceptable cell resolution as shown previously in Figure 6.1. Now, we know from the prior chapters that we expect a high density region to form behind the sphere due to the lensing effect. It may be justified to reduce the cell size in this region using additional mesh refinement. This is left as an exercise for the reader. Finally, I created a *surface element group* called `inlet` that captures all triangles on the inlet face. I then similarly made another group capturing all triangles on the sphere, called (no surprise here) `sphere`.

6.2.2 Mesh Formats

The created mesh is exported to a file. All unstructured mesh file formats store the mesh by outputting, in some manner, node *positions* and cell *connectiv-ity*. The connectivity block specifies which nodes belong to each element. The node ordering is important. For surface elements, it is used to set the ori-entation of normal vectors according to the right hand rule. As an example, consider the simple .dat format. It consists of a header specifying the number of nodes and the number of elements. The node positions are then saved on individual lines, with format node_id x y z. The elements then follow with syntax element_id element_type n1 n2 n3 .... The element_type field is 203 for surface triangles, and 304 is for tetrahedra,

```
2204 11821
1 -1.000000e+002 -1.000000e+002 4.000000e+002
2 -1.000000e+002 -1.000000e+002 0.000000e+000
...
146 203 175 140 28
147 203 26 185 175
...
2580 304 2183 771 808 61
2581 304 2183 808 771 1586
```

The downside of this format is that it lacks the ability to specify element or node *groups*. These groups are needed to assign boundary conditions. In our example, we need to know which triangles belong to the inlet for particle injection. We also need the inlet and sphere nodes to assign Dirichlet boundary conditions. While we could get around the .dat format limitation by exporting the surface groups as individual files, a cleaner approach is to use one of the more robust formats, such as Universal, or .unv. This file consists of multiple datasets with syntax [3]

```
-1
dataset_number
multiline_dataset_data...
-1
```

Specifically, the file generated by Salome starts with datasets 164 and 2420 that specify units and the coordinate system. We just ignore these, and assume that positions are given in millimeters and correspond to (x, y, z) Cartesian coordinates. Dataset 2411 then lists the node positions. Each node is saved using two lines, with line 1 starting with the node number. The rest of the information can be ignored. The second line contains the positions. Note that unlike the .dat format, this file does not contain a header specifying how many nodes to read. We simply continue reading until we find the dataset terminating line, -1.

```
      -1
    2411
          1            1            1           11
 -1.000000E+02  -1.000000E+02   4.000000E+02
          2            1            1           11
 -1.000000E+02  -1.000000E+02   0.000000E+00
          3            1            1           11
```

```
. . .
  -1
```

Element connectivity comes next in dataset 2412. Each element consists of multiple lines. The first entry on line 1 is the element number. The second entry is the element type, followed by physical properties, material, and color indexes, and finally the number of nodes. There are quite a few different element types, as can be seen in the format specification. To make the loader simpler, we determine the element type from the number of element nodes. We assume that there are no surface quadrangles, and that elements with four nodes are volumetric tetrahedrons. The next line specifies the connectivity: indexes of nodes making up this element. At least this is the case for "non-beam" elements. For these two-node elements, connectivity is stored on line 3, while line 2 contains orientation and cross-section entries. Salome outputs edges using this 3 line format. But since we don't actually need edge information, we simply skip these elements.

```
  -1
  2412
  11        11        2         1         7         2
  0         1         1
  11        2
. . .
  2259      41        2         1         7         3
  1035      1098      1096
. . .
  3880      111       2         1         7         4
  1460      577       1467      1454
. . .
  -1
```

Finally, card 2467 provides element groups. Just as the above datasets store multiple nodes or elements, this dataset stores multiple groups. Each group starts with a line containing the group number, followed by several properties we are not interested in. The last number in the eighth position provides the number of entities in the group. The next line then lists the group name, as set in Salome during the group construction. Finally, the elements are listed with two entries per line. The first value indicates entity type, which is 8 for all valid entries. Section starting with -1 closes out the last line for groups with an odd number of elements. The second value is the element index. This is the information we are after. The example below shows parts of this dataset for a file with two groups,

```
  -1
   2467
   1    0         0         0    0         0    0         211
inlet
```

8	1546	0	0	8	1547	0	0
8	1548	0	0	-1	0	0	0
2	0	0	0	0	0	0	426
sphere							
8	2154	0	0	8	2155	0	0
8	2156	0	0	8	2157	0	0

```
...
-1
```

6.2.3 Mesh Import

In order to actually store the mesh geometry, we need to define new data objects. We use `Node`, `Tri`, and `Tet` structures to hold the node, surface triangle, and volume tetrahedron information. These structures are defined in `VolumeMesh.h`.

```
// definition of a node
struct Node {
  Node(double x, double y, double z): pos{x,y,z} {}
  double3 pos;                          // node position
  enum NodeType {NORMAL, DIRICHLET, NEUMANN};
  NodeType type = NORMAL;
};

// definition of a triangular surface element
struct Tri {
  Tri (int n1, int n2, int n3): con{n1,n2,n3} {}
  int con[3];
  double area = 0;
  double3 normal;                       // normal vector
  double3 randomPos(const VolumeMesh &vm); // random triangle pos
};

// definition of a tetrahedral volume element
struct Tet {
  Tet (int n1, int n2, int n3, int n4): con{n1,n2,n3,n4} {}
  int con[4];
  double volume = 0;

  // data structures to hold precomputed 3x3 determinants
  double alpha[4], beta[4], gamma[4], delta[4];

  // cell connectivity
  int cell_con[4];
};
```

We also define a `Group` container for storing identification numbers of elements in a group,

```
struct Group {
  std::vector<int> elements;
};
```

These items are collected in a class called `VolumeMesh`,

```
class VolumeMesh {
public:
  std :: vector <Node> nodes ;
  std :: vector <Tri> tris ;
  std :: vector <Tet> tets ;
  std :: map<std :: string , Group> groups ;
};
```

The `map` container maps a string name to `Group` object. The actual file loading is accomplished using the `LoadUNV` function,

```
bool VolumeMesh :: loadUNV( string file_name) {
  ifstream in( file_name );
    if (! in ){ cerr <<"Could not open "<<file_name <<endl; return
      false ;}

  while (in ) {              // continue until end of file
    string line ;
    getline (in , line );      // read a line , should contain "  -1"
    if ( line != "    -1") { cerr <<"Data file mismatch"<<endl; return
      false ;}
    getline (in , line );      // read the next line
    stringstream ss ( line ); // wrap by stream parser
    int code;   ss>>code;    // read an integer from the line

    switch ( code ) {
      case 2411:  loadNodes ( in ); break ;
      case 2412:  loadElements ( in ); break ;
      case 2467:  loadGroups ( in ); break ;
      default :  skipDataset ( in ); break ;
    }
  }
  init ();                  // computes node volumes
  return true ;
}
```

For the sake of brevity, the source for individual loaders is not included here, but you will find it in `ch6/World.cpp`. These functions parse the data as described above. Since node and element numbering may not be sequential, the functions utilize *lookup tables* to remap node and element identification numbers. The volume mesh is passed as a parameter to the `World` constructor,

```
class World {
public:
  World ( VolumeMesh &vm) ;
  VolumeMesh vm;
};
```

6.2.4   Cell Volumes

As part of `VolumeMesh` construction, we also compute cell volumes. Volume of a tetrahedron is given by

$$V_{1234} = \pm\frac{1}{6}\begin{vmatrix} 1 & x_1 & y_1 & z_1 \\ 1 & x_2 & y_2 & z_2 \\ 1 & x_3 & y_3 & z_3 \\ 1 & x_4 & y_4 & z_4 \end{vmatrix} \tag{6.1}$$

This code is placed within a `init` function that is called by the `VolumeMesh` constructor,

```
void VolumeMesh::init() {
  for (Tet &tet:tets) {
    double M[4][4];
    // set first column to 1
    for (int i=0;i<4;i++) M[i][0] = 1;

    // loop over vertices and fill the matrix
    for (int v=0;v<4;v++) {
      for (int dim=0;dim<3;dim++) {
        M[0][dim+1] = nodes[tet.con[0]].pos[dim];
        M[1][dim+1] = nodes[tet.con[1]].pos[dim];
        M[2][dim+1] = nodes[tet.con[2]].pos[dim];
        M[3][dim+1] = nodes[tet.con[3]].pos[dim];
      }
    }
    tet.volume = (1.0/6.0)*utils::det4(M);
    // flip 0123 to 0321 if negative volume
    if (tet.volume<0) {int t=tet.con[1]; tet.con[1]=tet.con[3];
      tet.con[3]=t; tet.volume=-tet.volume;}
  }
}
```

We start by setting the first column to 1 and then copy the x, y, and z positions of the four nodes to columns 2, 3, and 4. The determinant is evaluated using a helper function `utils::det4`, which is placed inside of `FESolver.cpp`. The last line is important. Depending on the node ordering, the determinant in Equation 6.1 may evaluate to a negative number. In that case we simply flip the ordering such that a 0123 tetrahedron becomes 0321.

6.2.5   Field Object

Our code is written around the `Field` object that stores node or cell-centered data. Since there is no longer a three dimensional structure to the mesh, it does not make sense to retain three dimensional arrays in the underlying container. All data is now stored in a linear fashion, with the index corresponding to the node or cell identification number. Therefore, the `Field` object is reverted to a one-dimensional form,

```
template <typename T>
class Field_ {
public:
```

```
// constructor
Field_(VolumeMesh &vm)  :  ni{(int)vm.nodes.size()}, vm{vm} {
   data = new T[ni];      // allocate memory for a 1D array
   clear();
}
protected:
   int ni;        // allocated dimensions
   T *data;       // data held by this field
   VolumeMesh &vm; // volume mesh reference
   int ave_samples = 0;   // number of samples used for averaging
};
```

Similar changes are made to all functions, including the various overloaded operators. Note that the constructor argument is a reference to the **VolumeMesh**. The containing connectivity is needed by the scatter and gather functions. In order to avoid a circular dependence of class definitions, the **vec3** data type was moved into its own file, **Vec.h**.

6.2.6 Node Volumes

Now that we have cell volumes and the updated **Field** object, we can also set node volumes. Node volumes are stored in **Field node_vol** so that we can use the overloaded division operator to compute number density. The actual computation of node volumes is deferred to the **World** constructor, since that is where the node volume field is stored. We compute node volumes by effectively scattering one fourth of the cell volume to each cell node,

```
World::World(VolumeMesh &vm): vm{vm}, node_vol(vm),
   phi(vm),rho(vm), ef(vm) {
   // compute node volumes by averaging cell volumes
   for (Tet &tet: vm.tets) {
      for (int v=0;v<4;v++)
         node_vol[tet.con[v]] += 0.25*tet.volume;
   }
}
```

6.2.7 Output

Previously, we used VTK's **ImageData** format to store the mesh data. This format is designed for structured, Cartesian data. Obviously, this is no longer the case. VTK supports additional formats with an increasing complexity. The most complex is the **UnstructuredMesh**. As the name indicates, this is the format to use for unstructured meshes. Files of this type use the .vtu extension. Similarly to the **PolyData** format discussed previously in Chapter 4, the file consists of a data block containing node positions, and another data block containing cell connectivity. However, unlike the previous format, .vtu files support three dimensional cell types such as tetrehdra or various bricks. The .vtp format is suitable only for exporting *surface* meshes made up of triangles and other polygons.

This file type starts with a header

```
<?xml version="1.0"?>
<VTKFile type="UnstructuredGrid" version="0.1"
  byte_order="LittleEndian">
<UnstructuredGrid>
<Piece NumberOfPoints="2204" NumberOfVerts="0" NumberOfLines="0"
  NumberOfStrips="0" NumberOfCells="9252">
```

Next, a `DataArray` in the `Points` block specifies node positions,

```
<Points>
<DataArray type="Float32" NumberOfComponents="3" format="ascii">
-0.1 -0.1 0.4
-0.1 -0.1 0
...
</DataArray>
</Points>
```

Another block of type `Cells` specifies the cell information. This block contains three data arrays. The first one, called `connectivity`, contains a list of node numbers that make up the elements. For tetrahedrons, the first four entries correspond to the nodes of the first element, the second four to the nodes of the second element, and so on. But since an unstructured mesh may have cells of different types, we also need to inform the loader where each element ends. This is done with the `offsets` array. The entries are indexes to the start of the subsequent element. The first cell is composed of four nodes. Therefore, the first entry in the `offsets` array is 4, as `connectivity[4]` is the first node of the second element. Finally, different element types, such as quadrangles and tetrahedrons, may contain cells with the same number of nodes. We need to inform the loader that our cells are of type 10, which is the VTK code for a tetrahedron [51]. This information is stored in the `types` data array.

```
<Cells>
<DataArray type="Int32" Name="connectivity" format="ascii">
2182 770 807 60
2182 807 770 1585 ...
</DataArray>
<DataArray type="Int32" Name="offsets" format="ascii">
4 8 12 16 20 24 28 32 36 40 44 48 52 56 60 64 68 72 ...
</DataArray>
<DataArray type="UInt8" Name="types" format="ascii">
10 10 10 10 10 10 10 10 10 10 10 10 10 10 10 10 10 ...
</Cells>
```

The rest of the file follows a familiar syntax. The file contains `PointData` and `CellData` blocks, with each containing one or more data arrays storing the node or cell-centered scalar or vector quantities,

```
<CellData Vectors="ef">
</DataArray>
<DataArray type="Float64" Name="cell_volume" format="ascii">
8.09913e-07 9.45629e-07 ...
</DataArray>
</CellData>
```

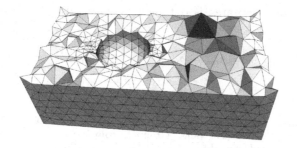

Figure 6.3: Initial mesh output, with coloring corresponding to cell volumes.

With these changes implemented in `Output::fields` in `Output.cpp`, we obtain the result shown in Figure 6.3. This plot was generated by clipping the domain along the $x = 0$ plane, and enabling "crinkle clip". Cells are shaded according to their volume.

6.3 PARTICLE INJECTION

Next we introduce particles. For simplicity, we implement only the cold beam source. The algorithm is similar to what we are familiar with from Chapter 3. However, instead of hard-coding the sampling plane, we use the surface group data from the loaded `.unv` file. Since the group is composed of individual triangles, we could inject the particles by looping through all injection group surface elements, and for each, computing the number of macroparticles to generate per $M_{ele} = \text{int}(nA_{ele}v\Delta t/w_{mp} + \mathcal{R})$. This approach does not scale well to cases in which the number of injection elements greatly exceeds the number of particles. Assume that the source is composed of 100 triangles, but we only inject 10 simulation particles per time step. Next, let's assume that, just due to luck, the random number $\mathcal{R} < 0.9$ on all 100 triangles. The source then generates no particles. On the other hand, if $\mathcal{R}$ were to evaluate to ≥ 0.9 for all triangles, the source would generate 100 particles instead of the expected 10. While *on average* we obtain the correct mass flow rate, the time step to time step noise may be unacceptably high. Therefore, I prefer to use an approach that first computes the total number of particles to generate using the group area,

$$M_{tot} = \text{int}(nA_{tot}v\Delta t/w_{mp} + \mathcal{R}) \tag{6.2}$$

These particles are then distributed to the surface elements, with probability

$$P = \frac{A_{ele}}{A_{tot}} \tag{6.3}$$

This relationship simply assures that larger triangles generate more particles as expected.

The total injection area is obtained by aggregating areas of individual triangles in the inlet group,

$$A_{tot} = \sum_j A_j \qquad j \in G_{inlet} \tag{6.4}$$

This code is added to the `ColdBeamSource` constructor,

```
ColdBeamSource :: ColdBeamSource (Species &species , World &world ,
    double v_drift , double den, Group &group) :
sp{species}, world{world}, v_drift{v_drift}, den{den},
    src_group{group} {
    A_tot = 0;
    for (int e:group.elements)    // compute total source area
        A_tot += world.vm.tris[e].area;
}
```

As can be seen here, the constructor call includes a reference to the injection group. The source is activated in `main` using

```
vector<ColdBeamSource> sources;
Group &group = world.vm.groups.at("inlet");  // exception if not
    found
sources.emplace_back(species[0], world,7000,1e10,group);  // ions
```

Note that the `map::at` function is used instead of the `[]` operator. This function throws an exception if the key (our desired group) is not found. The operator would instead add a new empty group under this key, which is not desired.

Area of a triangle is given by

$$A_{012} = \frac{1}{2} |(\vec{v}_1 - \vec{v}_0) \times (\vec{v}_2 - \vec{v}_0)| \tag{6.5}$$

where $\vec{v}_i$ is the position of the i-th vertex. The cross-product points in the surface normal direction, which we assume to point into the plasma domain. This orientation helps us determine the initial orientation of particle velocities. The normal vector is computed using the cross-product following the standard right hand convention. These calculations are added to `VolumeMesh::init`

```
for (Tri &tri:tris) {
    double3 v0 = nodes[tri.con[0]].pos;   // triangle node 0 position
    double3 v1 = nodes[tri.con[1]].pos;   // triangle node 1 position
    double3 v2 = nodes[tri.con[2]].pos;   // triangle node 2 position
    double3 c = cross(v1−v0,v2−v0);
    tri.area = 0.5*mag(c);                // triangle area
    tri.normal = unit(c);                 // triangle normal
}
```

A stochastic approach is used to distribute particles to surface elements. For each particle, we use a `while` loop to continues picking a random element from the injection group. The loop terminates once $\mathcal{R} \geq A_{ele}/A_{tot}$. This algorithm is demonstrated in the following code

```
void ColdBeamSource :: sample () {
  VolumeMesh &vm = world . vm;

  // inject particles
  for (int i=0;i<num_sim;i++) {
    Tri *tri;
    do {
      // pick random entry in the group
      int ge = (int)(rnd()*src_group.elements.size());
      tri = &vm.tris[src_group.elements[ge]];
    } while ( tri->area/A_tot<rnd() );

    double3 pos = tri->randomPos(vm);
    double3 vel = tri->normal*v_drift;
    sp.addParticle(pos, vel, sp.mpw0);
  }
}
```

Once an acceptable triangle is selected, we use a `randomPos` function to sample a random point on the triangle. For the cold beam, the velocity is simply the local normal vector scaled by the drift velocity magnitude, $\vec{v} = v_d \hat{n}$.

All that is missing at this point is the ability to sample a random point on a triangle. Here we use an algorithm from [47],

$$f_0 = 1 - \sqrt{\mathcal{R}_1} \tag{6.6}$$

$$f_1 = \sqrt{\mathcal{R}_1}(1 - \mathcal{R}_2) \tag{6.7}$$

$$f_2 = \left(\sqrt{\mathcal{R}_1}\right)\mathcal{R}_2 \tag{6.8}$$

$$\vec{x}_p = f_0\vec{x}_0 + f_1\vec{x}_1 + f_2\vec{x}_2 \tag{6.9}$$

The two random numbers $\mathcal{R}_1$ and $\mathcal{R}_2$ are sampled in range $[\epsilon, 1 - \epsilon]$, where ϵ is some small positive value. This approach prevents particles from being born on the edges. These equations are implemented in a member function of `Tri` called `randomPos`,

```
double3 Tri :: randomPos(const VolumeMesh &vm) {
  // triangle vertex positions
  double3 verts[3] = {vm.nodes[con[0]].pos, vm.nodes[con[1]].pos,
    vm.nodes[con[2]].pos};

  double eps = 0.0001;                    // avoid sampling on edges
  double r1 = eps + rnd()*(1-2*eps);
  double r2 = eps + rnd()*(1-2*eps);
  double fac0 = (1-sqrt(r1));
  double fac1 = sqrt(r1)*(1-r2);
  double fac2 = sqrt(r1)*r2;
  return fac0*verts[0] + fac1*verts[1] + fac2*verts[2];
}
```

6.4 MESH INTERPOLATION

Now that we have the particles, we need to update the `scatter` and `gather` functions to support unstructured meshes. In the previous chapters, we used

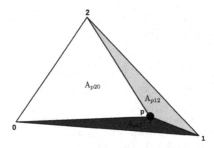

Figure 6.4: Illustration of shape factors for a triangular element.

volume or area ratios to calculate the weights associated with the cell nodes. The new algorithm is no different, except that we update the math for a tetrahedral geometry.

6.4.1 Shape Factors

Consider the 2D analogy in Figure 6.4. If a point is located inside a triangle, the three sub-areas generated using the point as one of the vertices add up to the area of the original triangle,

$$A_{p12} + A_{p20} + A_{p01} = A_{012} \tag{6.10}$$

or

$$\frac{A_{p12}}{A_{012}} + \frac{A_{p20}}{A_{012}} + \frac{A_{p01}}{A_{012}} \equiv L_0 + L_1 + L2 = 1 \tag{6.11}$$

These three ratios are known as *shape factors* or *basis functions*. The scenario for tetrahedral elements is identical, except that we consider volume ratios,

$$\frac{V_{p123}}{V_{0123}} + \frac{V_{p230}}{V_{0123}} + \frac{V_{p301}}{V_{0123}} + \frac{V_{p012}}{V_{0123}} = 1 \tag{6.12}$$

or

$$L_0 + L_1 + L_2 + L_3 = 1 \tag{6.13}$$

But since computer arithmetic is not precise, we need to check $L_0 + L_1 + L_2 + L_3 \approx 1$.

Volume of a tetrahedron is computed using Equation 6.1. Let the particle position $\vec{x}_p$ take place of one of the nodes. For instance, for V_{p123}, the 4×4 determinant evaluates to

$$6V_{p123} = 1\begin{vmatrix} x_1 & y_1 & z_1 \\ x_2 & y_2 & z_2 \\ x_3 & y_3 & z_3 \end{vmatrix} + x_p \begin{vmatrix} 1 & y_1 & z_1 \\ 1 & y_2 & z_2 \\ 1 & y_3 & z_3 \end{vmatrix} + $$
$$y_p \begin{vmatrix} 1 & x_1 & z_1 \\ 1 & x_2 & z_2 \\ 1 & x_3 & z_3 \end{vmatrix} + z_p \begin{vmatrix} 1 & x_1 & y_1 \\ 1 & x_2 & y_2 \\ 1 & x_3 & y_3 \end{vmatrix} \tag{6.14}$$

These 3×3 determinants depend only on cell vertices and can be precomputed for efficiency [50]. We then have

$$6V_{p123} = \alpha - x_p\beta + y_p\gamma + z_p\delta \tag{6.15}$$

We add this computation to the `VolumeMesh::init` function,

```
// precompute 3x3 determinants for LC computation
for (Tet &tet : tets) {
  double M[3][3];
  // loop over vertices
  for (int v=0;v<4;v++) {
    int v1,v2,v3;
    switch (v) {
      case 0: v1=1;v2=2;v3=3;break;     // Vp123
      case 1: v1=3;v2=2;v3=0;break;     // Vp320
      case 2: v1=3;v2=0;v3=1;break;     // Vp301
      case 3: v1=1;v2=0;v3=2;break;     // Vp102
    }

    double3 p1 = nodes[tet.con[v1]].pos;
    double3 p2 = nodes[tet.con[v2]].pos;
    double3 p3 = nodes[tet.con[v3]].pos;

    // alpha
    M[0][0] = p1[0]; M[0][1] = p1[1]; M[0][2] = p1[2];
    M[1][0] = p2[0]; M[1][1] = p2[1]; M[1][2] = p2[2];
    M[2][0] = p3[0]; M[2][1] = p3[1]; M[2][2] = p3[2];
    tet.alpha[v] = utils::det3(M);

    // beta
    M[0][0] = 1; M[0][1] = p1[1]; M[0][2] = p1[2];
    M[1][0] = 1; M[1][1] = p2[1]; M[1][2] = p2[2];
    M[2][0] = 1; M[2][1] = p3[1]; M[2][2] = p3[2];
    tet.beta[v] = utils::det3(M);

    // gamma
    M[0][0] = 1; M[0][1] = p1[0]; M[0][2] = p1[2];
    M[1][0] = 1; M[1][1] = p2[0]; M[1][2] = p2[2];
    M[2][0] = 1; M[2][1] = p3[0]; M[2][2] = p3[2];
    tet.gamma[v] = utils::det3(M);

    // delta
    M[0][0] = 1; M[0][1] = p1[0]; M[0][2] = p1[1];
    M[1][0] = 1; M[1][1] = p2[0]; M[1][2] = p2[1];
    M[2][0] = 1; M[2][1] = p3[0]; M[2][2] = p3[1];
    tet.delta[v] = utils::det3(M);
  }
}
```

Note that this computation needs to take place after cell volumes are calculated, since node ordering may change. These precomputed determinants are stored in each `Tet`'s `alpha`, `beta`, `gamma`, and `delta` arrays. Each array stores four coefficients, with the index corresponding to the node that is replaced by the particle position. For instance, α_2 is used to evaluate V_{01p3}.

6.4.2 Cell Search

The above shape functions can be used to interpolate data once we know which cell a particle is located in. But the shape functions can also be used to find the cell containing the particle. We simply loop over all cells until we find one for which Equation 6.11 holds. If a cell is not found, then the particle must be located outside the computational domain. While this brute force approach works, it is very slow for meshes containing a large number of cells. Therefore, a better approach is to implement a *neighbor traverse*.

We start by letting each particle carry its logical coordinate,

```
struct Particle {
    double3 pos;        // position
    double3 vel;        // velocity
    double mpw;         // macroparticle weight
    LCord lc;           // logical coordinate
};
```

The LCord type is defined in World.h as follows,

```
struct LCord {
    int cell_id = 0;           // cell id
    double L[4] = {0,0,0,0};   // shape factors
    LCord (int c=0):cell_id{c} {}
    explicit operator bool() {return cell_id>=0;}
};
```

This structure stores two sets of data: the cell identification c and the four interpolation shape factors $L[4]$. We use $c < 0$ to indicate an invalid, out of domain, position. To simplify this checking, we overload the bool() operator so that we can write

```
if (!lc) part.mpw = 0;
```

Assuming Δt is sufficiently small, the particle should not travel too far from its previous cell during a single push. The pre-push cell offers a good starting point in the search for the new cell. Consider the drawing in Figure 6.5. Triangles are shown for clarity, but the algorithm is identical for tetrahedrons. Prior to the push in (a), the three shape factors evaluate to $L_0 + L_1 + L_2 = 1$, with $L_i \in [0, 1]$. After the push, the three shape functions evaluated using the initial cell no longer add up to 1. Furthermore, L_0 lies completely outside the original cell, and thus $L_0 < 0$. This basis function shares the edge (or a face in 3D) with the neighbor cell actually containing the particle. This observation suggests an improved algorithm for performing the cell search:

1. Evaluate basis functions using the last known cell id number

2. Terminate if $\sum_i^4 L_i \approx 1$ and $L_i \in [0, 1]$0

3. Move to the cell sharing the face with the most negative shape function

4. Return to step 1

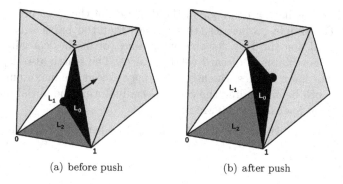

(a) before push   (b) after push

Figure 6.5: Neighbor traverse to find particle cell.

This recursive search is implemented in the **XtoL** function added to **VolumeMesh**,

```cpp
LCord VolumeMesh::XtoL(const double3 &pos, int e, bool search) {
    Tet &tet = tets[e];       // first try the current tetrahedron
    LCord lc(e);

    bool found = true;
    for (int i=0;i<4;i++) {          // loop over vertices
        lc.L[i] = (1.0/6.0)*(tet.alpha[i] - pos(0)*tet.beta[i] +
            pos(1)*tet.gamma[i] - pos(2)*tet.delta[i])/tet.volume;
        if (lc.L[i]<0 || lc.L[i]>1.0) found=false;
    }

    if (found) return lc;
    if (!search) return LCord(-1);

    // find the most negative weight is outside this tet
    int min_i = 0;
    double min_lc = lc.L[0];
    for (int i=1;i<4;i++)
        if (lc.L[i]<min_lc) {min_lc=lc.L[i]; min_i=i;}

    // is there a neighbor in this direction?
    if (tet.neighbor[min_i]>=0)
        return XtoL(pos,tet.neighbor[min_i]);

    // no neighbor - particle out of domain
    return LCord(-1);
}
```

This function is placed in **VolumeMesh** since it requires cell volumes and neighbor data. To simplify the calls, we also include a pass through wrapper in **World**,

```cpp
LCord XtoL(const double3 &pos, int e, bool search=true) {return
    vm.XtoL(pos,e,search);}
```

Finally, this method is not completely fool-proof. The search terminates if the cell does not have a neighbor in the direction of the most negative basis function. For simple geometries, this implies that the particle has left the computational domain. But, for concave meshes, or meshes containing small holes, this algorithm may discard valid particles. The search also depends on recursion. For large meshes, the algorithm may exceed the maximum allowed recursion depth before finding the cell, especially during the initial placement of a particle in `Species::addParticle`,

```
void Species::addParticle(double3 pos, double3 vel, double mpw) {
  LCord lc = world.XtoL(pos,0);   // or call world.XtoLbrute(pos)
  if (!lc) return;
  /*...*/
}
```

Therefore instead of starting the traverse in cell 0 as is done here, we may want to initialize cell ids with a brute force search,

```
LCord World::XtoLbrute(const double3 &pos) {
  for (size_t e = 0;e<vm.tets.size();e++) {
    LCord lc = vm.XtoL(pos,e,false);
    if (lc) return lc;          // uses overloaded bool operator
  }
  return LCord(-1);
}
```

This code uses the optional **search** argument to `XtoL` to disable the recursive search.

6.4.3 Cell Neighbors

The above traverse algorithm requires a *cell neighbor* table. The purpose of this table is to store the ids of the four neighbor cells sharing faces with each tetrahedron. We store this information in a `int neighbor[4]` member array of `Tet`. The connectivity is built in `VolumeMesh::init` using a rather inefficient algorithm. We start by setting all neighbors to -1. We then loop through all cells, $l \in [0, n_{tets} - 1]$. For each cell, we loop through the four faces, and for each, collect the list of the three node ids making up the face. We then loop through the remaining tetrahedrons, $m \in [l + 1, n_{tets} - 1]$, and search for one having a face with the same node ids. If we find a match, we set the neighbor information for both element l and m. The set faces are skipped in the subsequent search. Faces that remain set to -1 form the external boundary of the computational mesh. The algorithm is listed below.

```
// reset cell neighbors
for (Tet &tet:tets) {for (int i=0;i<4;i++) tet.neighbor[i]=-1;}

// set cell neighbors
for (size_t l=0;l<tets.size();l++) {
  Tet &tet = tets[l];
  int v1,v2,v3;
  for (int v=0;v<4;v++) {
    // skip if already set
```

```
if (tet.neighbor[v]>=0) continue;

switch(v) {  // hard-coded tet connectivity
    case  0:  v1=1;v2=2;v3=3;break;
    case  1:  v1=2;v2=3;v3=0;break;
    case  2:  v1=3;v2=0;v3=1;break;
    case  3:  v1=0;v2=1;v3=2;break;
}

// loop over rest, look for a tet with these three vertices
for (size_t m=l+1;m<tets.size();m++) {
    Tet &other = tets[m];
    bool matches[4] = {false, false, false, false};
    int count = 0;
    for (int k=0;k<4;k++) {
        if (other.con[k]==tet.con[v1] || other.con[k]==tet.con[v2]
            || other.con[k]==tet.con[v3]) {
            count++; matches[k]=true;}
    }
    // if three vertices match
    if (count==3) {
        tet.neighbor[v] = m;
        for (int k=0;k<4;k++) // set reverse lookup
            if (!matches[k]) other.neighbor[k] = l;
    }
}
    }
  }
}
```

Although we need to perform this calculation only during initialization, the required time can be prohibitive for large meshes due to the N^2 scaling. A more robust algorithm may involve the use of octrees or a uniform mesh to obtain a list of nearby elements.

6.4.4 Scatter and Gather

The scatter and gather functions simply use the shape factors to deposit a fractional value to the cell nodes, or to collect node-based data. The gather function evaluates

$$y = \sum_{i=0}^{3} L_i y_i \tag{6.16}$$

where y_i is some node-based data. These two functions exist as member functions of the Field object,

```
void scatter(const LCord &lc, T s) {
    Tet &tet = vm.tets[lc.cell_id];
    for (int v=0;v<4;v++) data[tet.con[v]] += lc.L[v]*s;
};

T gather(const LCord &lc) {
    T res;
    Tet &tet = vm.tets[lc.cell_id];
    for (int v=0;v<4;v++) res+=lc.L[v]*data[tet.con[v]];
```

```
  return res;
}
```

6.5 PARTICLE MOTION

Changes to the particle moving algorithm are quite minimal and mainly consist of the need to pass in the prior cell id to the XtoL function. We also update the particle's logical coordinate after it is moved to a new position. We use the overloaded bool operator to check for invalid lc, indicating the particle has left the domain. The particle removal step is identical to the Cartesian version,

```
void Species::advance() {
  // get the time step
  double dt = world.getDt();

  // loop over all particles
  for (Particle &part: particles) {
    // electric field at particle position
    double3 ef_part = world.ef.gather(part.lc);

    // update velocity from F=qE
    part.vel += ef_part*(dt*charge/mass);

    // update position from v=dx/dt
    part.pos += part.vel*dt;

    // update particle's lc
    part.lc = world.XtoL(part.pos, part.lc.cell_id);
    if (!part.lc) part.mpw = 0;  // kill particle
  }

  // particle removal step
}
```

Note that although we are gathering the electric field, we have not yet defined the field solver, and hence ef is a zero vector.

6.5.1 Number Density

Finally, the code to compute number density is identical to the Cartesian version,

```
void Species::computeNumberDensity() {
  den.clear();
  for (Particle &part: particles)
    den.scatter(part.lc, part.mpw);
  den /= world.node_vol;  // divide by node volume
}
```

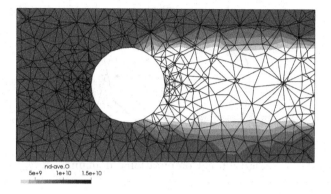

Figure 6.6: Number density of neutral particles on a unstructured mesh.

6.5.2 Neutral Results

We now have a fully functioning particle pushing code. Figure 6.6 shows the number density after 800 time steps. Since we have not yet implemented the potential solver, the ions effectively act as neutrals. We can notice that the source produces the correct number density, and that a wake forms behind the sphere. This plot can be compared to the first plot in Figure 4.1.

6.6 FINITE ELEMENT METHOD

We have all the pieces of a PIC code for an unstructured mesh with the exception of the Poisson solver. As you may remember, methods for solving differential equations include:

- **The Finite Difference Method (FDM):** This is the method that has been used in the majority of the examples developed so far.

- **The Finite Volume Method (FVM):** This method was used to introduce axisymmetric codes. It can be used to compute potentials on an unstructured mesh, however computing derivatives inside the cells is not straightforward.

- **The Finite Element Method (FEM):** This method is widely used on unstructured meshes.

In this section, we introduce the FEM method. Our implementation follows that of Hughes [35].

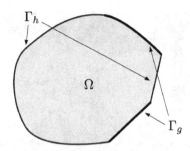

Figure 6.7: Computational domain.

6.6.1 Strong Form

Poisson's equation (which is a form of a steady state heat equation), can be written as

$$\nabla^2 u \equiv \frac{\partial^2 u}{\partial x^2} + \frac{\partial^2 u}{\partial y^2} + \frac{\partial^2 u}{\partial z^2} \equiv u_{,ii} = f \tag{6.17}$$

The third form used what is known as *index* or Einstein notation. Subscripts following a comma indicate a derivative, and repeated indexes imply a sum. We are interested in solving this equation on some spatial domain Ω. In order to obtain a unique solution, we also need to specify boundary conditions. We assume that we know the solution on a subset of the boundary denoted by Γ_g. These are the Dirichlet boundaries. We also assume that on another subset denoted by Γ_h we have the derivative of the solution. This is the Neumann boundary. Therefore, following the notation in [35], we can summarize the problem using *Strong form*,

(S)	Given $f : \Omega \to \mathbb{R}$, $g : \Gamma_g \to \mathbb{R}$ and $h : \Gamma_h \to \mathbb{R}$, find $u : \Omega \to \mathbb{R}$ such that $-ku_{,ii} = f$ on Ω $u = g$ on Γ_g $ku_{,i} = h$ on Γ_h

This is the standard form of a boundary value problem you may be familiar with from basic differential equations course. For us, $u = \phi$, $k = 1$, and $f = \rho/\epsilon_0$. The domain is drawn in Figure 6.7. Note, $\Gamma_g \notin \emptyset$, but Γ_h can be an empty set. Dirichlet boundaries are required for a unique solution, but Neumann boundaries are optional. The two boundary sets do not need to be continuous. We can have multiple discontinuous Γ_g and Γ_h regions.

6.6.2 Weak Form

Strong form is not the only way to formulate a boundary value problem. There is an alternate formulation known as *weak* (or *variational*) form. This form defines the solution using two types of functions:

- Candidate (or trial) solutions

- Weighting functions (or variations)

Solutions are required to satisfy the boundary condition $u = g$ on Γ_g. We also require that the derivatives of the solutions are square integrable,

$$\int_\Omega (u, i)^2 d\Omega < \infty \tag{6.18}$$

Functions satisfying this requirement are known as H^1 functions. Thus $u \in H^1$. Collection of trial solutions is given by

$$\mathcal{S} = \{u | u \in H^1, u = g \text{ on } \Gamma_g\} \tag{6.19}$$

The weighting functions are similar to the solutions, except they only satisfy the homogeneous version of the g boundary condition,

$$\mathcal{V} = \{w | w \in H^1, w = 0 \text{ on } \Gamma_g\} \tag{6.20}$$

The weak form is then given by

(W)	Given $f : \Omega \to \mathbb{R}$, $g : \Gamma_g \to \mathbb{R}$, and $h : \Gamma_h \to \mathbb{R}$, find $u \in \mathcal{S}$ such that for all $w \in \mathcal{V}$ $\int_\Omega w_{,i}(ku_{,i})d\Omega = \int_\Omega wf d\Omega + \int_{\Omega_h} whd\Gamma$

Unlike the strong form, the weak form defines the solution in terms of the derivative of u. The two solutions are identical, as shown in Section 1.4 in [35]. The proof shows that if u is a solution of (S), then it is also a solution of (W), and that if u is a solution of (W), then it is also a solution of (S). We now define several new operators to simplify the notation,

$$a(w, u) = \int_\Omega w_i k u_i d\Omega \tag{6.21}$$

$$(w, f) = \int_\Omega wf d\Omega \tag{6.22}$$

$$(w, f)\Gamma = \int_{\Gamma_h} wf d\Gamma \tag{6.23}$$

The second and third equations are similar, except that the last one is defined over a surface instead a volume.

6.6.3 Galerkin Form

The weak form is next approximated on a discretized domain Ω^h. The first step is to construct a finite-dimensional approximation of $\mathcal{S}$ and $\mathcal{V}$. Instead of continuous functions, we now have subsets $\mathcal{S}^h$ and $\mathcal{V}^h$, $\mathcal{S}^h \in \mathcal{S}$ and $\mathcal{V}^h \in \mathcal{V}$. The h subscript denotes a property associated with a mesh. Therefore if

$u^h \in \mathcal{S}^h$, $u^h \in \mathcal{S}$. The boundary conditions also hold, $u^h = g$ on Γ_g. Similarly, all members of $\mathcal{V}^h$ vanish on Γ_g. We can then let each member of $\mathcal{S}^h$ be represented by

$$u^h = v^h + g^h \tag{6.24}$$

where $v^h \in \mathcal{V}^h$ and g^h results in satisfactory representation of $u = g$ on Γ_g. We then have the following Galerkin form of the problem,

(G)	Given f,g, and h as before, find $u^h = v^h + g^g \in \mathcal{V}^h$, such that for all $w^h \in \mathcal{V}^h$ $a(w^h, v^h) = (w^h, f) + (w^h, h)_\Gamma - a(w^h, g^h)$

6.6.4 Formulation with Shape Functions

Now assume that our domain consists of $\eta = \{0, 1, \ldots, n_{np-1}\}$ nodal points. The nodes on which the Dirichlet g boundary condition is specified are given by η_g. Then $\eta - \eta_g$ is the subset of nodes on which the solution u^h is to be determined. We next let the members of $\mathcal{V}^h$ take the following form,

$$w^h(\vec{x}) = \sum_{A \in \eta - \eta_g} N_A(\vec{x}) c_A \tag{6.25}$$

and

$$v^h(\vec{x}) = \sum_{A \in \eta - \eta_g} N_A(\vec{x}) d_A \tag{6.26}$$

Here d_A is the unknown (plasma potential) that we are solving for. The nodal values of g are interpolated using shape functions,

$$g^h(\vec{x}) = \sum_{A \in \eta_g} N_a(\vec{x}) g_A \qquad g_A = g(\vec{x}_A) \tag{6.27}$$

We substitute these interpolation functions to the (G) equation to obtain

$$\sum_{B \in \eta - \eta_g} a(N_A, N_B) d_B = (N_A, f) + (N_A, h)_\Gamma - \sum_{B \in \eta_g} a(N_A, N_B) g_B \qquad ; A \in \eta - \eta_g \tag{6.28}$$

6.6.5 Matrix Form

We can simplify the above expression by defining

$$K_{AB} = a(N_a, N_B) \tag{6.29}$$

$$F_A = (N_A, f) + (N_a, h)_\Gamma - \sum_{B \in \eta_g} a(N_A, N_B) g_B \qquad ; A \in \eta - \eta_g \tag{6.30}$$

Equation 6.28 then becomes

$$\sum_{B \in \eta - \eta_g} K_{AB} d_b = F_a \tag{6.31}$$

which can be simplified further using matrix notation,

$$\boldsymbol{K} = [K_{AB}] = \begin{bmatrix} K_{0,0} & K_{0,1} & \cdots & K_{0,n-1} \\ K_{1,0} & K_{1,1} & \cdots & K_{1,n-1} \\ \vdots & \vdots & \ddots & \vdots \\ K_{n-1,0} & K_{n-1,1} & \cdots & K_{n-1,n-1} \end{bmatrix} \tag{6.32}$$

$$\vec{F} = \{F_A\} = \left\{ \begin{array}{c} F_0 \\ F_1 \\ \vdots \\ F_{n-1} \end{array} \right\} \tag{6.33}$$

and

$$\vec{d} = \{d_B\} = \left\{ \begin{array}{c} d_0 \\ d_1 \\ \vdots \\ d_{n-1} \end{array} \right\} \tag{6.34}$$

or

$$\boldsymbol{K}\vec{d} = \vec{F} \tag{6.35}$$

where $\boldsymbol{K}$ is known as the *stiffness matrix*, $\vec{F}$ is the force vector, $\vec{d}$ is the displacement vector. This terminology arises from FEM's historical use to model structural deformation. The *Matrix form* of the problem is written as

	Given the coefficient matrix $\boldsymbol{K}$ and vector $\vec{F}$, find $\vec{d}$ such that $\boldsymbol{K}\vec{d} = \vec{F}$
(M)	$\boldsymbol{K} = [K_{PQ}] \quad \vec{d} = \{d_q\} \quad \vec{F} = \{F_p\} \quad 0 \le P, Q < n_{eq}$
	$K_{PQ} = a(N_A, N_B) \quad P = ID(A) \quad Q = ID(B)$
	$F_P = (N_A, f) + (N_A, h)\Gamma - \sum_{B \in \eta_{eq}} a(N_A, N_B) g_B$

We are applying this equation only to the unknown nodes $n_{eq} = n_{np} - n_g$, and thus we need a look up table between the node number and the unknown equation number. We call this table ID following Hughes' notation,

$$ID(A) = \begin{cases} P & A \in \eta - \eta_g \\ 0 & A \in \eta_g \end{cases} \tag{6.36}$$

As an example, consider the mesh in Figure 6.8. Of the 6 nodes, 2 nodes are given known g values. 3 nodes form an h boundary on which the derivative needs to be prescribed. The remaining 1 node is a standard internal node.

6.6.6 Element View

The above formulation states the matrix problem using a *global view*. In this view, spatial position and interpolation functions N_A are given in physical $\vec{x}$ coordinates. This formulation is not very practical, and thus we instead switch to an *element view*. We let each element span some logical domain with coordinates $\vec{\xi} = \{\xi, \eta, \zeta\}$. The interpolations functions $N_{0,1,...}$ are now given in terms of these logical or *natural* coordinates. The integrals over the entire domain can now be rewritten as sums over elements,

$$\boldsymbol{K} = \sum_{e=0}^{n_{el}-1} \boldsymbol{K}^e \qquad \boldsymbol{K}^e = [K^e_{AB}] \tag{6.37}$$

$$\vec{F} = \sum_{e=0}^{n_{el}-1} \vec{F}^e \qquad \vec{F}^e = \{F^e_A\} \tag{6.38}$$

The main difference between the two definitions of K_{AB} and F_A is that in the element view (denoted by the superscript e), the integration is only over the element domain,

$$K^e_{AB} = a(N_A, N_B)^e = \int_{\Omega^e} \nabla N_A \cdot k \nabla N_B d\Omega \tag{6.39}$$

and

$$F^e_p = (N_A, f)^e + (N_A, h)^e_\Gamma - \sum_{B \in \eta_g} a(N_A, N_B)^e g_B \tag{6.40}$$

$$= \int_{\Omega^e} N_A f d\Omega + \int_{\Gamma^e_h} N_A h d\Gamma - \sum_{B \in \eta_g} a(N_A, N_B)^e g_B \tag{6.41}$$

The $a(N_A, N_B)^e$ terms in the summation are the components of the stiffness matrix. Γ^e_h is the part of the element boundary on Γ_h. It is non-empty only for elements on the domain boundary. We also have $P = ID(A)$, and $Q = ID(B)$.

We can write these equations for a single element,

$$k^e = [k^e_{ab}], \quad f^e = \{f^e_a\}, \quad 0 \le a, b < n_{en} \tag{6.42}$$

$$k^e_{ab} = a(N_a, N_b)^e = \int_{\Omega^e} (\nabla N_a)^T k \nabla N_b d\Omega \tag{6.43}$$

$$f^e_a = \int_{\Omega^e} N_a f d\Omega + \int_{\Omega^e} N_a h d\Gamma - \sum_{b=1}^{n_{en}} k^e_{ab} q^e_b \tag{6.44}$$

The difference between these equations and the ones defined previously in Equation 6.39 is that here a, b range only over n_{en} element nodes. In the case of a *linear* tetrahedron, we have four nodes, giving us a $[4 \times 4]$ matrix.

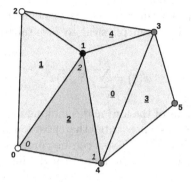

Figure 6.8: Illustration of indexing to relate local and global node numbers.

6.6.7 Matrix Assembly

Let's assume that we have computed the k^e matrix for some arbitrary element. These values now need to be *assembled* into the global stiffness matrix. The algorithm is demonstrated here with a triangular mesh, and hence each k^e is a 3×3 matrix. Let's consider the node ordering in Figure 6.8. Nodes 0 and 2, shown in white, are the Dirichlet boundary conditions, and nodes 4,5 and 3 are part of the Neumann boundary. We use the ID array to relate global node numbers to the global equation number,

A	0	1	2	3	4	5
ID(A)	-1	0	-1	1	2	3

We use -1 to denote global Dirichlet nodes that do not appear in the solution vector. We also need a *element node array* $IEN(e, a) = A$ to relate local and global node numbers. For instance, in the highlighted element 2, the local node 0 is global node 0, and node 1 is node 4,

a	0	1	2
IEN(2,a)	0	4	1

We already have this data - this is the `con` connectivity array defined for each `Tet`. Finally, we also need a *location matrix* $P = LM(e, a) = ID(IEN(e, a))$ to relate local node number to global unknown number,

a	0	1	2
LM(2,a)	-1	2	0

6.6.8 Quadrature

Only a "small" detail remains: how do we evaluate the k^e and $\vec{f}^e$ terms? As a reminder, the mathematical form is given by Equation 6.43. Let's now forget about the particular form under the integral and assume that we just want to integrate some arbitrary function $f(x, y, z)$ defined over Ω^e. Next assume that there exists mapping from physical to natural (or logical) coordinates $\vec{\xi} = (\xi, \eta, \zeta)$ such that

$$x(\vec{\xi}) = \sum_{a=1}^{4} N_a(\vec{\xi})\vec{x}_a^e \tag{6.45}$$

where $\vec{x}_a^e$ are the positions of the four nodes making up element e, and $\xi, \eta, \zeta \in [-1 : 1]$. This equation basically states that position in physical coordinates as a function of the logical coordinates can be computed by interpolating the positions of the nodes. We can then transform the integration to the $\vec{\xi}$ space,

$$\int_{\Omega^e} f(\vec{x})dxdydz = \int_{-1}^{1}\int_{-1}^{1}\int_{-1}^{1} f(x(\vec{\xi}), y(\vec{\xi}), z(\vec{\xi}))j(\vec{\xi})d\xi d\eta d\zeta \tag{6.46}$$

where $j(\vec{\xi}) = \partial x_i / \partial \xi_i$ is the *Jacobian*.

Let's now consider just the integral $\int_{-1}^{1} g(\xi)d\xi$. Integrating this one dimensional form is quite simple - we can use the trapezoidal rule,

$$\int_{-1}^{1} g(\xi)d\xi = \frac{1}{2}\Big(g(-1) + g(1)\Big)(1+1) + R$$
$$= g(-1) + g(1) + R \tag{6.47}$$

This is an example of numerical integration or *quadrature*. We can generalize the above formula as a sum

$$\int_{-1}^{1} g(\xi)d\xi = \sum_{l=1}^{n_{int}} g(\tilde{\xi}_l W_l + R \tag{6.48}$$

For the trapezoidal rule with two integration points $n_{int} = 2$, we have $\tilde{\xi}_1 = -1$, $\tilde{\xi}_2 = 1$, $W_1 = 1$, $W_2 = 1$. The residue (or error term) is $R = -(2/3)g_{,\xi\xi}(\tilde{\xi})$. This term is zero, giving us an exact numerical integration, if g varies linearly since then $g_{,\xi\xi}$ vanishes. This is the case in our code since we are using *linear* shape functions. As will be apparent soon, linear shape functions have a major drawback in that they offer only a constant derivative per element. In your own FEM-PIC implementation, you may be interested in utilizing higher order elements. For these elements, the trapezoidal integration produces *quadrature errors*. You can improve the computation by utilizing more control points. But where to put the extra points? The optimal configuration of sampling locations and the associated weights is given by *Gaussian quadrature*. The first three sets of coefficients are listed in Table 6.1, with higher order versions readily found online.

Using these coefficients, we evaluate a multi-dimensional integral as

$$\int_{-1}^{1}\int_{-1}^{1}\int_{-1}^{1} g(\xi, \eta, \zeta)d\xi d\eta d\zeta \approx \sum_{n=1}^{n_{int}}\sum_{m=1}^{n_{int}}\sum_{l=1}^{n_{int}} g(\tilde{\xi}_l, \tilde{\xi}_m, \tilde{\xi}_n)W_l W_m W_n \tag{6.49}$$

n_{int}	ξ_1	ξ_2	ξ_3	W_1	W_2	W_3	$O(R)$
1	0			2			2
2	$-\frac{1}{\sqrt{3}}$	$\frac{1}{\sqrt{3}}$		1	1		4
3	$-\sqrt{\frac{3}{5}}$	0	$\sqrt{\frac{3}{5}}$	$\frac{5}{9}$	$\frac{8}{9}$	$\frac{5}{9}$	6

Table 6.1: Gaussian quadrature coefficients for up to 3 integration points.

For arbitrary limits, we have

$$\int_a^b g(\xi)d\xi \approx \frac{b-a}{2}\sum_{l=1}^{n_{int}} g\left(\frac{b-a}{2}\tilde{\xi}_l + \frac{a+b}{2}\right)W_l \tag{6.50}$$

6.6.9 Shape Functions

Again, we want to compute

$$k_{ab}^e = a(N_a, N_b)^e = \int_{\Omega^e} N_{a,i}kN_{b,i}d\Omega \tag{6.51}$$

We can compute this derivative in the logical $\vec{\xi}$ space. Using chain rule, we write

$$\frac{\partial N_a}{\partial x} = \frac{\partial N_a}{\partial \xi}\frac{\partial \xi}{\partial x} + \frac{\partial N_a}{\partial \eta}\frac{\partial \eta}{\partial x} \tag{6.52}$$

or using index notation,

$$N_{a,x} = N_{a,\xi}\xi_{,x} + N_{a,\eta}\eta_{,x} \tag{6.53}$$

Similarly,

$$N_{a,y} = N_{a,\xi}\xi_{,y} + N_{a,\eta}\eta_{,y} \tag{6.54}$$

From this equation, we see that we need an expression for shape functions N_a in the logical space, $N_a = N_a(\xi)$. We also need an expression for the derivative of the mapping $\vec{\xi}_{,\vec{x}}$. We then evaluate

$$\int_\Omega f(\vec{x})d\Omega = \int_\xi f(\vec{\xi})\left|\vec{x}_{,\vec{\xi}}\right|d\vec{\xi} \tag{6.55}$$

So far, we managed to ignore the details of the shape functions, but cannot do so any longer. Let's consider triangular elements to simplify the visualization. Moving from a triangle to a tetrahedron is straightforward. Imagine that you have some arbitrary triangle in the $\{x, y\}$ space. We need a coordinate transformation to map every point to a new space defined in $\{\xi, \eta\}$. Here we take a different approach from [35]. Instead of mapping the triangle to a degenerate quadrilateral in $[-1 : 1] \times [-1 : 1]$, we map the triangle to $\{\xi, \eta\} \in [0 : 1]$.

One way to achieve this mapping is using *triangle coordinates*. These coordinates are the area-based shape factors L_i you are already familiar with from our prior discussion on particle to mesh interpolation. To maintain the existing notation, we denote these functions as N_i. Consider N_0. We have $N_0(\vec{x}_0) = 1$, and $N_1(\vec{x}_1) = N_1(\vec{x}_2) = 0$. In general,

$$N_i(\vec{x}_i) = \begin{cases} 1 & ; i = j \\ 0 & ; i \neq j \end{cases} \tag{6.56}$$

In other words, these shape functions transform physical coordinates to $[0, 1]$ space, which is exactly what we are after. The only issue is that there are three of them for only two spatial coordinates (in 2D). But since $N_0 + N_1 + N_2 = 1$, we can define

$$N_0 = \xi \tag{6.57}$$
$$N_1 = \eta \tag{6.58}$$
$$N_2 = 1 - \xi - \eta \tag{6.59}$$

Returning to Equations 6.53 and 6.54, we now have expressions for the $N_{a,\xi}$ and $N_{a,\eta}$ terms,

$$\begin{array}{ll} N_{0,\xi} = 1 & N_{0,\eta} = 0 \\ N_{1,\xi} = 0 & N_{1,\eta} = 1 \\ N_{2,\xi} = -1 & N_{2,\eta} = -1 \end{array} \tag{6.60}$$

6.6.10 Coordinate Transformation

Computing the other terms is bit more complicated since we do not have an analytical expression for $\vec{\xi}(\vec{x})$. Instead, we rewrite Equations 6.53 and 6.54 in matrix form,

$$\begin{Bmatrix} N_{a,x} \\ N_{a,y} \end{Bmatrix} = \begin{Bmatrix} N_{a,\xi} \\ N_{a,\eta} \end{Bmatrix} \begin{bmatrix} \xi_x & \xi_y \\ \eta_x & \eta_y \end{bmatrix} \tag{6.61}$$

Again, we already have the expression for the vector on the right, but we need an expression for the 2×2 matrix. Although we don't have an equation for $\vec{\xi}(\vec{x})$, we have one for $\vec{x}(\vec{\xi})$,

$$x(\xi, \eta) = \sum_{a=0}^{2} N_a(\xi, \eta) x_a^e \tag{6.62}$$

$$y(\xi, \eta) = \sum_{a=0}^{2} N_a(\xi, \eta) y_a^e \tag{6.63}$$

We can thus compute $\vec{x}, \vec{\xi}$,

$$\vec{x}_{,\vec{\xi}} = \begin{bmatrix} x_{,\xi} & x_{,\eta} \\ y_{,\xi} & y_{,\eta} \end{bmatrix} \tag{6.64}$$

The terms inside the matrix are evaluated from

$$x_{,\xi} = \sum_{a=0}^{2} N_{a,\xi} x_a^e \quad x_{,\eta} = \sum_{a=0}^{2} N_{a,\eta} x_a^e \tag{6.65}$$

$$y_{,\xi} = \sum_{a=0}^{2} N_{a,\xi} y_a^e \quad y_{,\eta} = \sum_{a=0}^{2} N_{a,\eta} y_a^e \tag{6.66}$$

$$\tag{6.67}$$

Note that we already have expressions for the derivatives of the shape functions that appear inside the summations, $N_{a,\xi}$ and $N_{a,\eta}$.

We have just written an expression for $\vec{x}_{,\vec{\xi}}$ in Equation 6.64. We are however interested in $\vec{\xi}_{,\vec{x}}$, which turns out to be the inverse,

$$\vec{\xi}_{,\vec{x}} = (\vec{x}_{,\vec{\xi}})^{-1} \tag{6.68}$$

This inverse can be computed easily for a 2×2 matrix, since for

$$A = \begin{bmatrix} a & b \\ c & d \end{bmatrix} \tag{6.69}$$

the inverse is given by

$$A^{-1} = \frac{1}{\det(A)} \begin{bmatrix} d & -b \\ -c & a \end{bmatrix} \tag{6.70}$$

6.6.11   Single Element Algorithm

Therefore, the algorithm for computing the $N_{a,x}$ and $N_{b,x}$ terms in an arbitrary element can be described by the following steps:

1. Compute $N_{a,\xi}$ at quadrature points l_ξ

2. Use $N_{a,\xi}$ to evaluate terms of $\vec{x}_{,\vec{\xi}}$

3. Compute $\vec{\xi}_{,\vec{x}} = \left(\vec{x}_{,\vec{\xi}}\right)^{-1}$

4. Use $N_{a,\xi}\xi_{,x}$ to evaluate $N_{a,x}$

5. Repeat steps 1-4 with b

The quadrature is then computed using limits transformation,

$$\int_0^1 \int_0^1 g(\vec{\xi}) d\vec{\xi} \approx \left(\frac{1-0}{2}\right)\left(\frac{1-0}{2}\right) \sum_n^{n_{int}} \sum_m^{n_{int}} g(\xi_m', \eta_n') W_m W_n \tag{6.71}$$

where

$$\xi_m' = \left(\frac{1-0}{2}\right)\xi_m + \left(\frac{1+0}{2}\right) \tag{6.72}$$

and

$$g(x) = N_{a,x} \cdot k N_{b,x} \tag{6.73}$$

This transformation arises since Gaussian quadrature is defined for $[-1:1]$.

6.6.12 Derivatives

At this point, you should notice that the derivatives of the shape functions are constant. As such, we do not need to recompute $N_{a,x}$ at each quadrature point as they are all the same. Elements like this are known as C^0 elements. The number in the superscript is the order of the derivative. With linear basis functions, we obtain 0-th order (constant) derivatives. In PIC simulations, we are actually interested in the derivative of the solution, $\vec{E} = -\nabla\phi$. Using the FEM formulation, the derivative is evaluated from

$$E(\vec{x}) = - \sum_{a=0}^{n_{en}-1} N_{a,x}(\vec{x})\phi_a^e \tag{6.74}$$

Since the shape function derivative is constant, we obtain only a constant (not spatially varying) electric field per element. The electric field is a step-wise function. This is not ideal, as particles experience a jump as they cross a cell boundary. For this reason, it is recommended to use second order elements in your implementation. Such a formulation is beyond the scope of this book, but Hughes provides definitions for the higher order basis functions.

6.6.13 Tetrahedral Element

Modifying the algorithm to a three-dimensional tetrahedral element is quite straightforward. The primary difference is that we now have an additional spatial coordinate, and thus need to perform the mapping $\{x, y, z\} \to \{\xi, \eta, \zeta\}$. We again use the normalized volumes as the natural tetrahedral coordinates,

$$N_0 = \xi \qquad N_1 = \eta \qquad N_2 = \zeta \qquad N_3 = 1 - \xi - \eta - \zeta \tag{6.75}$$

where

$$N_0(x) \equiv L_0(x) = \frac{V_{x123}}{V_{0123}} \tag{6.76}$$

We again obtain constant derivatives,

$$
\begin{array}{lll}
N_{0,\xi} = 1 & N_{0,\eta} = 0 & N_{0,\zeta} = 0 \\
N_{1,\xi} = 0 & N_{1,\eta} = 1 & N_{1,\zeta} = 0 \\
N_{2,\xi} = 0 & N_{2,\eta} = 0 & N_{2,\zeta} = 1 \\
N_{3,\xi} = -1 & N_{3,\eta} = -1 & N_{3,\zeta} = -1
\end{array} \tag{6.77}
$$

6.7 PUTTING IT ALL TOGETHER

6.7.1 Initialization

The above equations are implemented in class FESolver in FESolver.cpp. We start by defining the constructor,

```
FESolver::FESolver(World &world, int max_it, double tol):
    world{world}, vm{world.vm}, max_solver_it(max_it),
    tolerance(tol) {
    neq = 0;                          // count the number of uknowns
    for (size_t i=0;i<vm.nodes.size();i++)
        if (vm.nodes[i].type==NORMAL ||
            vm.nodes[i].type==OPEN) neq++;   // h nodes
    cout<<"There are "<<neq<<" unknowns"<<endl;

    // allocate neq*neq K matrix
    K = new double*[neq];
    for (int i=0;i<neq;i++) K[i] = new double[neq];
    cout<<"Allocated "<<neq<<"x"<<neq<<" stiffness matrix"<<endl;

    // allocate neq*neq J matrix
    J = new double*[neq];
    for (int i=0;i<neq;i++) J[i] = new double[neq];
    cout<<"Allocated "<<neq<<"x"<<neq<<" Jacobian matrix"<<endl;

    // allocate F0 and F1 vectors
    F0.reserve(neq);
    F1.reserve(neq);
    cout<<"Allocated two "<<neq<<"x1 force vectors"<<endl;

    n_nodes = vm.nodes.size();
    n_elements = vm.tets.size();

    // allocate ID vector
    ID.reserve(n_nodes);
    cout<<"Allocated "<<n_nodes<<"x1 ID vector"<<endl;

    // allocate location matrix, n_elements*4
    LM = new int*[n_elements];
    for (int e=0;e<n_elements;e++) LM[e] = new int[4];
    cout<<"Allocated "<<n_elements<<"x4 location matrix"<<endl;

    // allocate NX matrix
    NX = new double**[n_elements];
    for (int e=0;e<n_elements;e++) {
      NX[e] = new double*[4];
        for (int a=0;a<4;a++) NX[e][a] = new double[3];
    }
    cout<<"Allocated "<<n_elements<<"x4x3 NX matrix"<<endl;

    // solution array, initialized to zero
    d.reserve(neq);

    // allocate memory for g and uh arrays
    g.reserve(n_nodes);
    uh.reserve(n_nodes);
```

```
cout<<"Allocated "<<n_nodes<<"x1 g and uh vector"<<endl;
detJ.reserve(n_elements);
cout<<"Allocated "<<n_elements<<"x1 detJ vector"<<endl;

// electric field
ef.reserve(n_elements);

// set up the ID array note valid values are 0 to neq-1 and -1
   indicates "g" node
int P=0;
for (int n=0;n<n_nodes;n++)
   if (vm.nodes[n].type==NORMAL ||
      vm.nodes[n].type==OPEN) {ID[n]=P; P++;}
   else ID[n]=-1;   // Dirichlet node

// now set up the LM matrix
for (int e=0;e<n_elements;e++)
   for (int a=0;a<4;a++)   // tetrahedra
      LM[e][a] = ID[vm.tets[e].con[a]];
cout<<"Built ID and LM matrix"<<endl;

// set quadrature points
l[0]=-sqrt(1.0/3.0); l[1]=sqrt(1.0/3.0);
W[0]=1; W[1]=1;
n_int = 2;

// initialize solver "g" array
for (int n=0;n<n_nodes;n++) {
   if (vm.nodes[n].type == INLET) g[n]=0;       // phi_inlet
   else if (vm.nodes[n].type == SPHERE) g[n]=-100; // phi_sphere
   else g[n] = 0;                                // default
}

// compute NX matrix
computeNX();

// sample assembly code
startAssembly();
preAssembly();          // form K and F0
}
```

This code mainly consists of allocating memory for the K matrix and various helper vectors. For simplicity we use a full matrix to store K, with the matrix allocated using the **new** operator. The allocated matrices are deleted in the destructor. The one-dimensional vectors are stored as std::vector<double>. We also set the ID and LM lookup tables, and also set the quadrature points. We are using the scheme with two integration points.

6.7.2 Matrix Assembly

Next, we call startAssembly to set all matrix and vector components to zero.

```
void FESolver::startAssembly() {
   for (int i=0;i<neq;i++)
      for (int j=0;j<neq;j++) K[i][j] = 0;
```

```
  for (int i=0;i<neq;i++) {
    F0[i]=0;
    F1[i]=0;
  }
}
```

After this, we call **preAssembly** to build the matrix and the linear parts (f_h and f_g as described below) of the forcing vector $\vec{F}$,

```
void FESolver::preAssembly() {
  for (int e=0;e<n_elements;e++) {
    Tet &tet = vm.tets[e];
    double ke[4][4];

    for (int a=0;a<4;a++)
      for (int b=0;b<4;b++) {
        ke[a][b] = 0;                        // reset

        // perform quadrature
        for (int k=0;k<n_int;k++)
          for (int j=0;j<n_int;j++)
            for (int i=0;i<n_int;i++) {
              double nax[3],nbx[3];

              double xi = 0.5*(l[i]+1);    // not used
              double eta = 0.5*(l[j]+1);   // not used
              double zeta = 0.5*(l[k]+1);  // not used
              double3 nax = getNax(e,a,xi,eta,zeta);
              double3 nbx = getNax(e,b,xi,eta,zeta);

              // dot product
              double dot=0;
              for (int d=0;d<3;d++) dot+=nax[d]*nbx[d];

              // set ke matrix components
              ke[a][b] += dot*detJ[e]*W[i]*W[j]*W[k];
            }
      }

    // we now have the ke matrix
    addKe(e,ke);

    // force vector
    double fe[4];

    for (int a=0;a<4;a++) {
      // second term int(na*h), zero since only h=0 supported
      double fh=0;

      // third term, -sum(kab*qb)
      double fg = 0;
      for (int b=0;b<4;b++) {
        int n = tet.con[b];
        double gb = g[n];
        fg-=ke[a][b]*gb;
      }
```

```
    // combine
    fe[a] = fh + fg;
    }
  addFe(F0, e, fe);
  } // for element
}
```

This code loops over all elements. For each element, we loop over the 16 node combinations of a and b. and for each, we evaluate the integral

$$\sum_k \sum_j \sum_i g(i,j,k) W_i W_j W_k \tag{6.78}$$

where

$$g(i,j,k) = N_{a,x}(\xi_i, \eta_j, \zeta_k) \cdot N_{b,x}(\xi_i, \eta_j, \zeta_k) \tag{6.79}$$

We set ke[a][b] to this sum.

We also need to evaluate the force vector. It consists of three terms,

$$f_a^e = \int_{\Omega^e} N_a f d\Omega + \int_{\Omega^e} N_a h d\Gamma - \sum_{b=0}^{n_{en}-1} k_{ab}^e g_b^e \tag{6.80}$$

I call these terms f_f, f_h, and f_g. The second and third items are built by preAssembly. The first term is the interpolation of the forcing function. We replace it with quadrature,

$$\left(\frac{1}{8}\right) \sum_k^{n_{int}} \sum_j^{n_{int}} \sum_k^{n_{int}} N_a\left(\xi_i', \eta_j', \zeta_k'\right) \left|\vec{x}_\xi(\xi_i', \eta_j', \zeta_k')\right| W_i W_j W_k \tag{6.81}$$

Note that the $\vec{x}_{,\xi}$ terms are constant due linear elements. The $(1/8)$ ratio arises from the change of integration limits. We also have

$$\xi' = \frac{x^+ - x^-}{2}\xi + \frac{x^+ + x^-}{2} = \frac{1}{2}(\xi+1) \tag{6.82}$$

This is the part that captures charge density. The second term encompasses the Neumann boundary condition. The code was simplified to support only the zero version, $\partial\phi/\partial n = h = 0$, and hence this term drops out. The last term inserts the Dirichlet boundaries. We keep this part. The g array is initialized prior to assembly using the NodeType type properties stored for each volume mesh node.

6.7.3 Shape Function Derivatives

The assembly code utilizes getNax to evaluate $N_{a,x}(\vec{\xi})$. These values are constant in each cell due to the use of linear elements, and we can thus precompute them. The values are stored in NX array and this function simply returns the appropriate member,

```
double3 FESolver::getNax(int e, int a, double xi, double eta,
  double zeta) {
  return {NX[e][a][0],NX[e][a][1],NX[e][a][2]};
}
```

The derivatives are computed in a call to `computeNX` in the constructor,

```
void FESolver::computeNX() {
  double na_xi[4][3] = {{1,0,0}, {0,1,0}, {0,0,1}, {-1,-1,-1}};

  for (int e=0;e<n_elements;e++) {
    Tet &tet = vm.tets[e];
    double x[4][3]; // node positions
    for (int a=0;a<4;a++) {
      double3 pos = vm.nodes[tet.con[a]].pos;
      for (int d=0;d<3;d++) x[a][d] = pos[d];   // copy
    }

    // build x_xi matrix
    double x_xi[3][3];
    for (int i=0;i<3;i++)                        // x/y/z
      for (int j=0;j<3;j++) {                    // xi/eta/zeta
        x_xi[i][j] = 0;
        for (int a=0; a<4; a++)                  // tet node
          x_xi[i][j] += na_xi[a][j]*x[a][i];
      }

    // save det(x_xi)
    detJ[e] = utils::det3(x_xi);

    // compute matrix inverse
    double xi_x[3][3];
    inverse(x_xi,xi_x);

    // evaluate na_x
    for (int a=0;a<4;a++)
      for (int d=0;d<3;d++) {
        NX[e][a][d]=0;
        for (int k=0;k<3;k++)
          NX[e][a][d]+=na_xi[a][k]*xi_x[k][d];
      }
  }
}
```

We use the analytical forms of the $N_{i,\xi}$ derivatives as they are constant in each cell. We then populate the $\vec{x}_{,\xi}$ matrix for each element and store its inverse in `x_xi`. We then compute the derivatives of the shape functions. The inverse is computed using Cramer's rule. For a 3×3 matrix,

$$A^{-1} = \frac{1}{\det(A)} C^T \qquad (6.83)$$

where C is a matrix of *cofactors*. Cofactors are the determinants of *minors* with alternating signs that are also utilized when computing the cross-product.

6.7.4 Nonlinear Solver

We use the Newton-Rhapson scheme to solve the nonlinear problem arising from the Boltzmann electron term. Again, we want to find the root of

$$\vec{G}(\phi) = \mathbf{K}\vec{phi} - \vec{F} \qquad (6.84)$$

The derivative is given by

$$G_{,\phi}(\phi) = K - F_{,\phi} \qquad (6.85)$$

As we saw above, the forcing vector is composed of three terms. Only the first term is a function of potential, and as such, it is the only term that survives the derivative,

$$f^e_{a,\phi} = \int_{\Omega^e} N_a f_{,\phi} \qquad (6.86)$$

In our case,

$$f = \frac{e}{\epsilon_0}\left[Z_i n_i + n_0 \exp\left(\frac{\phi - \phi_0}{kT_e}\right)\right] \qquad (6.87)$$

so

$$f_{,\phi} = \frac{e}{\epsilon_0} n_0 \exp\left(\frac{\phi - \phi_0}{kT_e}\right)\frac{1}{kT_e} \qquad (6.88)$$

This f_f term is evaluated using the following function,

```cpp
void FESolver :: buildF1Vector (dvector &ion_den) {
  dvector f(neq);
  // compute the RHS term on all unknown nodes
  for (int n=0;n<n_nodes;n++)  {
    int A = ID[n];
    if (A<0) continue;                        // skip known nodes
    f[A] = (QE/EPS_0)*(ion_den[n]+n0*exp((d[A]-phi0)/Te0));
  }

  for (int e=0;e<n_elements;e++) {            // loop over elements
    double fe[4];
    for (int a=0;a<4;a++) {
      double ff=0;
      int A = LM[e][a];
      if (A>=0) {                             // if unknown node
        // perform quadrature
        for (int k=0;k<n_int;k++)
          for (int j=0;j<n_int;j++)
            for (int i=0;i<n_int;i++) {
              double xi = 0.5*(l[i]+1);       // change of limits
              double eta = 0.5*(l[j]+1);
              double zeta = 0.5*(l[k]+1);
              double Na=evalNa(a,xi,eta,zeta);
              ff += f[A]*Na*detJ[e]*W[i]*W[j]*W[k];
            }
        ff*=(1.0/8.0);                        // change of limits scaling
        fe[a] = ff;
      }
    }
```

```
    }
    addFe(F1, e, fe);
  }
}
```

The `buildF1Vector` function is called at every iteration of the non-linear Newton Raphson solver,

```
void FESolver::solveNonLinear(dvector &rhoi) {
  dvector y(neq);     // allocate space for y, initialized to zero
  dvector G(neq);

  bool converged = false;
  double L2;
  for (int it=0;it<10;it++) {
    buildF1Vector(rhoi); // build the ff part of the force vector

    // form G=K*d-F
    matVecMultiply(G,K,d, neq);   // G=K*d
    vecVecSubtract(G,G,F0, neq);  // G=G-F giving us G=K*d-F
    vecVecSubtract(G,G,F1, neq);

    buildJmatrix();               // assemble the Jacobian
    solveLinear(J,y,G);           // solver the linear system

    // now that we have y, update solution
    for (int n=0;n<neq;n++) d[n]-=y[n];

    // compute residue
    double sum=0;
    for (int u=0;u<neq;u++) sum+=y[u]*y[u];

    L2 = sqrt(sum)/neq;
    if (L2<1e-2) {
      cout<<" NR converged in "<<it+1<<" iterations with
        L2="<<setprecision(3)<<L2<<endl;
      converged=true;
      break;
    }
  }
  if (!converged) cerr<<"NR failed to converge, L2 = "<<L2<<endl;
}
```

The Jacobian matrix is built using the following function,

```
void FESolver::buildJmatrix() {
  // first compute exponential term
  dvector fp_term(neq);
  dvector FP(neq);

  for (int n=0;n<neq;n++)
    fp_term[n] = -QE/EPS_0*n0*exp((d[n]-phi0)/Te0)*(1/Te0);

  // now set J=K
  for (int i=0;i<neq;i++)
    for (int j=0;j<neq;j++)
      J[i][j] = K[i][j];
```

```
// build fprime vector
double fe[4];

for (int e=0;e<n_elements;e++) {
  for (int a=0;a<4;a++) {
    double ff=0;
    int A = LM[e][a];
    if (A>=0) {                    // perform quadrature on unknown
      nodes
      for (int k=0;k<n_int;k++)
        for (int j=0;j<n_int;j++)
          for (int i=0;i<n_int;i++) {
            double xi = 0.5*(l[i]+1);    // change of limits
            double eta = 0.5*(l[j]+1);
            double zeta = 0.5*(l[k]+1);
            double Na=evalNa(a,xi,eta,zeta);
            ff += fp_term[A]*Na*detJ[e]*W[i]*W[j]*W[k];
          }
      ff*=(1.0/8.0);               // change of limits scaling
    }
    fe[a] = ff;
  }

  // assembly
  for (int a=0;a<4;a++) {          // tetrahedra
    int P = LM[e][a];
    if (P<0) continue;             // skip g nodes

    FP[P] += fe[a];
  }
}

// subtract diagonal term
for (int u=0;u<neq;u++)    J[u][u]-=FP[u];
}
```

6.7.5 Solution Assembly

The non-linear solver is called from FESolver::solve,

```
void FESolver::solve() {
  // copy data to a dvector
  dvector rhoi(vm.nodes.size());
  for (size_t i=0;i<vm.nodes.size();i++)
    rhoi[i] = world.rho[i];

  // solve the system
  solveNonLinear(rhoi);

  // combine d and g to phi
  for (int n=0;n<n_nodes;n++) {
    // zero on non-g nodes
    uh[n] = g[n];

    // is this a non-g node?
    int A=ID[n];
```

```
        if (A>=0) uh[n] += d[A];
        world.phi[n] = uh[n];
    }
}
```

After the solver completes, we still have one step left, and that is to combine the solution on the "unknown" nodes with the Dirichlet boundaries, $u^h = v^h + g$. We store this data locally and also copy it to the World phi field.

6.7.6 Electric Field

And finally, we add code for computing the electric field in each cell. This function is effectively the gather operation but evaluated using derivatives of the basis function.

```
void FESolver::computeEF() {
    for (int e=0;e<n_elements;e++) {
        Tet &tet = vm.tets[e];
        for (int d=0;d<3;d++) ef[e][d]=0;

        for (int a=0;a<4;a++) {
            int A = tet.con[a];
            double3 nx = getNax(e,a,0.5,0.5,0.5);
            // minus sign since negative gradient
            for (int d=0;d<3;d++) ef[e][d]-=nx[d]*uh[A];
        }
    }
}
```

For linear elements, these functions are constant in each cell. We thus precompute the field and store the components in a std::vector<double3> array called ef. We also add a function to evaluate the field at a particular logical coordinate. This function accepts the logical coordinate as an argument for generality, however, only the cell_id property is used to retrieve the appropriate $\vec{E}$.

```
double3 FESolver::evalEf(LCord &lc) {return ef[lc.cell_id];}
```

6.7.7 Results

Results obtained with the unstructured grid are visualized in Figure 6.9. These results are compared to those obtained in Chapter 3 on a structured grid. Both potential and ion density are quite similar to the sugarcubed solution. The simulation time varies significantly, however. The FEM solver required almost four times longer to complete the 400 iterations than the uniform Cartesian version.

6.8 SUMMARY

In this chapter we developed a FEM-PIC code that simulates the flow of ions past a sphere on an unstructured grid. We started by learning how to

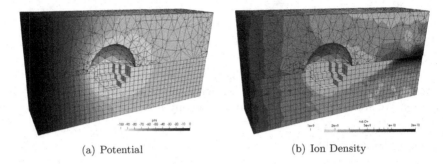

(a) Potential   (b) Ion Density

Figure 6.9: Comparison of results on the unstructured mesh to those obtained in Chapter 3.

import the mesh geometry. We then discussed methods for scattering particle positions to the grid and saw how to find the containing cell using neighbor traverse. We reviewed the Finite Element Method (FEM) and developed an FEM solver for the Poisson's equation. We also learned about shape functions, and saw that linear elements yield constant electric field per cell. Simulation results on the unstructured grid were then compared to the Cartesian version.

EXERCISES

6.1 *Particle Push Performance.* Compare neutral (no field solve) simulation run times with the brute force cell search, brute force with precomputed determinants (Equation 6.15), and the neighbor traverse. Depending on the number of elements, you should see at least two orders of magnitude speed up with the traverse method.

6.2 *Cartesian Comparison.* Compare particle push timing on the unstructured grid to the Cartesian version from Chapter 3.

6.3 *Two element FEM.* Derive the FEM equations for the one-dimensional form $u_{,xx} + f = 0$ on a system described by two elements. Let the domain span $x \in [0,1]$, and let $u_{,x}(0) = 0$, $u(1) = g$. Solve this system by manually evaluating the coefficients of the stiffness matrix.

6.4 *Three element FEM.* Now extend the formulation to three elements. Start by writing down expressions for N_0, N_1, N_2, and N_3. Next compute the integrals $K_{00} \ldots K_{22}$ and $F_0 \ldots F_2$. Solve the system for $d_0 \ldots d_2$ and assemble the solution by including the right-side Dirichlet condition. Compare the numerical and theoretical solutions.

6.5 *Matrix Assembly.* Write code to perform the assembly for the above exercise automatically by considering the element view. Start by creating an empty $n_{el} \times n_{el}$ $\boldsymbol{K}$ matrix and an empty $n_{el} \times 1$ $\vec{F}$ vector. Then loop through the elements and for each compute the 2×2 $\boldsymbol{k}^e$ and 2×1 $\vec{f}^e$ matrix and vector. Don't forget to add the boundary values to $\vec{f}^e$. Add these values to the appropriate place in the global matrix. Then solve the $\boldsymbol{K}\vec{d} = \vec{F}$ system and evaluate $\vec{u}^h$. Since we are using linear shape functions and the plotting program connects points with linear segments, we can just store the nodal values. Test your program with a varying number of nodes, and random node spacing.

CHAPTER 7

Electromagnetics

7.1 INTRODUCTION

T HE CODES developed so far considered only the electric field. In this chapter, we learn how to incorporate magnetic field effects. We start by introducing the Boris algorithm for pushing particles in a magnetic field. We then develop a magnetic field solver for a *magnetostatic problem*. Finally, we develop a two-dimensional electromagnetic PIC code.

7.2 CLASSIFICATION

One of the many ways to classify plasma dynamics is by considering the role of the magnetic field. There are three cases to consider. First, the field strength could be negligible such that neither ions nor electrons are *magnetized*. This is the assumption we have included implicitly in our codes so far. Second, magnetic field may be present, but be time invariant, $\partial B/\partial t = 0$. In this case, we need to modify the velocity integrator to take into account the full Lorentz force equation,

$$\vec{F} = q\left(\vec{E} + \vec{v} \times \vec{B}\right) \tag{7.1}$$

The electric field can still be computed per the electrostatic assumption by solving Poisson's equation. Finally, the magnetic field may be time varying, either due to external sources, or due to strong plasma currents. The electrostatic assumption no longer holds, and we now need to solve the electric and magnetic field by considering Maxwell's equation (Section 1.3.2), specifically Faraday's law of induction and Ampere's circuit law.

7.3 MAGNETIZATION

We now start with the second scenario from above. In the presence of magnetic field, charged particles orbit about the field lines with *cyclotron frequency*

$$\omega_c = \frac{|q|B}{m} \tag{7.2}$$

The orbit size is given by the *Larmor radius*,

$$r_L \equiv \frac{v_\perp}{\omega_c} = \frac{mv_\perp}{|q|B} \tag{7.3}$$

Here $v_\perp$ is the tangential velocity along the orbit. We can notice that the orbital radius decreases with particle mass, and also with field strength. If the orbital radius is much larger than some characteristic problem dimension, $r_L \gg L$, we can assume that the magnetic field plays an insignificant role. Such a particle is said to be *unmagnetized*. This is the case in the problems considered so far. Due to the m dependence, it is not uncommon to find cases in which lighter species, such as electrons, are magnetized, but the heavier ions are not. This setup is in fact exploited in Hall Effect thrusters that form the bulk of my current research. These plasma propulsion devices operate by trapping electrons in a closed $\vec{E} \times \vec{B}$ drift about the centerline. The heavier ions pass through the thruster without being impeded by the magnetic field.

In situations like these, we next need to decide how to treat the electrons. Obviously, one option is to include them directly as particles. This fully-kinetic approach is very inefficient since a tiny time step must be used to resolve the cyclotron motion. Therefore, we may want to disregard the orbital dynamics, and consider only the movement of the *guiding center*. This is known as the *gyrokinetic method*. Finally, we can use a hybrid approach similar to what was discussed in Chapter 3. There is however a very important caveat to keep in mind. The Boltzmann relationship, Equation 3.11, holds only along a single magnetic field line. A scattering event, such as collision or a turbulent field fluctuation, is necessary to bump a charged particle to a new field line. We can relate the guiding center drift velocity across the field lines to the applied electric field using a proportionality parameter μ known as *mobility*, $v = \mu E$. Since electron motion in the direction parallel to the field line is not restricted, we have $\mu_\parallel \gg \mu_\perp$. It is then customary to decouple the electron dynamics into a motion along and across the magnetic field. In the parallel direction, we can assume that temperature is constant along each field line λ, $T_e = T_e(\lambda)$. This assumption then leads to a quasi one-dimensional fluid model for the electron population known as *thermalized potential* [45]. Details of this approach are beyond the scope of this book, but one popular implementation is described in [29].

7.3.1 Boris Push

However, let's assume that we actually want to perform a fully kinetic simulation which resolves the orbital motion about magnetic field lines. As can be seen by the $\vec{v} \times \vec{B}$ term in Equation 7.1, the magnetic field always acts in the direction perpendicular to the velocity vector. The field thus only affects the direction, but not the magnitude, of particle velocity. If we were to simply modify our Leapfrog integrator to evaluate

$$\vec{v}^{k+0.5} = \frac{q}{m} \left(\vec{v}^{k-0.5} \times \vec{B} \right) \tag{7.4}$$

for a case with $\vec{E} = 0$ and a constant $\vec{B} = 0.01\hat{k}$ T field, we obtain the result shown by the dashed line in Figure 7.1. We expect the particle to complete a closed orbit, but instead the trajectory spirals outward. This increase in orbital radius indicates a non-physical energy gain.

In his 1970 paper [16], Boris outlined an algorithm for integrating particle velocity in the presence of a magnetic field. This algorithm has since become the de-facto standard for pushing magnetized particles. We are effectively solving

$$\frac{\vec{v}^{k+0.5} - \vec{v}^{k-0.5}}{\Delta t} = \frac{q}{m}\left[\vec{E} + \frac{\vec{v}^{k+0.5} + \vec{v}^{k-0.5}}{2} \times \vec{B}\right] \tag{7.5}$$

Boris noticed that we can eliminate $\vec{E}$ field by defining

$$\vec{v}^{k-0.5} = \vec{v}^- - \frac{q}{m}\vec{E}\frac{\Delta t}{2} \tag{7.6}$$

$$\vec{v}^{k+0.5} = \vec{v}^+ + \frac{q}{m}\vec{E}\frac{\Delta t}{2} \tag{7.7}$$

With these substitutions, equation 7.5 becomes

$$\frac{\vec{v}^+ - \vec{v}^-}{\Delta t} = \frac{1}{2}\frac{q}{m}(\vec{v}^+ + \vec{v}^-) \times \vec{B} \tag{7.8}$$

which represents a pure rotation. Next, from geometry we see that the velocity rotates through angle

$$\tan\left(\frac{\theta}{2}\right) = -\frac{q}{m}B\frac{\Delta t}{2} \tag{7.9}$$

The vector form of the rotation vector is

$$\vec{t} = -\hat{b}\tan\left(\frac{\theta}{2}\right) \equiv \frac{q}{m}\vec{B}\frac{\Delta t}{2} \tag{7.10}$$

and

$$\vec{v}' = \vec{v}^- + \vec{v}^- \times \vec{t} \tag{7.11}$$

This equation rotates the velocity through $\Delta t/2$. We then follow this by another half-step rotation but with magnitude scaled to maintain the velocity magnitude,

$$\vec{v}^+ = \vec{v}^- + \vec{v}' \times \vec{s} \tag{7.12}$$

where

$$\vec{s} = \frac{2\vec{t}}{1 + t^2} \tag{7.13}$$

The Boris method for integrating the Lorentz force thus consists of the following four steps:

1. Perform half acceleration to compute $\vec{v}^-$ from $\vec{v}^{k-0.5}$ per Equation 7.6

2. Perform half rotation per Equation 7.11 to compute $\vec{v}'$

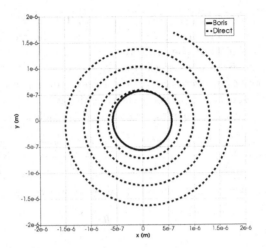

Figure 7.1: Orbit of an electron computed using the incorrect direct approach and with the Boris method.

3. Perform second half rotation per Equation 7.12 to compute $\vec{v}^{+}$

4. Perform second half acceleration per Equation 7.7 to compute $\vec{v}^{k+0.5}$

The result from this integration method is shown by the dark line in Figure 7.1. Now as expected, the electron completes a closed trajectory. This example can be found in ch7/lorentz_integrator.cpp. The velocity integration algorithm is copied below,

```cpp
// direct integration, yields a non-physical energy gain
void updateVelocityDirect(Particle &part, double dt, const double3
   &E, const double3 &B) {
   double qm = part.q/part.m;           // q/m
   double3 vxB = cross(part.v,B);       // compute cross-product
   part.v += qm*dt*(E+vxB);             // advance velocity -
      incorrect
}

// Boris method, conserves energy
void updateVelocityBoris(Particle &part, double dt, const double3
   &E, const double3 &B) {
   double qm = part.q/part.m;           // q/m
   // v minus, first half acceleration
   double3 v_minus = part.v + qm*E*dt/2;

   // v prime, first half rotation
   double3 t = qm*B*dt/2;
   double3 v_prime = v_minus + cross(v_minus, t);

   // v_plus, second half rotation
   double3 s = 2.0*t/(1+dot(t,t));
   double3 v_plus = v_minus + cross(v_prime, s);
```

```
// v[k+0.5], second half acceleration
part.v = v_plus + qm*E*dt/2;
}
```

The relevant part of the integrator is given below,

```
// rewind velocity per leapfrog
updateVelocityDirect(part1,-0.5*dt,E,B);
updateVelocityBoris(part2,-0.5*dt,E,B);

// main loop
for (int ts = 0;ts<=300;ts++) {
  // output data
  out<<ts<<" "<<part1.x<<" "<<part2.x<<"\n";

  updateVelocityDirect(part1,dt,E,B);
  updateVelocityBoris(part2,dt,E,B);
  part1.x += part1.v*dt;
  part2.x += part2.v*dt;
}
```

7.4 MAGNETOSTATICS

We now move away from the particle world, and learn how to solve the magnetic field. Just as we can assume that the magnetic field is time invariant in order to derive the governing equations of electrostatics, we can also assume a steady electric field $\partial \vec{E}/\partial t = 0$. This is the case in vacuum before plasma is introduced, or at steady state in which there is negligible temporal variation in charge density. The time derivative of electric field then drops out of the Ampere's law,

$$\nabla \times \vec{B} = \mu_0 \left(\vec{j} + \epsilon_0 \frac{\partial \vec{E}}{\partial t} \right) \tag{7.14}$$

and by including Gauss' Law for Magnetism, we obtain the two governing *microscopic* equations of *magnetostatics*,

$$\nabla \times \vec{B} = \mu_0 \vec{j} \tag{7.15}$$

$$\nabla \cdot \vec{B} = 0 \tag{7.16}$$

where $\mu_0 \approx 1.25663 \times 10^{-6}$ (V · s)/(A · m) is the *vacuum permeability*.

Just as we discussed with the case of electric displacement field in Section 4.9, sum of molecular magnetic moments gives rise to an average *macroscopic magnetization*. This term is also known magnetic moment density and is given the symbol $\vec{M}$. It can be shown that magnetization contributes to effective current density $j_M = \nabla \times \vec{M}$. Ampere's law then becomes

$$\nabla \times \vec{B} = \mu_0(\vec{j} + \nabla \times \vec{M}) \tag{7.17}$$

Combining the two cross-products allows us to define the *H field*,

$$\vec{H} = \frac{1}{\mu_0}\vec{B} - \vec{M} \tag{7.18}$$

We then have

$$\nabla\vec{H} = \vec{j} \tag{7.19}$$

and

$$\nabla \cdot \vec{B} = 0 \tag{7.20}$$

7.4.1 Vector Potential

There are two routes we can take in order to actually solve the magnetic field. Helmholtz decomposition tells us that any vector can be separated into an irrotational (curl-free) and a solenoidal (divergence-free) part,

$$\vec{F} = -\nabla\phi + \nabla \times \vec{A} \tag{7.21}$$

Gauss' magnetism law $\nabla \cdot \vec{B} = 0$ tells us that the magnetic field consists only of the solenoidal component. Hence

$$\vec{B} = \nabla \times \vec{A} \tag{7.22}$$

and

$$\nabla \times \vec{H}(\nabla \times \vec{A}) = \vec{j} \tag{7.23}$$

since $\vec{H} = \vec{H}(\vec{B})$. In the case of linear material, this complicated equation can be simplified by letting $\vec{B} = \mu\vec{H}$, similarly to the previous treatment of dielectrics. Then

$$\nabla \times \left(\frac{1}{\mu}\nabla \times \vec{A}\right) = \vec{j} \tag{7.24}$$

Furthermore, if μ is constant over some region of space, then in that piece we have

$$\nabla(\nabla \cdot \vec{A}) - \nabla^2\vec{A} = \mu\vec{j} \tag{7.25}$$

Since we only care about $\nabla \times \vec{A}$, we can let $\vec{A} = \vec{A} + \nabla\psi$, which is known as *gauge transformation*. It allows us to select a convenient form for $\nabla \cdot A$. The choice of zero is known as the *Coulomb gauge*. Then, finally we arrive at a vector form of Poisson's equation,

$$\nabla^2\vec{A} = -\mu\vec{j} \tag{7.26}$$

7.4.2 Scalar Potential

If $\vec{j} = 0$, the second equation of magnetostatics becomes

$$\nabla \times \vec{H} = 0 \tag{7.27}$$

which allows us to define a magnetic potential ϕ_M such that $\vec{H} = -\nabla \phi_M$ (this should look familiar from the derivation of the electrostatic method). We then obtain $\nabla \cdot (\mu \nabla \phi_M) = 0$ for linear material. If μ is constant, we obtain the Laplace equation

$$\nabla^2 \phi_M = 0 \tag{7.28}$$

If the region contains ferromagnets, we will have some finite magnetization $\vec{M}$. Then

$$\nabla \cdot \vec{B} = \mu_0 \nabla \cdot (\vec{H} + \vec{M}) = 0 \tag{7.29}$$

or

$$\nabla^2 \phi_m = -\rho_m \tag{7.30}$$

where

$$\rho_m = -\nabla \cdot \vec{M} \tag{7.31}$$

7.4.3 Magnetized Sphere

We now demonstrate this method by computing magnetic field for a uniformly magnetized sphere with magnetization $\vec{M} = M_0 \hat{e}_z$. The analytical solution for this configuration is given by Equation 5.104 in Jackson [36],

$$\phi_M(r, \theta) = \frac{1}{3} M_0 a^2 \left(\frac{\min(r, a)}{\max^2(r, a)} \right) \cos \theta \tag{7.32}$$

where a is the sphere radius.

We start by modifying the flow past a sphere code from Chapter 3 by including several new fields to the `World` object,

```
Field phi_m;            // magnetic potential
Field3 B;               // magnetic field
Field3 H;               // H field
Field3 M;               // magnetization
Field phi_m_theory;     // analytical solution from Jackson
```

Source files for this section are found in the **ch7/MS** directory. Initialization is added to the `World` constructor. We also add a function to set the magnetization vector,

```
void World::magnetizeSphere(const double3 &M0) {
  for (int i=0;i<ni;i++)
    for (int j=0;j<nj;j++)
      for (int k=0;k<nk;k++) {
        if (inSphere(pos(i,j,k)))
          M[i][j][k] = M0;
      }
}
```

We next add a new `MagneticSolver` class. Since the magnetic field is based on a solution of a Poisson's equation, we re-use much of the existing functionality in `PotentialSolver`. This class acts mainly as a wrapper that calls previously defined functions. Specifically, we take advantage of the `SolveGSLinear` function developed in Section 3.6.4. This function was already written to operate on an arbitrary matrix system passed via function arguments. It however utilized member variables to control the iteration count and convergence tolerance. These parameters are moved to the argument list, and the function is flagged `static` so that it can be called from outside the class,

```
// PotentialSolver.h
bool static solveGSLinear(Matrix &A, dvector &x, dvector &b, int
  max_it, double tolerance);
```

The `ComputeEF` function is also modified to use a static function that evaluates $-\nabla\phi$,

```
void PotentialSolver::computeEF() {
  computeNegGradient(world.ef, vec::deflate(world.phi), world);
}
```

where `computeNegGradient` is the former `computeEF` function, but generalized to operate on a vector field given by the function argument. This approach helps us avoid code duplication in the algorithms needed to evaluate $\vec{E}$ and $\vec{B}$.

In order to set the right hand side for Equation 7.30, we also need a function to compute divergence of a vector field. Divergence is computed using the same set of first derivative equations used for the gradient,

$$\nabla \cdot \vec{f} = \nabla_x f_x + \nabla_y f_y + \nabla_z f_z \tag{7.33}$$

giving us

```
dvector MagneticSolver::computeDivergence(Field3 &M) {
  dvector div(A.nu);

  for (int i=0;i<world.ni;i++)
    for (int j=0;j<world.nj;j++)
      for (int k=0;k<world.nk;k++) {
        double divX = 0, divY=0, divZ=0;

        // x component
        if (i==0)
          divX = (-3*M[i][j][k][0] + 4*M[i+1][j][k][0] -
            M[i+1][j][k][0]) / (2*dx);      // forward
        else if (i==world.ni-1)
          divX = (M[i-2][j][k][0] - 4*M[i-1][j][k][0] +
            3*M[i][j][k][0]) / (2*dx);      // backward
        else
          divX = (M[i+1][j][k][0] - M[i-1][j][k][0]) / (2*dx);
          // central

        /* ... */
        int u = k*world.ni*world.nj+j*world.ni+i;
        div[u] = divX+divY+divZ;
      }
  return div;
}
```

Finally, the code used to construct the **A** coefficient matrix is modified to take into account the difference in boundary conditions between the electric and magnetic potentials. For this solver, we assign $\phi_m = 0$ at the center of the sphere. The Neumann condition $\nabla_n \phi_m = 0$ is applied on all domain boundaries. The modified code reads as follows

```
void MagneticSolver::buildMatrix() {
    /* ... */
    // get sphere center
    double3 orig_lc = world.XtoL(world.getSphereOrig());
    sphere_orig_u = world.U((int)orig_lc[0], (int)orig_lc[1],
        (int)orig_lc[2]);

    // solve potential
    for (int k=0;k<nk;k++)
        for (int j=0;j<nj;j++)
            for (int i=0;i<ni;i++) {
                int u = world.U(i,j,k);
                A.clearRow(u);

                // apply Dirichlet boundary at sphere center
                if (u==sphere_orig_u) {
                    node_type[u] = DIRICHLET;
                    A(u,u)=1;
                    continue;
                }
                // Neumann boundaries
                node_type[u] = NEUMANN;          // set default
                /*...*/
```

The actual solution algorithm reads as follows

```
bool MagneticSolver::solve() {
    dvector b_m = computeDivergence(world.M);     // -rho_m=div(M)
    b_m[sphere_orig_u] = 0;                        // Dirichlet
        solution at sphere center
    bool conv = PotentialSolver::solveGSLinear(A, phi_m, b_m,
        max_solver_it, tolerance);

    PotentialSolver::computeNegGradient(world.H,phi_m,world);
    world.B = Const::MU_0*(world.H+world.M);

    vec::inflate(phi_m,world.phi_m);  // set 3D solution for
        visualization
    analyticalSol();                    // also compute analytical
        solution

    return conv;
}
```

We start by computing $b \equiv -\rho_m = \nabla \cdot \vec{M}$. We explicitly set $b = 0$ on the single Dirichlet node, although the divergence evaluates to zero here anyway due to the uniform magnetization. We then solve $\nabla^2 \phi_m = \vec{b}$ and use ϕ_m to evaluate the $\vec{H}$ field per $\vec{H} = -\nabla \phi_m$. Finally, we compute the $\vec{B}$ field per Equation 7.18, $\vec{B} = \mu_0(\vec{H} + \vec{M})$. As one additional step, we inflate the local one-dimensional ϕ_m into a three dimensional field for visualization. We

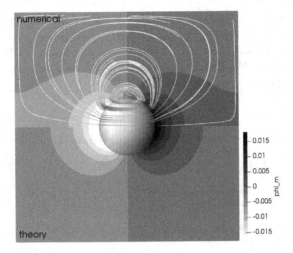

Figure 7.2: Magnetic field around a uniformly magnetized sphere with comparison to the analytical solution.

also call a `analyticalSol` function to evaluate Equation 7.32 to aid in code testing.

The solver is called from **main** using

```
world.magnetizeSphere({0,0,1});              // set sphere properties
MagneticSolver bsolver(world,20000,1e-4);
bsolver.solve();
```

The resulting ϕ_m is visualized in Figure 7.2. The top half shows the solution from the solver, while the bottom half is the analytical solution. Streamlines of the $\vec{B}$ field are also shown over the top half. We can see that the two solutions are similar, however, we can clearly see effect of the boundaries in the numerical solution. The magnetic field streamlines are closed, but are "squished" into the simulation domain bounding box. Since there is not a simple solution for taking into account the boundary effect, magnetic field solvers tend to operate on oversized domains, such that the boundaries are far removed from region of interest.

7.5 ELECTROMAGNETIC PIC

Finally, we briefly discuss the steps needed to implement an electromagnetic PIC (EM-PIC) code. There are various approaches available to us. For instance, Chapter 6 in [15] discusses a scheme for a periodic system with electromagnetic waves propagating in only a single direction. The solution is obtained with the help of Fourier Transforms. We instead consider a general algorithm for a two-dimensional problem. Returning to the Maxwell's equations from Section 1.3.2, we can rewrite Faraday's and Ampere's Laws in a form that isolates the time derivative,

$$\frac{\partial \vec{B}}{\partial t} = -\nabla \times \vec{E} \tag{7.34}$$

$$\frac{\partial \vec{E}}{\partial t} = c^2 \nabla \times \vec{B} - \frac{1}{\epsilon_0} \vec{j} \tag{7.35}$$

Here the identity $c^2 = (\mu_0 \epsilon_0)^{-1}$ is used. From this formulation we can see that new values of magnetic and electric field can be obtained from the curl of the electric and magnetic field. Including the two Gauss' laws,

$$\nabla \cdot \vec{E} = \frac{\rho}{\epsilon_0} \tag{7.36}$$

$$\nabla \cdot \vec{B} = 0 \tag{7.37}$$

we have the four governing equations of electromagnetics.

7.5.1 Curl

Clearly, we need an algorithm to evaluate curl, which is given by definition as

$$\nabla \times \vec{F} = \begin{vmatrix} \hat{i} & \hat{j} & \hat{k} \\ \frac{\partial}{\partial x} & \frac{\partial}{\partial y} & \frac{\partial}{\partial z} \\ F_x & F_y & F_z \end{vmatrix} \tag{7.38}$$

We can take one of two approaches perform this computation numerically. The first option is to rewrite the derivative terms using Finite Difference. This is in fact what we end up doing here. Alternatively, we can utilize the Kelvin-Stokes theorem,

$$\int_S \nabla \times \vec{F} \cdot d\vec{S} = \oint_C \vec{F} \cdot d\vec{l} \tag{7.39}$$

This formulation is analogous to the Finite Volume alternative to Finite Difference that was discussed in Chapter 5. It is introduced here since it useful for cases utilizing non-rectilinear (deformed) meshes. For a Cartesian grid, the two schemes yield an identical discretization.

Just as was the case for FVM, we start by defining a small region of space in which we assume that curl remains constant. We then approximate the right hand side of Equation 7.39 by defining a plane cutting through the control volume centroid and summing up contributions over the edges. Two of the three possible orientations are shown in Figure 7.3. Note that for clarity, only a subset of the mesh control volume is shown. The rectangle with the mesh nodes shown by dark circles corresponds to a slice through the cell centroid with z index $k + 0.5$.

For the $\hat{i}$ component, we have

$$(\nabla \times \vec{F}) \cdot \hat{i} \Delta A = \sum_{n=1}^{4} \vec{F}_n \cdot \vec{dl}_n \hat{e}_n \tag{7.40}$$

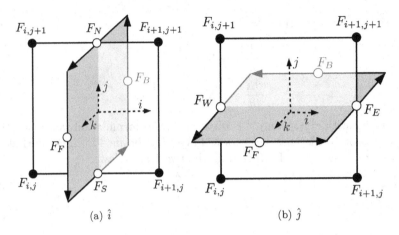

Figure 7.3: Control regions used to compute two components of $\nabla \times \vec{F}$.

The orientation of the $\vec{dl}$ line segments follows the right hand rule. By labeling $\vec{F}$ on the four edges as $\vec{F}_S$, $\vec{F}_E$, $\vec{F}_N$, and $\vec{F}_W$, and using Δx, Δy, and Δz for edge lengths, we obtain

$$(\nabla \times \vec{F}) \cdot \hat{i} \Delta y \Delta z = -(F_z)_S \Delta z + (F_y)_B \Delta y + (F_z)_N \Delta z - (F_y)_F \Delta y \quad (7.41)$$

or

$$(\nabla \times \vec{F}) \cdot \hat{i} = \frac{(F_z)_N - (F_z)_S}{\Delta y} - \frac{(F_y)_F - (F_y)_B}{\Delta z} \quad (7.42)$$

which can be seen to be the central difference discretization of $\partial(F_z)/\partial y - \partial(F_y)/\partial z$. The edge centered values of $\vec{F}$ are obtained by averaging the nearby neighbors. For instance,

$$\vec{F}_S = \vec{F}_{i+0.5,j,k+0.5}$$
$$= 0.25 \left(\vec{F}_{i,j,k} + \vec{F}_{i+1,j,k} + \vec{F}_{i,j,k+1} + \vec{F}_{i+1,j,k+1} \right) \quad (7.43)$$

7.5.2 Staggered Grid

For a vector field defined on nodes of a control volume, Equation 7.39 yields an expression for curl at the cell centers. We then obtain a *staggered* (or Yee) grid. It is visualized in Figure 7.4. The dark circles correspond to nodes on which the electric field is known. The curl $\nabla \times \vec{E}$ is known on the open circles. Collectively, these cell-centers form the nodes of a *magnetic field grid*. Cell centers of the magnetic grid correspond to the nodal positions of the electric field mesh. Hence, the computed $\nabla \times \vec{B}$ curl can then be used to update the electric field solution per Equation 7.35 everywhere except on the boundaries.

We use a similar approach to compute the curl of $\vec{B}$, with the important

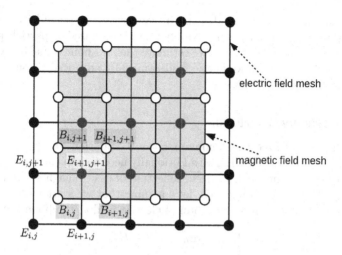

Figure 7.4: Staggered grid.

distinction that we now expect to get a two component vector since for $\vec{B} = B_z(x, y)\hat{k}$, we have

$$\nabla \times \vec{B} = (\partial B_z/\partial y)\hat{i} - (\partial B_z/\partial x)\hat{j} \qquad (7.44)$$

We again define two closed control regions following the right hand rule about $\hat{i}$ and $\hat{j}$. Dot products on the "front" and "back" $z-$ and $z+$ edges vanish, since the field is assumed to consist solely of the $\hat{z}$ component. After a bit of algebra, we arrive at

$$(\nabla \times \vec{B}_z) = \left(\frac{(B_z)_N - (B_z)_S}{\Delta y}\right)\hat{i} - \left(\frac{(B_z)_E - (B_z)_W}{\Delta x}\right)\hat{j} \qquad (7.45)$$

These are just the Finite Difference expressions for the first derivative. Just as was the case with Finite Volume, we recover the standard Finite Difference equations for a Cartesian mesh. Finally, note that because of the staggered grid indexing, $(\nabla \times \vec{B})_{i,j}$ on the staggered grid correspond to $\vec{E}_{i+1,j+1}$.

7.5.3 Time Integration

When advancing field quantities, we would like to integrate the forcing term at the half-way time point, similarly to the Leapfrog scheme used for particle velocities. In other words, we need

$$\frac{\vec{B}^{k+0.5} - \vec{B}^{k-0.5}}{\Delta t} = -\nabla \times \vec{E}^k \qquad (7.46)$$

and

$$\frac{\vec{E}^{k+1} - \vec{E}^k}{\Delta t} = c^2 \nabla \times \vec{B}^{k+0.5} - \frac{1}{\epsilon_0}\vec{j}^{k+0.5} \qquad (7.47)$$

Current density $\vec{j} = qn\vec{v}$ already contains velocity evaluated at half-step due to the Leapfrog integration scheme. Number density is based on particle positions which are defined at full times. We estimate the mid-point value by averaging, $n^{k+0.5} = (n^k + n^{k+1})/2$. Magnetic field is initially advanced through $\Delta t/2$ to bring the values to $k + 0.5$.

7.5.4 Boundary Conditions

In ES-PIC, we utilized the Neumann $E_{\hat{n}} = 0$ condition to represent the far-stream boundary. In EM-PIC, we additionally need to implement an absorbing boundary to prevent a non-physical reflection of electromagnetic waves at the grid edge. Following Tajima [55], let's assume that we start with the second order finite difference representation of electric field at the boundary

$$E_{-1} - 3E_0 + 3E_1 - E_2 = 0 \tag{7.48}$$

and let the solution on the grid point n be

$$E_n = Ae^{inb} + Be^{-inb} \tag{7.49}$$

where A and B are the amplitudes of the incoming and reflected wave, and i is the imaginary number. Substituting into the boundary condition equation, we obtain

$$B = Ae^{-2ib}\left(\frac{1 - e^{ib}}{1 - e^{-ib}}\right)^3 \tag{7.50}$$

or $|B| = |A|$ and no absorption is achieved. Clearly, this approach does not work. We could perhaps implement a stretched mesh at the boundaries (or over many ghost cells) but that also leads to an incomplete absorption.

Tajima instead suggests implementing a masking function

$$\bar{E}(x) = f(x)E(x) \tag{7.51}$$

where

$$f(x) = \begin{cases} -d^{-2}x^2 + 2d^{-1}x & ; 0 \le x < d \\ 1 & ; d \le x \le L - d \\ -d^2x^2 + 2d^{-2}(L - d) - d^{-2}(L - d)^2 + 1 & ; L - d < x \le L \end{cases} \tag{7.52}$$

This mask generates a parabolically decaying E field in the "ramp" (or moat) outside the particle domain. The expressions are obtained by computing coefficients of $ax^2 + bx + c = f$ with $f(0) = 0$, $f(d) = 1$, and $(df/dx)(d) = 0$.

7.5.5 Gauss' Law

Ampere's and Faraday's laws, Equations 7.35 and 7.34 specify how the electromagnetic fields evolve in time. These equations however do not take into account boundary conditions, magnetic materials, or stationary charges. This is where Gauss' Law and Gauss' Law for Magnetism come in. We use these two expressions to set the initial electric and magnetic field. Then in the absence of integration errors, both Gauss' laws hold throughout the simulation. To illustrate this, consider the ordinary Gauss' law that we are familiar with from our discussion of electrostatics

$$\nabla \cdot \vec{E} = \frac{\rho}{\epsilon_0} \tag{7.53}$$

If we take a time derivative of both sides and substitute Equation 7.35 for the $\partial \vec{E}/\partial t$ term, we obtain

$$\nabla \cdot \left(c^2 \nabla \times \vec{B} - \frac{1}{\epsilon_0} \vec{j} \right) = \frac{1}{\epsilon_0} \frac{\partial \rho}{\partial t} \tag{7.54}$$

The first term on left is identically zero. Therefore,

$$\frac{\partial \rho}{\partial t} + \nabla \cdot \vec{j} = 0 \tag{7.55}$$

which is simply the charge conservation statement.

Gauss' law can also be used to correct the solution as needed either due to a build up of numerical errors or to take into account a change in boundary conditions. Let's assume that $\vec{E}^*$ is some additional error field that needs to be added to the computed electric field to satisfy Equation 7.36. We have

$$\nabla \cdot (\vec{E} + \vec{E}^*) = \frac{\rho}{\epsilon_0} \tag{7.56}$$

which can be rewritten as

$$\nabla \cdot \vec{E}^* = \frac{\rho}{\epsilon_0} - \nabla \cdot \vec{E} \tag{7.57}$$

Since $\nabla \cdot (\nabla \times \vec{A}) = 0$ by definition, the above formulation subtracts out the solenoidal part of $\vec{E}$. We can then let

$$\vec{E}^* = -\nabla \phi^* \tag{7.58}$$

to obtain Poisson's equation,

$$\nabla^2 \phi^* = -\frac{\rho}{\epsilon_0} + \nabla \cdot \vec{E} \tag{7.59}$$

After ϕ^* is obtained, we use Equation 7.58 to compute $\vec{E}^*$.

7.5.6 Putting It All Together

The overall EM-PIC field update algorithm then consists of the following steps:

1. Use Equations 7.36 and Equation 7.37 to compute $\vec{E}^0$ and $\vec{B}^{-0.5}$

2. Starting with particle data $\vec{x}^k$ and $\vec{v}^{k-0.5}$, compute $\vec{x}^{k+1}$ and $\vec{v}^{k+0.5}$

3. Compute $\vec{j}^{k+0.5}$ using $\vec{v}^{k+0.5}$ and $(\vec{x}^k + \vec{x}^{k+1})/2$

4. Advance $\vec{B}^{k-0.5}$ to $\vec{B}^{k+0.5}$ using Equation 7.46

5. Advance $\vec{E}^k$ to $\vec{E}^{k+1}$ using Equation 7.47 with $\vec{j}^{k+0.5}$

6. Return to step 2 until exit criterion is reached

7.5.7 Approximations

Electromagnetic waves propagate at the speed of light, and hence the time step should be set according to $\Delta t = \Delta x/c$. The resulting tiny time step makes the simulation impractical to engineering problems in which we are generally interested in resolving low frequency modes or dynamics of the relatively slow ions. Therefore, it is common to apply approximations to the governing equations to speed up the simulations. In the well-known *Darwin* (or magnetoinductive) model, we drop the transverse displacement current in Ampere's equation,

$$\nabla \times \vec{B} = \vec{j}_T = \frac{1}{\epsilon_0}\vec{j} + \frac{\partial E_L}{\partial t} \tag{7.60}$$

$$\nabla \times \vec{E} = -\frac{\partial \vec{B}}{\partial t} \tag{7.61}$$

where the L and T subscripts correspond to the longitudinal and transverse components. The transverse electric field is obtained from $\nabla^2 E_T = (1/c)(\partial j_T/\partial t)$. A solver based on this approach is described in [31].

7.5.8 Relativistic Push

Electromagnetic problems often deal with cases in which particle, especially electron, velocities start approaching the speed of light. Our velocity update is based on the Newtonian $\vec{a} = \vec{F}/m$. For near-luminal speeds, we instead have to use the relativistic version,

$$\vec{a} = \frac{1}{m_0 \gamma(v)}\left(\vec{F} - \frac{(\vec{v} \cdot \vec{F})\vec{v}}{c^2}\right) \tag{7.62}$$

Here

$$\gamma(v) = \left(1 - \frac{v^2}{c^2}\right)^{-1/2} \tag{7.63}$$

is the *Lorentz factor* and m_0 is the rest mass. Sometimes the last term in Equation 7.62 is dropped to obtain

$$\vec{a} = \frac{a}{m_0 \gamma(v)}(\vec{E} + \vec{v} \times \vec{B}) \tag{7.64}$$

While this simplified form assures that particle velocities do not exceed the speed of light, all three terms should be retained to obtain the correct asymptotic behavior.

7.5.9 Example

The algorithm from Section 7.5.6 is demonstrated using a modified version of the 2D flow past a cylinder from Chapter 5. You will find this code in the ch7/EM-PIC directory. In this example, we simply inject a beam of positive protons and negative electrons, and let the solver compute the self-induced electric and magnetic field. In order to keep the simulation run time reasonable, the example is run with low current densities. The sphere is included, but is reduced in size, and we also set ϕ_{sphere} to only -0.001 V. This small value is used so the tiny contribution from the self-induced field can be resolved. The simulation time step is set such that information moving at the speed of light travels one cell length per time step, $\Delta t = \Delta x / c$. In order to avoid having to simulate too many time steps, electrons are injected into the domain with $w = 5 \times 10^6$ m/s. This corresponds to approximately 71 eV. At this velocity and Δt, it takes each electron 60 time steps to traverse the cell. Protons are loaded with $w = 10^6$ m/s. Due to the difference in velocities, the total current is non-zero for the initial transient period marked by protons traversing the domain. Density of both particles is 2×10^9 m^{-3}. Due to the large injection velocity and small field magnitudes, the effect of electromagnetic fields on the particle trajectories is negligible.

As the first modification, we add support for computing current density. The approach here is similar to computing ρ: we let each species compute its own j_s, and we then use a World function to aggregate the sum,

$$\vec{j} = \sum_s \vec{j}_s \qquad (7.65)$$

The code is listed below

```
void World::computeCurrentDensity(vector<Species> &species) {
    j = 0;                  // clear current density
    for (Species &sp:species) {
        if (sp.charge==0) continue;   // skip neutrals
        j += sp.j;
    }
}
```

The species current density is computed in Species::computeGasProperties. Unlike in the electrostatic case in which the mesh velocity was used only for visualization, here we need to evaluate $\vec{v}$ at each time step. Furthermore, since temperature is not needed (at least in this example), we can obtain velocity by direct averaging,

$$\vec{v} = \left(\sum_p w_{mp,p} \vec{v}_p \right) / \sum_p w_{mp,p} \qquad (7.66)$$

As noted previously, velocities are already defined at half-times. The gas density at the half-step is obtained with the help of a secondary `den_old` array. After new density corresponding to n^{k+1} is computed, we compute current from

$$\vec{j}^{k+0.5} = 0.5e\left(n^k + n^{k+1}\right)\vec{v}^{k+0.5} \tag{7.67}$$

The new density is then copied down to the `den_old` field. The resulting code is

```
void Species::computeGasProperties() {
  den_old = den;  // save old density for averaging
  den.clear();
  vel.clear();
  for (Particle &part:particles) {
    double2 lc = world.XtoL(part.pos);
    den.scatter(lc, part.mpw);
    vel.scatter(lc, part.mpw*part.vel);
  }

  // divide by count to get average vel
  vel/=den;

  // divide by node volume to get density
  den /= world.node_vol;

  // set current density
  j = charge*0.5*(den+den_old)*vel;
}
```

We next replace the `PotentialSolver` class with `EMSolver`,

```
class EMSolver {
public:
  // constructor
  EMSolver(World &world, int max_it, double tol): world(world),
    A(world.ni*world.nj), max_solver_it(max_it), tolerance(tol) {
    buildMatrix();
    init();
  }

  // builds the "A" matrix for linear potential solver
  void buildMatrix();

  void init();        // solves initial field
  void advance();     // advance fields

  Field3 curl(Field3 &f, bool inner=false);
  Field div(Field3 &f);
  Field3 grad(Field &f) {return grad(f,world);}
  static Field3 grad(Field &f, World &world);

  static bool solveGSLinear(Matrix &A, dvector &x, dvector &b, int
    max_solver_it, double tolerance);
};
```

The structure should look very familiar. The main change from the electrostatic version includes the renaming of `solve` to `advance` and the addition

of a new `init` function that can be used to set the initial fields from Gauss'
laws. We also add functions for computing vector operations such as the curl.
These operations could be added to `Field` but since they are used only by
this particular class, they are added here as member functions.

The `init` function computes the initial electric field due to the sphere. We
also rewind $\vec{B}$ by $-0.5\Delta t$, since we would like the two fields to leapfrog each
other,

```
void EMSolver :: init () {
  Field b3 = −(1/Const :: EPS_0)*world.rho ;

  // using solver operation on a 1D vector
  dvector b = vec :: deflate (b3) ;
  dvector phi_s = vec :: deflate (world.phi) ;
  // correct Dirichlet values
  for (int u=0;u<A.nu;u++)
    if (node_type[u]==NodeType :: DIRICHLET) b[u] = phi_s [u] ;

    linearSolveGS (A, phi_s ,b) ;
    world.phi = vec :: inflate (phi_s , world.ni , world.nj) ;

    world.E = −grad (world.phi) ;
    applyBoundaries () ;

    // Faraday law
    Field3 curlE = curl (world.E) ;
    world.B −= 0.5* world.getDt ()* curlE ;   // rewind by 0.5*dt
}
```

The `advance` function simply evaluates Equations 7.34 and 7.35. We use
the member `curl` functions to compute $\nabla \times \vec{E}$ and $\nabla \times \vec{B}$. As written here,
these functions allocate a new field on each use. This is inefficient, but this
approach was used for clarity. In a production code, you would want to pre-
allocate these fields in the constructor, and then pass a reference to the curl
operator.

```
void EMSolver :: advance () {
  double dt = world.getDt () ;

  // Faraday 's law
  Field3 curlE = curl (world.E) ;
  world.B −= dt*curlE ;     // B(k+0.5) = B(k−0.5) − curl [E(k)]*dt ;

  // Ampere 's law
  Field3 curlB = curl (world.B, true) ;
  world.E += dt *( Const :: C*Const :: C*curlB −(1/Const :: EPS_0)*world.j ) ;

  applyBoundaries () ;
}
```

The `applyBoundaries` function implements the electric field ramp per
Section 7.5.4,

```
void EMSolver :: applyBoundaries () {
  double d = 2 ;
```

```
for (int i=0;i<world.ni;i++)
  for (int j=0;j<world.nj;j++) {
    double f=1,g=1;
    int L=world.ni-1;
    int M=world.nj-1;

    if (i<d)
      f = -i*i/(d*d)+2*i/d;
    else if (i>L-d) // if (x>(L-d))
      f = -i*i/(d*d) + 2*(L-d)*i/(d*d) - (L-d)*(L-d)/(d*d)+1;
    else f=1;

    if (j<d)
      g = -j*j/(d*d)+2*j/d;
    else if (j>M-d) // if (x>(L-d))
      g = -j*j/(d*d) + 2*(M-d)*j/(d*d) - (M-d)*(M-d)/(d*d)+1;
    else g= 1;
    world.E[i][j][0]*=f;
    world.E[i][j][1]*=g;
  }
}
```

The *main loop* is also modified to include calls to these new functions. These modifications are fairly minimal.

```
int main(int argc, char *args[])
{
  // initialize domain
  World world(41,41);
  world.setExtents({0,-0.02},{0.04,0.02});
  double dt = world.getDh()[0]/Const::C;         // dx = 0.5*c*dt
  world.setTime(dt,4000);
  cout<<"dt = "<<dt<<endl;

  // set objects
  double phi_circle = -0.001;
  world.addCircle({0.02,0},0.0015,phi_circle);
  world.addInlet();

  // set up particle species
  species.push_back(Species("e-", ME, -1*QE, 1e-0, world));
  species.push_back(Species("H+", AMU, 1*QE, 1e-0, world));
  Species &eles = species[0];
  Species &ions = species[1];

  // setup injection sources
  const double nde = 2e9;                        // mean ion density
  vector<ColdBeamSource> sources;
  sources.emplace_back(eles,world,5e6,nde);      // electron source
  sources.emplace_back(ions,world,1e6,nde);      // ion source

  // initialize EM solver
  EMSolver solver(world,10000,1e-3);

  // main loop
  while(world.advanceTime()) {
    // inject particles
```

```
for (auto source: sources)
  source.sample();

// move particles
for (Species &sp: species) {
  sp.advance();
  sp.computeGasProperties();
}

// compute charge and current density
world.computeChargeDensity(species);
world.computeCurrentDensity(species);

// integrate E/M fields
solver.advance();

// update averages at steady state
if (world.steadyState(species)) {
  for (Species &sp: species)
    sp.updateAverages();
}

// screen and file output
Output::screenOutput(world, species);
Output::diagOutput(world, species);

// periodically write out results
if (world.getTs()%100==0 || world.isLastTimeStep())
    Output::fields(world, species);
}

// show run time
cout<<"Simulation took "<<world.getWallTime()<<" seconds"<<endl;
return 0;            // indicate normal exit
}
```

The resulting $\vec{E}$ and $\vec{B}$ fields are plotted in Figure 7.5. In this image, the electric field slice is warped by the magnetic field. The first image shows the solution at the beginning of the simulation, while the second plot was generated after 7,000 time steps. Due to the non-zero j, the system continues to charge which is demonstrated by an on-going increase in the electric field magnitude.

7.6 SUMMARY

In this chapter we discussed approaches for including electromagnetic effects. We started by discussing changes needed in the particle integrator to recover the orbital motion of charged particles about magnetic field lines. We then learned how to integrate Maxwell's equations to compute the time evolution of the electric and magnetic field. Here along the way we learned about staggered grids, and developed numerical algorithms for curl and divergence.

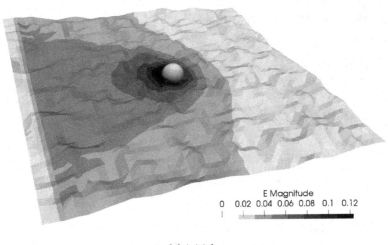

(a) initial

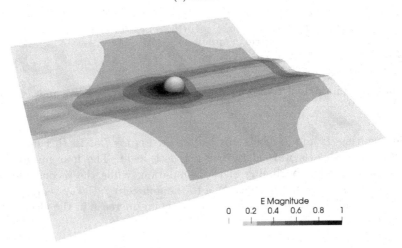

(b) 7,000 time steps

Figure 7.5: Electric and magnetic fields at the start of the simulation and after 7,000 time steps. The slice, plotting electric field magnitude, is warped by the magnetic field.

EXERCISES

7.1 $E \times B$ *drift.* In the presence of both electric and magnetic fields, charged particles drift with velocity given by

$$\vec{v}_E = \frac{\vec{E} \times \vec{B}}{B^2} \tag{P7.1}$$

The magnitude of this velocity is $v_d = E/B$. Note that this drift applies to the *guiding center* about which the particle orbits. Develop a program that simulates the orbit of an electron ($m = m_e$, $q = -e$) and a positron ($m = m_e$, $q = +e$) with starting position $(x, y) = (0, 0)$ and velocity $(v_x, v_y) = (10, 10^5)$ in $\vec{E} = (100, 0, 0)$ V/m and $\vec{B} = (0, 0, 0.1)$ T. Plot the two orbits using appropriately small Δt. You should observe that both particles drift in the same direction despite having different charge. Then repeat the simulation but double the mass of the first particle. Notice that v_d does not change although the orbital Larmor radius increases. Numerically compute the drift velocity and compare to the theory.

7.2 *Magnetic Confinement.* Chen provides a good overview of the *magnetic mirror* effect. Consider a magnetic field primarily in the z direction, but which strongly increases in magnitude at some left and right bound. The strong $\partial B/\partial z$ gradient that forms here traps charged particles within this "magnetic bottle". The confinement is however not perfect, with the "loss cone" described on pages 33 and 34 in [24]. For this exercise, write a tracing code for a single charged particle in a magnetic field. You may want to consider just a one dimensional analytical model for the magnetic field such as $B = \exp(-2(z - 1)^2) + \exp(-2(z + 1)^2)$ in $z \in [-1, 1]$. In this case, it may be necessary to incorporate a "grad-B" force, $\vec{F}_{gB} = -\mu \nabla B$, where $\mu = (mv_\perp^2)/(2B)$ is the *magnetic moment*. This force is needed since the one-dimensional magnetic field does not recover the convergence of magnetic field lines in the "neck". Run this algorithm with a large number of particles with uniformly sampled $v_\perp$ and $v_\parallel$ velocity components. Determine which particles pass through the mirror. Output all sampled particles along with a code indicating their confinement status. Compare the resulting population to the theoretical loss cone angle.

7.3 *Relativistic Push.* Write a simple velocity integrator for an initially stationary electron in a constant $\vec{E} = 10^4$ V/m electric field. Compute and plot velocity for 10,000 $\Delta t = 1.5 \times 10^{-7}$ s time steps with the Newtonian, and 2 and 3 term relativistic form for acceleration. You should observe that in the first case, particle velocity continues to increase linearly, exceeding c. With the two term model, velocity reaches c which is also non-physical. The three term form recovers the correct asymptotic behavior.

CHAPTER 8

Eulerian Methods

8.1 INTRODUCTION

$\mathbb{S}$ O FAR we have focused on *Lagrangian* methods in which plasma species are represented by particles that move about the computational domain. In this chapter, we discuss the alternate *Eulerian* methods based on solving governing differential equations on a stationary computational grid. We start by reviewing the governing equations of magnetohydrodynamics. We then illustrate several solution methods using the one-dimensional advection diffusion model equation. Next we review additional electron models that can be used in hybrid PIC simulations instead of the Boltzmann relationship. We close the chapter by developing a 1D-1V Vlasov solver for two-stream instability.

8.2 MAGNETOHYDRODYNAMICS

Regardless of whether we take the Lagrangian or the Eulerian approach, the primary objective of plasma simulations is to compute the time evolution of plasma species densities, velocities, and temperatures. In general, any plasma simulation method consists of the following two steps:

1. Use gas state $(\rho, \vec{v}, T)$ to compute electromagnetic fields

2. Use the $\vec{E}$ and $\vec{B}$ fields to advance gas properties by Δt

8.2.1 Conservation Equations

The primary difference between the two simulation approaches lies in the implementation of the first step. In the Lagrangian description, we let the simulation consist of discrete particles that move about the domain. Their positions and velocities are used to compute the new macroscopic flow parameters. We use Newtonian equations of motion to integrate their velocity and position. On the other hand, with the Eulerian approach we let the simulation consist of a stationary grid. We then use *conservation equations* to control the in- and out-flow of properties of interest in each cell. These equations assure that physical properties, such as mass, momentum, and energy, remain conserved. Any standard plasma or gas dynamics text, such as Chen [24] or Anderson [8], contains a detailed derivation for magnetohydrodynamics (MHD) or computational fluid dynamics (CFD). The primary difference between MHD and CFD is that CFD equations are developed for neutral gas, and thus do not consider electromagnetic effects.

Here for brevity, only the final form is summarized. The first relationship of interest is one for *conservation of mass*. For species j, we have

$$\frac{\partial n_j}{\partial t} + \nabla \cdot (n\vec{u}) = \dot{n}_j \tag{8.1}$$

This expression simply states that the rate of change of number density in some control volume is due to the in- and out-flux of the material through the volume boundaries. The term on the right side captures volumetric mass creation by chemical reactions. If ionization or recombination is not present, this term vanishes. By multiplying number density n by species charge and summing over all species, $\rho = \sum_j q_j n_j$, we obtain relationship for current density,

$$\frac{\partial \rho}{\partial t} + \nabla \cdot \vec{j} = 0 \tag{8.2}$$

The right side is now always zero since the number of created positive and negative charges balances. At steady state, we further see that

$$\nabla \cdot \vec{j} = 0 \tag{8.3}$$

which, by employing the Divergence theorem, indicates that at steady state, there is no net current through domain boundaries, $\oint_S \vec{j} \cdot \hat{n} dA = 0$.

Next, the *momentum conservation equation* states

$$m_j n_j \left(\frac{\partial \vec{v}_j}{\partial t} + \vec{v}_j \cdot \nabla \vec{v}_j \right) = n_j \vec{F} - \nabla : \vec{\vec{\Pi}}_j + \vec{R} \tag{8.4}$$

This is simply the fluid version of Newton's second law, $m(d\vec{v}/dt) = \vec{F}$ that takes into account hydrodynamic pressure and resisistive drag arising from collisions. For a fluid acted solely on by the Lorentz force with isotropic pressure and no resistive drag, Equation 8.4 simplifies to

$$m_j n_j \left[\frac{\partial \vec{v}_j}{\partial t} + (\vec{v}_j \cdot \nabla)(\vec{v}_j) \right] = q n_j \left(\vec{E} + \vec{v}_j \vec{B} \right) - \nabla p_j \tag{8.5}$$

Finally, we need some closure mechanism to relate pressure p and density n. Typically we use the ideal gas law, $p = nkT$. Computing temperature then involves including an equation for *conservation of energy*. Starting with formulation in [29] and [42], we have

$$\frac{\partial}{\partial t} \left(\frac{3}{2} n_j k T_j \right) + \nabla \cdot \left(\frac{3}{2} n_j \vec{v}_j k T_j + \vec{q}_j \right) + p_j \nabla \cdot \vec{v}_j = (S_e)_j - (S_i)_j \tag{8.6}$$

The system is closed by defining the heat conduction vector as $\vec{q}_j = -\kappa_e \nabla T_j$. The $(S_i)_j$ term captures inter-species energy transfer due to inelastic processes such as ionization. For the case of electrons ionizing background neutrals, this term can be written as $(S_i)_e = n_e n_a C_i \epsilon_i$, where C_i is the ionization rate, and ϵ_i is the ionization energy. It is negligible in low density plasmas. The elastic term can be written for electrons as [29]

$$(S_e)_e = \sigma_s m_e \nu_{se} n_j \left[(\vec{v}_s - \vec{v}_e)^2 + \frac{2}{m_s} \frac{3}{2} k (T_s - T_e) \right] \tag{8.7}$$

In the case of a two-species plasma with no neutrals, we have

$$(S_e)_e = m_e \nu_{ei} n_e (\vec{v}_i - \vec{v}_e)^2 + \nu_{ei} n_e \frac{m_e}{m_i} 3k(T_i - T_e) \tag{8.8}$$

The first term can be rewritten with the help of $\vec{j} = en\vec{v}$ and by assuming quasi-neutrality as

$$\frac{m_e \nu_{ei}}{e^2 n_e} (\vec{j}_i - \vec{j}_e)^2 \equiv \frac{\vec{j}^2}{\sigma} \tag{8.9}$$

From the momentum equation, Equation 8.85, we also have $\vec{j}/\sigma = \vec{E} + 1/(n_e e)\nabla p_e$. Therefore, and by once again assuming neutrality, the elastic term can be written as

$$S_e = \vec{j} \cdot \vec{E} + (\vec{v}_i - \vec{v}_e) \cdot \nabla p_e + \nu_{ei} n_e \frac{m_e}{m_i} 3(T_i - T_e) \tag{8.10}$$

The first term on the right hand side is known as *Joule heating*. Often, the drag-induced energy transfer term is simplified by assuming $\vec{v}_e \gg \vec{v}_i$, allowing the ion term to drop out. The thermal energy transfer given by the second term is also sometimes ignored, but here we retain it for completeness. With the substitution into Equation 8.6, we obtain for electrons

$$\frac{\partial}{\partial t} \left(\frac{3}{2} n_e k T_e \right) + \nabla \cdot \left(\frac{5}{2} n_e \vec{v}_e k T_e \right) = \nabla \cdot (\kappa_e \nabla T_e) + \vec{j} \cdot \vec{E} +$$
$$\vec{v}_i \cdot \nabla p_e + 3 \frac{m_e}{m_i} n u_{ei} n_e k (T_i - T_e) - n_e n_a C_i \epsilon_i \tag{8.11}$$

Given that electrons respond rapidly to ion disturbances, it is not uncommon to assume that electrons establish a temporary thermodynamic equilibrium between each ion move. This assumption then allows us to drop the time derivative, and compute electron temperature per

$$\nabla^2 T_e = -\nabla \ln(\kappa_e) \cdot \nabla T_e + \frac{1}{\kappa_e} \left[\nabla \cdot \left(\frac{5}{2} n_e \vec{v}_e k T_e \right) - \vec{j} \cdot \vec{E} \right.$$
$$\left. -\vec{v}_i \cdot \nabla p_e - 3 \frac{m_e}{m_i} \nu_{ei} n_e k (T_i - T_e) + n_e n_a C_i \epsilon_i \right] \tag{8.12}$$

In many plasma discharges, such as those encountered in electric propulsion applications, $T_i \ll T_e$ and we can assume that ion temperature remains time invariant. Alternatively, ion pressure can be computed from density per

$$p = Cn^\gamma \tag{8.13}$$

where C is set such that $p = p_0$ for $n = n_0$,

$$p_0 \equiv n_0 kT_0 = C(n_0)^\gamma \tag{8.14}$$

yielding

$$C = \frac{kT_0}{(n_0)^{\gamma-1}} \tag{8.15}$$

The term γ is the specific heat ratio, with $\gamma = 5/3$ frequently used for plasmas.

8.2.2 Single Fluid MHD Equations

MHD equations can be applied individually to each species, or we can treat ions and electrons together as if they form a single "plasma fluid". Let's assume that we have only two species: singly charged ions and electrons. The momentum conservation equations can then be written as

$$m_i n_i \frac{D\vec{v}_i}{Dt} = q_i n_i \left(\vec{E} + \vec{v}_i \times \vec{B} \right) + n_i m_i \vec{g} - \nabla p_i - \nabla \cdot p_i + \vec{R}_{ie} \tag{8.16}$$

$$m_e n_e \frac{D\vec{v}_e}{Dt} = q_e n_e \left(\vec{E} + \vec{v}_e \times \vec{B} \right) + n_e m_e \vec{g} - \nabla p_e - \nabla \cdot \pi_e + \vec{R}_{ei} \tag{8.17}$$

Here $D/Dt \equiv \partial/\partial t + \vec{v} \cdot \nabla$ is the total (or material) derivative. The pressure tensor has been split into the isotropic and anisotropic terms, ∇p and $\nabla \cdot \pi$. We next make some simplifications. First, we assume quasineutrality, $n_i = n_e = n$. This simplification makes this model applicable only to regions outside the sheath. Furthermore, for singly charged ions, $q_i = e$ while $q_e = -e$. To conserve momentum, $\vec{R}_{ie} = -\vec{R}_{ei}$ since momentum lost in collisions by one species has to be recovered by the other one. Finally, we ignore the anisotropic stress term since viscous effects are generally negligible in plasma. We then obtain

$$m_i n_i \frac{D\vec{v}_i}{Dt} = en \left(\vec{E} + \vec{v}_i \times \vec{B} \right) + nm_i \vec{g} - \nabla p_i - \vec{R}_{ei} \tag{8.18}$$

$$m_e n_e \frac{D\vec{v}_e}{Dt} = en \left(\vec{E} + \vec{v}_e \times \vec{B} \right) + nm_e \vec{g} - \nabla p_e - \vec{R}_{ei} \tag{8.19}$$

$$\tag{8.20}$$

Adding these two equations we obtain,

$$n \frac{D}{Dt}(m_i \vec{v}_i + m_e \vec{v}_e) = en(\vec{v}_i - \vec{v}_e) \times \vec{B} + n(m_i + m_e)\vec{g} - \nabla p \tag{8.21}$$

Note that the electric field term $\vec{E}$ drops out. The total pressure $p = p_i + p_e$ is set from Dalton's law. Next, total mass density is given by $\rho = n(m_i + m_e)$.

We can also define a mass-averaged fluid velocity as

$$\vec{v} = \frac{m_i \vec{v}_i + m_e \vec{v}_e}{m_i + m_e} \tag{8.22}$$

Finally, current density is $\vec{j} = en(\vec{v}_i - \vec{v}_e)$. With these definitions, Equation 8.21 is rewritten as

$$\rho \frac{D\vec{v}}{Dt} = \vec{j} \times \vec{B} + \rho \vec{g} - \nabla p \tag{8.23}$$

This relationship governs mass convection in the single-fluid MHD model.

While the electric field does not appear in the above equation, it is needed to advance the magnetic field. Instead of adding Equations 8.18 and 8.20 as was done above, we can multiply the ion equation by electron mass, and vice versa. We then subtract the second equation from the first and obtain

$$\frac{m_e m_i}{e} \frac{D\vec{j}}{Dt} = en(m_i + m_e)\vec{E} + en(m_e \vec{v}_i - m_i \vec{v}_e) \times \vec{B}$$
$$- m_e \nabla p_i - m_i \nabla p_e - (m_e + m_i)\vec{R}_{ei} \tag{8.24}$$

The second group on the RHS forms the *Hall term*. The part in parentheses looks similar to the relationship for fluid velocity, except that the masses are reversed. With a bit of algebra and the prior definition of mass-averaged fluid velocity and current density, this term is rewritten as

$$m_e \vec{v}_i - m_i \vec{v}_e = m_i \vec{v}_i + m_e \vec{v}_e + m_i(\vec{v}_e - \vec{v}_i) + m_e(\vec{v}_i - \vec{v}_e) \tag{8.25}$$

$$= (m_i + m_e)\vec{v} - \frac{(m_i - m_e)}{ne}\vec{j} \tag{8.26}$$

The Hall term thus becomes

$$en(m_e \vec{v}_i - m_i \vec{v}_e) \times \vec{B} = e\rho\vec{v} \times \vec{B} - (m_i - m_e)\vec{j} \times \vec{B} \tag{8.27}$$

Next, following the derivation in Chen [24], let's consider the friction term. In a fully ionized two-species plasma,

$$\vec{R}_{ei} = m_e n_e(\vec{v}_i - \vec{v}_e)\nu_{ei} \tag{8.28}$$

where

$$\nu_{ei} = n_i \overline{\sigma v_{ei}} \tag{8.29}$$

is the collision frequency. At the same time, we can expect this friction term to also arise from Coulomb forces,

$$F = -\frac{e^2}{4\pi\epsilon_0 r^2} \tag{8.30}$$

so the friction term should also be a function of e^2. Because of quasineutrality, $n_e = n_i$, we can also define

$$\vec{R}_{ei} = \eta e^2 n^2(\vec{v}_i - \vec{v}_e) = \eta en\vec{j} \tag{8.31}$$

By comparing these two forms, we see that

$$\eta = \frac{\nu_{ei} m_e}{n e^2} \tag{8.32}$$

Also note that in Equation 8.24, we actually have

$$(m_e + m_i)\vec{R}_{ei} = en(m_e + m_i)\eta\vec{j} = e\rho\eta\vec{j} \tag{8.33}$$

Let's now consider a case with no magnetic field and cold gas so T and $p = 0$. The electron momentum equation then simplifies to

$$0 = -en_e\vec{E} + \vec{R}_{ei} \tag{8.34}$$

Here we assumed that we are viewing the system from the frame of reference of ions, so that $\partial j_e / \partial t$ term drops out. The advective term is also ignored since it is generally small. Using the definition for $\vec{R}_{ei}$, we obtain

$$\vec{E} = \eta\vec{j} \tag{8.35}$$

which is just Ohm's law, $V = RI$ or $I = V/R$. Alternatively, we can write

$$\vec{j} = \frac{1}{\eta}\vec{E} = \sigma\vec{E} \tag{8.36}$$

Therefore, η is the plasma resistivity and $\sigma = 1/\eta$ is conductivity. With this definition, and by assuming $m_i \gg m_e$ so that $(m_i + m_e) \approx m_i$ and $m_e/m_i \approx 0$, Equation 8.24 is rewritten as

$$\vec{E} + \vec{v} \times \vec{B} - \eta\vec{j} = \frac{1}{en}(\vec{j} \times \vec{B} - \nabla p_e) \tag{8.37}$$

This is relationship is known as *Generalized Ohm's Law*, and is frequently encountered in plasma simulations. By coupling it with current conservation $\nabla \cdot \vec{j} = 0$, it offers us a method for computing electric field.

8.3   ADVECTION-DIFFUSION EQUATION

The governing equations of CFD and MHD tend to have a form similar to

$$\frac{\partial \rho\phi}{\partial t} + \nabla \cdot (\rho\vec{u}\phi) = \nabla \cdot (D\nabla\phi) + S \tag{8.38}$$

Fluid simulations of plasma then "simply" involve time marching the governing equations. Unfortunately, this is not completely trivial since the choice of discretization scheme affects the solution stability. Furthermore, some schemes introduce numerical diffusion, which can lead to incorrect results especially in flows containing sharp gradients. Various *shock capturing* methods have been devised to address these problems. These methods are beyond the scope of this book.

Equation 8.38 is called the *advection-diffusion* (or convection-diffusion) equation. It governs evolution of some scalar quantity ϕ due to advective mass flux $(\rho \vec{u})$ of the surrounding fluid, and diffusion with coefficient D. Imagine that a drop of dye is placed a river, the advective term captures the dye being carried downstream by the flow, while the diffusive term captures the radial spreading of the drop. The A-D equation consists of four terms. First, it contains the time derivative term. This term is used by the solver to advance the solution to the next time step. Next comes the advective term. This term requires special treatment to avoid numerical errors. The diffusion term on the right hand side reduces to the Laplacian operator for a spatially uniform diffusion coefficient D. Finally, the source S term captures generation of the quantity ϕ within the volume from chemical reactions (in mass conservation) or heat sources (in energy conservation equation).

8.3.1 Steady State Form

We now demonstrate few solution schemes for this model equation in one dimension. Let's start with the steady state form in which $\partial \phi / \partial t = 0$. We also ignore the source term. For a one-dimensional problem, we obtain

$$\frac{\partial(\rho u \phi)}{\partial x} = \frac{\partial}{\partial x} \left(D \frac{\partial \phi}{\partial x} \right) \tag{8.39}$$

This elliptic equation holds inside the computational volume, which, for 1D, is simply a line segment with length L. On the boundaries, we prescribe the Dirichlet boundary $\phi(0) = \phi_0$ and $\phi(L) = \phi_L$. We next apply the Finite Difference method to rewrite the derivatives. Starting with the diffusive term, we have

$$-D \left[\frac{\partial^2 \phi}{\partial x^2} \right]_i \approx -D \left[\frac{\phi_{i-1} - 2\phi_i + \phi_{i+1}}{(\Delta x)^2} \right] \tag{8.40}$$

Following the notation in [28], we rewrite the equation as

$$-D \left[\frac{\partial^2 \phi}{\partial x^2} \right]_i \approx A_w^d \phi_{i-1} + A_p^d \phi_i + A_e^d \phi_{i+1} \tag{8.41}$$

where the coefficients for the west, east, and central points are given by

$$A_w^d = -D/(\Delta x)^2 \tag{8.42}$$

$$A_e^d = -D/(\Delta x)^2 \tag{8.43}$$

$$A_p^d = 2D/(\Delta x)^2 \tag{8.44}$$

Next, we need to discretize the advective term. Discretization of this term is not trivial. While it is customary to use the central difference scheme (CDS) for the diffusive term, there are different approaches available for the advective term. We consider two options: CDS, and Upwind Difference Scheme (UDS). This term controls the transport of the quantity ϕ due to mass flux $\rho \vec{u}$. The central difference discretization implies that the value of i-th node should be computed using nodes $i + 1$ and $i - 1$. If the flow is moving from left to right, $u > 0$, it is not reasonable to expect that ϕ_{i+1} should influence ϕ_i. This is where the UDS scheme comes in. This scheme takes into account the direction of gas flow. If $u > 0$, ϕ is advected from left to right, and the term is written using backward difference (BDS),

$$\left[\frac{\partial(\rho u \phi)}{\partial x}\right]_{i,BDS} = \frac{(\rho u \phi)_i - (\rho u \phi)_{i-1}}{\Delta x} \tag{8.45}$$

On the other hand, if $u < 0$, information is moving from right to left. We then use the forward difference scheme (FDS),

$$\left[\frac{\partial(\rho u \phi)}{\partial x}\right]_{i,FDS} = \frac{(\rho u \phi)_{i+1} - (\rho u \phi)_i}{\Delta x} \tag{8.46}$$

Alternatively, we can use the central difference scheme,

$$\left[\frac{\partial(\rho u \phi)}{\partial x}\right]_{i,CDS} = \frac{(\rho u \phi)_{i+1} - (\rho u \phi)_{i-1}}{2\Delta x} \tag{8.47}$$

We again rewrite the derivative term using coefficients,

$$\left[\frac{\partial(\rho u \phi)}{\partial x}\right]_i = A_w^a \phi_{i-1} + A_p^a \phi_i + A_e^a \phi_{i+1} \tag{8.48}$$

For the UDS scheme, we write

$$A_w^a = -\frac{\max(\rho u, 0)}{\Delta x} \tag{8.49}$$

$$A_w^a = \frac{\min(\rho u, 0)}{\Delta x} \tag{8.50}$$

$$A_p^a = -(A_w^a + A_e^a) \tag{8.51}$$

Similarly, for CDS we have

$$A_w^a = -\frac{\rho u}{2\Delta x} \tag{8.52}$$

$$A_e^a = -\frac{\rho u}{2\Delta x} \tag{8.53}$$

$$A_p^a = -(A_w^a + A_e^a) = 0 \tag{8.54}$$

Combining the diffusive and advective coefficients, we have on each internal node

$$A_w \phi_{i-1} + A_p \phi_i + A_e \phi_{i+1} = 0 \tag{8.55}$$

where $A_w = A_w^a + A_w^d$ with similar equations for e and p. The resulting tri-diagonal $\mathbf{A}\vec{\phi} = \vec{b}$ matrix system is then solved using the direct solver from

Section 1.4.3.1. Here $\vec{b} = 0$ since there is no source term. The numerical result can then be compared to the analytical solution [28]

$$\phi = \phi_0 = \frac{e^{xP_e/L} - 1}{e^{P_e} - 1}(\phi_L - \phi_0) \tag{8.56}$$

where $P_e = \rho u L/D$ is the *Peclet number*.

The two approaches are compared in `ad1-steady.cpp`, which is copied below.

```cpp
#include <iostream>
#include <vector>
#include <fstream>
#include <math.h>

using namespace std;
using dvector = vector<double>;

// simple tri diagonal matrix
struct TriD {
  TriD(int ni) {
    a.reserve(ni);
    b.reserve(ni);
    c.reserve(ni);
  }
  dvector a,b,c;    // left, main, and right diagonals
};

// solves Ax=b using the Thomas algorithm
dvector solveDirect(TriD &A, dvector &rhs);

int main() {
  // simulation inputs
  double rho = 1;
  double u = 1;
  double D = 0.02;
  double phi0 = 0;
  double phiL = 1;

  // set domain parameters
  double L = 1;
  int ni = 11;

  // write the analytical solution on a fine mesh
  ofstream out_th("theory.csv");
  out_th<<"x,phi_th"<<endl;
  for (int i=0;i<501;i++) {
    double x = L*i/500.0;
    double Pe = rho*u*L/D;
    double phi_true = phi0 + ((exp(x*Pe/L)-1)/
      (exp(Pe)-1)*(phiL-phi0));
    out_th<<x<<","<<phi_true<<"\n";
  }

  // set matrix
  TriD A(ni);
```

```cpp
dvector b(ni);

// dirichlet condition on left and right
A.b[0] = 1;
b[0] = phi0;
A.b[ni-1] = 1;
b[ni-1] = phiL;

// assuming uniform spacing
double dx = L/(ni-1);

// diffusive term
double AdW = -D/(dx*dx);
double AdE = -D/(dx*dx);
double AdP = -(AdE + AdW);

// upwind scheme for the convective derivative
double AcE = min(rho*u,0.)/dx;
double AcW = -max(rho*u,0.)/dx;
double AcP = -(AcE+AcW);

// contribution from both terms
double Aw = AdW + AcW;
double Ap = AdP + AcP;
double Ae = AdE + AcE;

// set internal nodes
for (int i=1;i<ni-1;i++) {
  A.a[i] = Aw;   // A[i,i-1]
  A.b[i] = Ap;   // A[i,i]
  A.c[i] = Ae;   // A[i,i+1]
}

// obtain the solution
dvector phi_uds = solveDirect(A, b);

// repeat for CDS
AcE = rho*u/(2*dx);
AcW = -rho*u/(2*dx);
AcP = -(AcW+AcE);

// contribution from both terms
Aw = AdW + AcW;
Ap = AdP + AcP;
Ae = AdE + AcE;

// set internal nodes
for (int i=1;i<ni-1;i++) {
  A.a[i] = Aw;
  A.b[i] = Ap;
  A.c[i] = Ae;
}

// obtain the solution
dvector phi_cds = solveDirect(A,b);

// output results
```

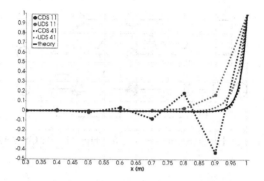

Figure 8.1: Comparison of upwind and central difference schemes for the steady problem.

```
ofstream out ("numerical.csv");
out<<"x,phi_uds,phi_cds"<<endl;
for (int i=0;i<ni;i++) {
    double x = L*i/(ni-1.0);
    out<<x<<","<<phi_uds[i]<<","<<phi_cds[i]<<"\n";
}
return 0;
}
```

The code starts by outputting the analytical solution on a fine 500 node mesh. The code then instantiates an object of type TriD, which is a simple representation of a tri-diagonal matrix. We then set the coefficients for the UDS scheme, and the solution is obtained by calling solveDirect, which is basically a simplified (since the coefficient matrix is passed in) form of solvePotentialDirect from Section 1.4.3.1. The process then repeats, but with coefficients for the CDS scheme. Figure 8.1 compares results for $L = 1$ with 11 and 41 nodes, respectively. We can see that for the lower node count, the UDS scheme produces the qualitatively correct behavior, however the solution exhibits excessive numerical diffusion. CDS produces oscillations, which for the case of continuity equations, would imply a non-physical negative density. Using the finer mesh, however, CDS produces a result that follows the analytical solution very closely. The numerical diffusion, while reduced, is still present and noticeable, for the UDS scheme even with the finer mesh.

8.3.2 Unsteady Problem

Let's now return to the original unsteady 1D advection-diffusion equation,

$$\frac{\partial \rho \phi}{\partial t} + \frac{\partial (\rho u \phi)}{\partial x} - \frac{\partial}{\partial x} \left(D \frac{\partial \phi}{\partial x} \right) = 0 \tag{8.57}$$

The addition of the time derivative term makes this PDE parabolic. As $t \to \infty$, the solution should approach the steady solution from the previous section.

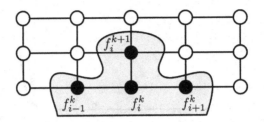

Figure 8.2: Integration stencil for the 1D FTCS scheme.

Just as with the convective term, there are multiple ways to discretize the time derivative. However, we require that ϕ^{k+1} only depends on $k+1$ or prior information since time cannot flow backwards. The simplest strategy is to use the central difference for the spatial derivatives and forward difference for the time derivative. This scheme is known as *Forward Time Central Difference* (FTCS),

$$\frac{\phi^{k+1} - \phi^k}{\Delta t} + \frac{(\rho u \phi)^k_{i+1} - (\rho u \phi)^k_{i-1}}{\Delta x} + \frac{(\rho u \phi)^k_{i-1} - 2(\rho u \phi)^k_i + 2(\rho u \phi)^k_{i+1}}{\Delta^2 x} = 0$$
(8.58)

The computational stencil for the 1D FTCS scheme is shown in Figure 8.2. The scheme can be generalized as

$$\frac{\vec{\phi}^{k+1} - \vec{\phi}^k}{\Delta t} = -\mathbf{A}\vec{\phi}^k$$
(8.59)

where $\mathbf{A}$ is a coefficient matrix. The basic integration algorithm becomes:

1. Set initial conditions

2. March solution to new time per $\phi^{k+1} = \phi^k - \Delta t (A\phi^k)$

3. Return to step 2 while $k < k_{max}$

The FTCS method is demonstrated in `ad1-ftcs.cpp`. The relevant changes from the steady version are listed below.

```
// initial value
dvector phi(ni);
phi[0] = b[0];
phi[ni-1] = b[ni-1];

// temporary vector for phi[k+1]
dvector phi_new(ni);

// iterate using forward time
double dt = 1e-2;

// integrate solution in time
for (int it=0;it<100;it++) {
```

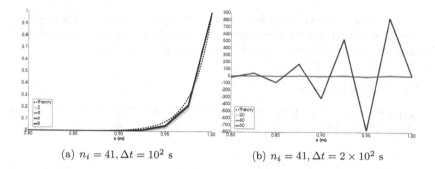

(a) $n_i = 41, \Delta t = 10^2$ s   (b) $n_i = 41, \Delta t = 2 \times 10^2$ s

Figure 8.3: Solution obtained with the FTCS method for different values of Δx and Δt.

```
// set only the non−boundary nodes
for (int i=1;i<ni−1;i++) {
  // compute (A*phi) for node i
  double R = A.a[i]*phi[i−1] +
             A.b[i]*phi[i] +
             A.c[i]*phi[i+1];
  phi_new[i] = phi[i] − dt*R;
}

// copy down non−boundary nodes
for (int i=1;i<ni−1;i++)
  phi[i] = phi_new[i];

// create file name results/ftcs_xxxx.csv
stringstream ss;
ss<<"results/ftcs_"<<setw(4)<<setfill('0')<<it<<".csv";
ofstream out(ss.str());    // open output file
out<<"x,phi"<<endl;        // write header
for (int i=0;i<ni;i++)
  out<<L*i/(ni−1.0)<<","<<phi[i]<<"\n";
}  // file is closed here
```

Figure 8.3(a) plots the solution near the right boundary obtained with $n_i = 41$ mesh nodes after 20, 40, 60, and 80 $\Delta t = 10^{-2}$ s time steps. We can see that the solution is indeed converging to the previously discussed analytical steady state form. It is interesting to note that this solution was obtained without ever solving the $A\vec{\phi} = \vec{b}$ system. This feature of parabolic systems is in fact exploited by some Poisson equation solvers. Instead of solving the $\nabla^2\phi = b$ system, these algorithms time march $\partial\phi/\partial t = b - \nabla^2\phi$ in pseudo-time until the term on left vanishes.

Now, let's say that we decide to speed up the convergence by utilizing twice as large time steps, $\Delta t = 2 \times 10^{-2}$ s. This result is shown in Figure 8.3(b). The two results should be identical, but clearly this is not the case. This solution is demonstrating non-physical oscillations. Similarly, let's assume that we decide to refine the solution by using a finer mesh. Using the original $\Delta t = 10^{-2}$ s but increasing ni to 81 such that Δx is decreased by two, we again obtain a divergent solution. By trial and error, we may find out that Δt needs to be reduced by a factor of four to maintain convergence with $\Delta x/2$.

8.3.3 Stability Analysis

In order to explain this behavior, we need to perform *Von Neumann stability analysis*. We start by rewriting the previous equation in terms of ϕ coefficients,

$$\phi_i^{k=1} = (1 - 2d)\phi_i^k + \left(d - \frac{a}{2}\right)\phi_{i+1}^k + \left(d + \frac{a}{2}\right)\phi_{i-1}^k \tag{8.60}$$

with

$$d = \frac{d\Delta t}{\rho(\Delta x)^2} \qquad a = \frac{u\Delta t}{\Delta x} \tag{8.61}$$

This allow us to write the method as

$$\phi^{k+1} = \mathbf{A}\phi^k \tag{8.62}$$

with $A_w = (d + a/2)$, $A_p = (1 - 2d)$, and $A_e = (d - a/2)$.

Von Neumann hypothesized that boundary conditions are rarely the source of instability and can be ignored. Then, every point of the solutions follows

$$\phi^{k+1} = \mathbf{A}\phi^k = \mathbf{A}(\mathbf{A}\phi^{k-1})\ldots = \mathbf{A}^{k+1}\phi^0 \tag{8.63}$$

In other words, the solution after some number of time steps is achieved by a repetitive multiplication of the initial solution by the matrix A. If we let the initial $\phi^0 = \phi + \epsilon$ where the two terms are some correct initial value and some initial error, we clearly see that we need the coefficient matrix A to be such that $\mathbf{A}^{k+1}\epsilon < \mathbf{A}^k\epsilon$.

A general linear system of the form $\vec{f}'(t) + \mathbf{A}\vec{f}(t) = 0$ has a solution in the form $\vec{f} = \sum_l c_l e^{\lambda_l t}\vec{\eta}_l$, where the terms on the right hand side include a constant that relates to the initial condition (which can be ignored), an exponential term containing an eigenvalue λ_l, and the corresponding eigenvector $\vec{\eta}_l$. In order for the error to decay, we need $\lambda_l < 1$ for all l to prevent the exponential term from increasing in magnitude. For our one-dimensional problem with no boundary conditions, the system matrix contains constant terms on all three diagonals. This special matrix is called *Toeplitz matrix* and has eigenvalues that can be written in the form of sines and cosines. In general, we write

$$\phi_j^k = \lambda^k e^{i\alpha j} \tag{8.64}$$

where $i = \sqrt{-1}$ and α is some arbitrary wave number. If we substitute this expression for ϕ_i into Equation 8.60, we obtain

$$\lambda = 1 + 2d(\cos\alpha - 1) + i2a\sin\alpha \tag{8.65}$$

Since the magnitude of a complex number is the sum of the squares of the real and the imaginary part, we also have

$$\lambda^2 = [1 + 2d(\cos\alpha - 1)]^2 + 4a^2 \sin^2\alpha \qquad (8.66)$$

We can now consider some special cases. If $d = 0$ (no diffusion), $\lambda^2 > 1$ for any value of α and this system is *unconditionally unstable*. The FTCS scheme diverges regardless of node spacing or the integration time step. On the other hand if $a = 0$ (no advection), then $\lambda^2 = [1 + 2d(\cos\alpha - 1)]^2$. This quadratic equation evaluates to one at $\alpha = 1$ and has some maximum value at $\alpha = -1$. We can check graphically (or determine analytically) that we need $d < 0.5$ in order for $\lambda^2 < 1$ everywhere. Using this stability requirement in the full equation, we also obtain $c < 2d$. The stability of the FTCS method is thus

$$\Delta t < \frac{\rho(\Delta x)^2}{2D} \qquad (8.67)$$

and

$$\frac{\rho u \Delta x}{D} \equiv P_{e,cell} < 2 \qquad (8.68)$$

Here $P_{e,cell}$ is the Peclet number computed using cell dimension.

8.3.4 Alternative Time Integration Methods

The Δt condition in Equation 8.67 agrees with the previous observation that time step had to be decreased by a factor of four when Δx was decreased by half. We are generally interested in long term solutions and hence prefer to use large Δt. This stability requirement makes the FTCS method non-ideal for simulations utilizing fine spatial meshes.

In general, we are interested in solving the following system

$$\frac{d\phi(t)}{dt} = f(t, \phi(t)) \qquad (8.69)$$

which can be discretized as $\phi^{k+1} = \phi^k + f(\phi^k)\Delta t$. This is the Forward Euler method used previously. But there are other approaches. We could use the Backward Euler method in which the forcing vector on the right hand side is evaluated at time $k + 1$,

$$\phi^{k+1} = \phi^k + f(\phi^{k+1})\Delta t \qquad (8.70)$$

Backward Euler can also be written in matrix form as

$$(\mathbf{I} - \mathbf{A})\phi^{k+1} = \phi^k \qquad (8.71)$$

Alternatively, f could be evaluated at $k+0.5$. This is the midpoint (or Crank-Nicolson) method

$$\phi^{k+1} = \phi^k + f(\phi^{k+0.5})\Delta t \qquad (8.72)$$

The $\phi^{k+0.5}$ term is obtained by averaging ϕ^k and ϕ^{k+1} to obtain

$$\phi^{k+1} = \phi^k + \left(\frac{1}{2}\right) \left[f(\phi^k + f(\phi^{k+1}))\right] \Delta t \tag{8.73}$$

Alternatively,

$$\left(\mathbf{I} - \frac{1}{2}\mathbf{A}\right) \phi^{k+1} = \left(\mathbf{I} + \frac{1}{2}\mathbf{A}\right) \phi^k \tag{8.74}$$

This popular method combines the stability of the Euler Backward method with the second order accuracy of the trapezoidal method.

Unlike the explicit FTCS method, the two methods given by Equations 8.71 and 8.74 are implicit since advancing from time k to $k+1$ requires the solution at time $k+1$. As was discussed in Chapter 1, implicit methods are *unconditionally stable*. This characteristic does not imply that Δt can be arbitrarily large for correct results. The criterion merely indicates that the solution remains bounded instead of growing to infinity. The downside of implicit methods is that they require solution of a matrix system

It is not unreasonable for the total computational time to actually increase with an implicit method despite using fewer, larger Δt time steps. Therefore, additional schemes exist that attempt to recover the stability of implicit methods without requiring the time-consuming matrix solve. These schemes can be divided into two classes: *predictor-corrector* and *multipoint* methods.

With the predictor-corrector method, we first use the standard forward integration to obtain a prediction for the new value,

$$\phi^* = \phi^k + f(\phi^k)\Delta t \tag{8.75}$$

This value is then corrected from

$$\phi^{k+1} = \phi^k + \frac{1}{2}\left(f(\phi^k) + f(\phi^*)\right)\Delta t \tag{8.76}$$

This method is second order accurate but retains the stability requirements similar to the forward Euler explicit scheme.

Alternatively, instead of using the predicted future value, we can perform the integration using one or more "old" values from previous time steps. An example of this multi-point scheme is the second order *Adams-Bashford* method given by

$$\phi^{k+1} = \phi^k + \frac{\Delta t}{2} \left[3f(\phi^k) - f(\phi^{k-1})\right] \tag{8.77}$$

Solution at time $k+1$ is computed using solutions at k and $k-1$ time steps. This method introduces memory overhead since the solution vector from the prior time step needs to be retained. Multipoint methods can also be written in an implicit fashion. For instance, the third-order *Adams-Moulton* method computes ϕ^{k+1} using ϕ^{k-1}, ϕ^k, and ϕ^{k+1},

$$\phi^{k+1} = \phi^k + \frac{\Delta t}{12} \left[5f(\phi^{k+1}) + 8f(\phi^k) - f(\phi^{k-1})\right] \tag{8.78}$$

Finally, the *Runge-Kutta* and *Leapfrog* methods that we have been using to integrate particle velocities can also be applied to a generic differential equation. The fourth order Runge-Kutta method (RK4) is

$$\phi^a = \phi^k + \frac{\Delta t}{2} f(\phi^k) \tag{8.79}$$

$$\phi^b = \phi^k + \frac{\Delta t}{2} f(\phi^a) \tag{8.80}$$

$$\phi^* = \phi^k + \Delta t f(\phi^b) \tag{8.81}$$

$$\phi^{k+1} = \phi^k + \frac{\Delta t}{6} \left[f(\phi^k) + 2f(\phi^a) + 2f(\phi^b) + f(\phi^*) \right] \tag{8.82}$$

The Leapfrog method is instead written as

$$\phi^{k+1} = \phi^{k-1} + f(\phi^k) 2\Delta t \tag{8.83}$$

The forcing vector f is evaluated at time k and is used to advance the solution from $k-1$ to $k+1$. This method actually turns out be unconditionally unstable for the A-D equation, but it can be stabilized with some manipulation. One such trick is to replace the ϕ_i^k term with the average $(\phi_i^{k+1} + \phi_i^{k-1})/2$. This is known as *Dufort-Frankel* Method.

8.3.5 Example

We now demonstrate the use of fluid models in plasma simulation codes. Here for simplicity, we only model a neutral population. Let's assume that the sphere contains some molecular "spherium" contaminant that is continuously outgassing into the environment. Instead of using particles, we let the density be governed by the diffusion equation,

$$\frac{\partial n}{\partial t} = D\nabla^2 n = 0 \tag{8.84}$$

Note that since the advection term is removed, this equation models a scenario in which momentum transfer from collisional coupling with other species is negligible. In other words, $n_{sp} \gg n_{ions}$. Here we also assumed that D is constant so that it can be taken out of the gradient.

The example is based on the code developed in Chapter 4. In line with the algorithm on page 273, we start by generalizing the species integration scheme. We have already used a `Species::advance` function to advance particle velocities and positions through a time step. This allowed us to write

```
for (Species &sp : species)
    sp.advance();
```

We would like to keep this same general approach, but have the `advance` function perform the integration of Equation 8.84 for "fluid species". Therefore, following the same familiar approach from Section 4.4, we let `species` be a container for a generic base class of type `Species`,

```
vector<unique_ptr<Species>> species;
```

We let this class contain functionality common to any material species, regardless of how the integration scheme is implemented. We have

```
class Species {
public:
    Species(std::string name, double mass, double charge, World
        &world) : name(name), mass(mass), charge(charge),
        den(world.nn), T(world.nn), vel(world.nn), den_ave(world.nn),
        world(world) { }
    ~Species() {};

    virtual void advance() = 0;  // integrates state by dt
    virtual std::string printSelf() = 0;  // screen outpu

    double3 sampleIsotropicVel(double T);  // samples random
        isotropic velocity
    double sampleVth(double T);            // returns random thermal
        velocity

    // updates number density
    void updateAverages() {den_ave.updateAverage(den);}

    // hooks to be overridden as needed
    virtual void computeGasProperties() {}
    virtual void clearSamples() {}
    virtual double getMass() {return 0;}  // total mass
    virtual double3 getMomentum() {return {0,0,0};}  // total
        momentum
    virtual double getKE() {return 0;}  // total kinetic energy

    const std::string name;  // species name
    const double mass;       // mass in kg
    const double charge;     // charge in Coulomb

    Field den;       // number density
    Field T;         // temperature
    Field3 vel;      // stream velocity
    Field den_ave;   // averaged number density

protected:
    World &world;
};
```

This class contains two pure virtual functions, **advance** and **printSelf**. The latter simply returns a string to be output to the screen at every time step. Since fluid species do not contain particles, it does not make sense to output a macroparticle count for these materials. This overloaded function is called from **Output::screenOutput** as

```
void Output::screenOutput(World &world,
    vector<unique_ptr<Species>> &species) {
    cout<<"ts: "<<world.getTs();
    for (auto &sp:species)
        cout<<" "<<sp->printSelf();
    cout<<endl;
```

```
}
```

The functionality specific to particle-based species is then delegated to the derived `KineticSpecies` class,

```cpp
class KineticSpecies : public Species {
public:
    KineticSpecies(std::string name, double mass, double charge,
        double mpw0, World &world) : Species(name, mass, charge,
        world), mpw0(mpw0), mpc(world.ni-1,world.nj-1,world.nk-1),
        n_sum(world.nn),nv_sum(world.nn), nuu_sum(world.nn),
        nvv_sum(world.nn), nww_sum(world.nn) { }

    std::string printSelf();
    double getMass();
    double3 getMomentum();
    double getKE();

    // moves all particles using electric field ef
    void advance();
    /* ... */
}
```

The particle-based `Species::advance` function from prior chapters now becomes `KineticSpecies:: advance`. We similarly declare a new class of type `FluidSpecies` that also derives from the `Species` base class,

```cpp
class FluidSpecies : public Species {
public:
    FluidSpecies(std::string name, double mass, double charge,
        double den0, double D, World &world):
        Species{name, mass, charge, world}, den0{den0}, D{D},
        den_new(world.nn) {}

    void advance();
    std::string printSelf();

protected:
    double den0;       // sphere density
    double D;          // diffusion coefficient
    Field den_new;     // temporary solution buffer
};
```

We can note that we are using constructor chaining to initialize common data objects, such as `Field den`. With this change, we modify the initialization of our `species` container in `Main.cpp` to read

```cpp
vector<unique_ptr<Species>> species;
    species.emplace_back(new KineticSpecies("O+", 16*AMU, QE, 5e3,
        world));
species.emplace_back(new FluidSpecies("Sph", 100*AMU, 0, 1e16,
    100, world));
Species &ions = *species[0];
```

The reference to `ions` is used to initialize the ion injection source. We subsequently perform a code-wide replacement of loops over species to take into account that this variable is now a vector of unique_ptr. For instance, the integration loop becomes

```
for (auto &sp:species)
  sp->advance();
```

Previously, this code block also included calls to computeNumberDensity, sampleMoments, and computeMPC. As these functions are specific to kinetic species, a call to them was added to the end of KineticSpecies::advance.

All that remains at this point is implementing the advance function. This function uses the FTCS scheme to integrate Equation 8.84. The only "difficulty" is dealing with boundaries. We assume that the sphere acts as an infinitely large source of the contaminant, such that $n = n_0$ holds for the duration of the simulation for any node on the sphere. On the external boundaries, we use the Neumann condition $\partial n / \partial \hat{n} = 0$. The algorithm thus consists of looping through all the nodes and checking if the Dirichlet or the Neumann boundary is appropriate. If not, the FTCS scheme is used to compute the new value $n_{i,j,k}^{k+1}$. After we complete looping over the entire mesh, we "copy down" the solution, $n^{k+1} \to n^k$. The code is given below

```
void FluidSpecies::advance() {
  double dt = world.getDt();
  double3 dh = world.getDh();

  for (int i=0;i<world.ni;i++)
    for (int j=0;j<world.nj;j++)
      for (int k=0;k<world.nk;k++) {
        if (world.inSphere(world.pos(i,j,k))) den_new[i][j][k] =
            den0;
        else if(i==0) den_new[i][j][k] = den[i+1][j][k];
        else if(i==world.ni-1) den_new[i][j][k] = den[i-1][j][k];
        else if(j==0) den_new[i][j][k] = den[i][j+1][k];
        else if(j==world.nj-1) den_new[i][j][k] = den[i][j-1][k];
        else if(k==0) den_new[i][j][k] = den[i][j][k+1];
        else if(k==world.nk-1) den_new[i][j][k] = den[i][j][k-1];
        else {
          double lap_x = (den[i-1][j][k] - 2*den[i][j][k] +
              den[i+1][j][k]) / (dh[0]*dh[0]);
          double lap_y = (den[i][j-1][k] - 2*den[i][j][k] +
              den[i][j+1][k]) / (dh[1]*dh[1]);
          double lap_z = (den[i][j][k-1] - 2*den[i][j][k] +
              den[i][j][k+1]) / (dh[2]*dh[2]);
          double lap = lap_x + lap_y + lap_z;

          den_new[i][j][k] = den[i][j][k] + dt*D*lap;
        }
      }
  den = den_new;   // copy down
}
```

We also add code for the printSelf function. We use it to output the average number density,

```
std::string FluidSpecies::printSelf() {
```

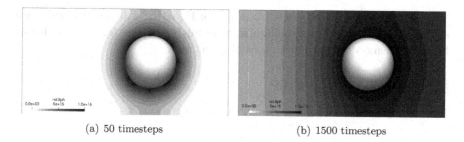

(a) 50 timesteps   (b) 1500 timesteps

Figure 8.4: Material number density evolution from diffusion equation.

```
std::stringstream ss;
ss<<name<<": "<<std::setprecision(4)<<den.mean();
return ss.str();
}
```

The simulation output now looks like

```
ts: 0 O+: 56 Sph: 2.843e+14
ts: 1 O+: 112 Sph: 3.109e+14
ts: 2 O+: 168 Sph: 3.322e+14
...
ts: 9999 O+: 6532 Sph: 9.898e+15
ts: 10000 O+: 6527 Sph: 9.898e+15
Simulation took 134.777 seconds
```

As the simulation progresses, the averaged number density of the "spherium" contaminant keeps increasing. Eventually, the density in the entire domain becomes uniform and equal to the prescribed 10^{16} m^{-3} value on the sphere. We can observe this graphically in Figure 8.4. The first image shows the spherium number density shortly after the simulation starts. The material is found mainly in a thin layer around the sphere. After 1,500 time steps, the concentration of the material has increased, and the region where $n = n_0$ has grown in size. A similar picture generated at the conclusion of the simulation would show an almost entirely flat field.

8.4 ELECTRON MODELS

Fluid models are also widely used in *hybrid* Particle in Cell simulations. In Chapter 3, we represented electrons by the Boltzmann relationship $n_e = n_0 \exp((\phi - \phi_0)/kT_{e,0})$. This relationship arises from the momentum conservation equation for the electrons, Equation 8.4. With $\vec{F}$ given by the Lorentz force and assuming that the magnetic field is negligible, we have for plasma containing only ions and electrons,

$$0 = -n_e e \vec{E} - \nabla p_e + m_e n_e (\vec{v}_i - \vec{v}_e) \nu_{ie} \qquad (8.85)$$

The total derivative on the left hand side of Equation 8.4 was dropped since electrons are expected to respond instantaneously and electron density should be varying smoothly with large spatial gradients.

8.4.1 Polytropic Relationship

Next, if we ignore the resistive drag, we obtain,

$$0 = e\vec{E} + \frac{1}{n_e}\nabla p_e \tag{8.86}$$

It should be noted that the above assumption implies $\vec{v}_e = \vec{v}_i$ since generally $\nu_{ie} \neq 0$. In the case of neutral plasma, $n_e = n_i = n$, this assumption also implies that the total current $\vec{j} = e(n_i\vec{v}_i - n_e\vec{v}_e) = en(\vec{v}_i - \vec{v}_e) = 0$. With electron pressure given by the ideal gas law $p_e = kn_eT_e$ and using the electrostatic assumption $\vec{E} = -\nabla\phi$, we next write

$$e\nabla\phi = \frac{k}{n_e}\nabla\left(n_eT_e\right) \tag{8.87}$$

In Chapter 3, we assumed that electrons are isothermal, $T_e = T_{e,0}$. The T_e term can then be taken out of the derivative and we arrive at the Boltzmann relationship from Equation 3.11. But instead of using constant temperature, we can assume that electron pressure follows the polytropic relationship, $p_e = Cn^\gamma$ where γ is the specific heat ratio [50]. Again, starting with Equation 8.85, we write

$$e\vec{E} = -\frac{1}{n_e}C\gamma n_e^{\gamma-1}\nabla n_e \tag{8.88}$$

Combining the densities and using $\vec{E} = -\nabla\phi$,

$$e\nabla\phi = C\gamma n_e^{\gamma-2}\nabla n_e \tag{8.89}$$

We next integrate both sides

$$\int_{\phi_0}^{\phi} e\,d\phi = C\gamma\int_{n_0}^{n_e} n_e^{\gamma-1}dn_e \tag{8.90}$$

to obtain

$$e(\phi - \phi_0) = C\frac{\gamma}{\gamma-1}\left[n_e^{\gamma-1} - n_0^{\gamma-1}\right] \tag{8.91}$$

This is the final form except for the constant C. We desire $p_0 = p(n_0)$ or $n_0kT_{e,0} = Cn_0^\gamma$, leading to $C = kT_{e,0}n_0^{1-\gamma}$. With this substitution we obtain

$$\phi = \phi_0 + \frac{\gamma}{\gamma-1}\frac{kT_{e,0}}{e}\left[\left(\frac{n_e}{n_0}\right)^{\gamma-1} - 1\right] \tag{8.92}$$

By letting $n_e = n_i = n$, we obtain the polytropic alternative to the QN scheme from Equation 3.44. The above model can also again be inverted and solved for n_e,

$$n_e = n_0 \left[\frac{e(\phi - \phi_0)}{kT_{e,0}} \frac{\gamma - 1}{\gamma} + 1 \right]^{\frac{1}{\gamma - 1}}$$

(8.93)

This expression can then be used in Poisson's equation. As an aside, from the initial relationship for pressure as a function of density, we can also derive a relationship for electron temperature, $nkT_e = kT_{e,0}n_0^{1-\gamma}n^\gamma$, or

$$\frac{T_e}{T_{e,0}} = \left(\frac{n}{n_0} \right)^{\gamma - 1}$$

(8.94)

8.4.2 Detailed Electron Model

The polytropic and Boltzmann relationships were derived by assuming that the resistive term in Equation 8.85 vanishes. However, if plasma is assumed to be neutral, then the $n_e(\vec{v}_i - \vec{v}_e)$ term can be written as $\vec{j}/e$ where $\vec{j} = \vec{j}_i + \vec{j}_e$ is the total current density. Then,

$$0 = -\vec{E} - \frac{1}{n_e e} \nabla p_e + \frac{m_e \nu_{ei}}{n_e e^2} \vec{j}$$

(8.95)

The coefficient on $\vec{j}$ is the plasma resistivity $\eta = 1/\sigma$. We therefore have the generalized Ohm's law,

$$\vec{j} = \sigma \vec{E} + \frac{\sigma}{n_e e} \nabla p_e$$

(8.96)

Comparing this equation to the one encountered during the derivation of the Boltzmann relationship, we see that this is a more general case with $\vec{j} \neq 0$. If $\vec{j} = 0$, the Boltzmann relationship is recovered. Next, in neutral plasma, the charge conservation equation $\partial \rho / \partial t + \nabla \cdot \vec{j} = 0$ reduces to $\nabla \cdot \vec{j} = 0$. With our expression for $\vec{j}$ and also, using $\vec{E} = -\nabla \phi$, we can write

$$\nabla \cdot \left[-\sigma \nabla \phi + \frac{\sigma k_B}{en} \nabla(nT_e) \right] = 0$$

(8.97)

This equation forms the foundation of the "Detailed Electron Model" proposed by Boyd [20]. With the help of chain rule, this model can be seen to result in a Poisson-like elliptic equation, however, the RHS is now given by terms containing gradients of pressure and conductivity.

This scheme can be applied directly to a case with constant temperature. However, Boyd further expanded the model by including a relationship for temperature based on a solution of the energy conservation equation 8.6. The energy equation requires that electron velocity is known. The electron continuity expression reads

$$\frac{\partial n_e}{\partial t} + \nabla \cdot (n_e \vec{v}_e) = n_e n_a C_i$$

(8.98)

If the flow is assumed to be irrotational ($\nabla \times \vec{v}_e = 0$), then a stream function $n_e\vec{v}_e = \nabla\psi$ can be defined. The first term on the LHS vanishes at steady state, yielding

$$\nabla^2\psi = n_e n_a C_i \qquad (8.99)$$

With the neutral plasma assumption already made in other parts of this model, we obtain an expression for electron velocity

$$\vec{v}_e = \nabla\psi/n_i \qquad (8.100)$$

Temperature is obtained from energy conservation, Equation 8.6. Implementation of these models is left as a chapter exercise.

8.5 VLASOV SOLVERS

Eulerian methods are not limited to the MHD formulation in which the gas can be assumed to be in continuum. In Chapter 1 we encountered another PDE known as the Boltzmann equation 1.8

$$\frac{\partial f}{\partial t} + \vec{v} \cdot \nabla_x f + \frac{\vec{F}}{m} \cdot \nabla_v f = (\dot{f})_c \qquad (8.101)$$

If collisions are ignored and acceleration arises solely from the Lorentz force, we have the *Vlasov equation*.

$$\frac{\partial f}{\partial t} + \vec{v} \cdot \nabla_x f + \frac{q\left(\vec{E} + \vec{v} \times \vec{B}\right)}{m} \cdot \nabla_v f = 0 \qquad (8.102)$$

For the case with no magnetic field, the equation reduces further to

$$\frac{\partial f}{\partial t} + \vec{v} \cdot \nabla_x f + \frac{q\vec{E}}{m} \cdot \nabla_v f = 0 \qquad (8.103)$$

Here $f = f(\vec{x}, \vec{v}, t)$ is value of the distribution function evaluated at some time t for some spatial and velocity coordinate $(\vec{x}, \vec{v})$. In general, we do not have an analytical equation for f and thus need to represent it as a discretized form in the velocity phase space. For a 1D1V problem, we have a standard histogram in which one axis represents the velocity, and the other axis stores the count of particles with that velocity. As was discussed in Chapter 1, storing the discretized VDF leads to excessive computational requirements for multidimensional problems. For instance, let's say that we want each velocity component to capture $\pm 3v_{th}$. For 1 eV electrons, the range of velocities is $\approx \pm 1.25 \times 10^6$ m/s. Even limiting the smallest resolvable velocity increment to 10 km/s, we need 250 bins for each dimension. For a 3D problem, this leads to the need to store $250 \times 250 \times 250$ or over 15 million values per physical cell. On the other hand, storing 250 unique values for a one dimensional problem is quite feasible. Therefore, Vlasov codes can actually be a good alternative to the kinetic PIC method for 1D problems. They can be used to characterize numerical noise arising from the particle simulation. We now demonstrate how to develop a simple 1D Vlasov solver for case with no magnetic field.

Following the work of Cheng [25], we start with a normalized form of the VDF for an electron with mass $m = 1$ and charge $q = -1$. Per the naming convention from the paper, v is the x component of velocity. We have

$$\frac{\partial}{\partial t} f(x, v, t) + v \frac{\partial}{\partial x} f(x, v, t) - E(x, t) \frac{\partial}{\partial v} f(x, v, t) = 0 \tag{8.104}$$

Instead of trying to integrate this equation directly, we split the integration into two half-time steps,

$$\frac{\partial}{\partial t} f + v \frac{\partial}{\partial x} f = 0 \tag{8.105}$$

$$\frac{\partial}{\partial t} f - E(x, t) \frac{\partial}{\partial v} f = 0 \tag{8.106}$$

From the first equation, we have

$$\frac{f(x, v, \Delta t) - f(x, v, 0)}{\Delta t} = -v \frac{\partial}{\partial x} f \tag{8.107}$$

$$f(x, v, \Delta t) = f(x, v, 0) - v \Delta t \frac{\partial}{\partial x} f \tag{8.108}$$

$$f(x, v, \Delta t) = f(x - v \Delta t, v, 0) \tag{8.109}$$

Similarly, from the second equation

$$f(x, v, \Delta t) = f(x, v + E(\bar{x}, \tau) \Delta t, 0) \tag{8.110}$$

The authors next show that for the second order convergence, we need to evaluate $E(x)$ at $x - v(\Delta t/2)$. The following scheme, consisting of three shifts, is then proposed to evaluate f^{n+1} from f^n,

1. $f^*(x, v) = f^n(x - v \Delta t/2, v)$

2. $f^{**}(x, v) = f^*(x, v + E(x) \Delta t)$

3. $f^{n+1}(x, v) = f^{**}(x - v \Delta t, v)$

8.5.1 Two-Stream Instability Example

To demonstrate the integration scheme, let's consider spatially uniform Maxwellian plasma. We start by initializing a two-dimensional grid to store the one-dimensional variation of the one-dimensional VDF. This grid is shown in Figure 8.5. The i index corresponds to spatial location while the j index is the velocity. We let $v \in [0, L]$ and $v \in [-v_{max}, +v_{max}]$. We then have

$$\Delta x = L/(n_i - 1) \tag{8.111}$$

$$\Delta v = 2v_{max}/(n_j - 1) \tag{8.112}$$

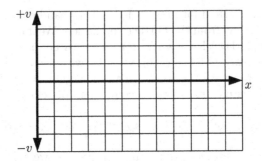

Figure 8.5: Phase-space grid used for a 1D-1V Vlasov problem.

and

$$x = i\Delta x \tag{8.113}$$

$$v = -v_{max} + j\Delta v \tag{8.114}$$

These data structures are allocated in `main.cpp` using the **Field** 2D container from Chapter 5 and `vector<double>` aliased as **dvector**,

```
const double pi = acos(-1.0);        // pi

World world;
world.setLimits(10,5);
world.setNodes(101,51);
world.periodic = true;
int ni = world.ni;
int nj = world.nj;
double dx = world.dx;
double dv = world.dv;
double dt = 1/8.0;

Field f(ni,nj);      // f
Field fs(ni,nj);     // fs
Field fss(ni,nj);    // fss
dvector ne(ni);      // number density
dvector b(ni);       // solver RHS, -rho=(ne-1) since ni=1 is assumed
dvector E(ni);       // electric field
dvector phi(ni);     // potential

// map is a list of keys and corresponding values for output
map<string,Field*> scalars2D;
map<string,dvector*> scalars1D;
scalars2D["f"] = &f;
scalars1D["ne"] = &ne;
scalars1D["E"] = &E;
```

We then load the Maxwellian population by evaluating

$$\hat{f}_m = \frac{1}{v_{th}\sqrt{\pi}} \exp\left(-\frac{v^2}{v_{th}^2}\right) \tag{8.115}$$

on each grid node. We load a normalized version with $v_{th} = \sqrt{2}$. This code is listed below,

```
// set initial distribution
for (int i=0;i<ni;i++)
  for (int j=0;j<nj;j++) {
    double x = world.getX(i);
    double v = world.getV(j);

    double vth2 = 0.02;
    double vs1 = 1.6;
    double vs2 = -1.4;
    f[i][j] = 0.5/sqrt(vth2*pi)*exp(-(v-vs1)*(v-vs1)/vth2);
    f[i][j] += 0.5/sqrt(vth2*pi)*exp(-(v-vs2)*(v-vs2)/vth2)*
               (1+0.02*cos(3*pi*x/world.L));
  }
```

This code loads two cold beams with velocities vs1 and vs2. The second beam is loaded with a slight sinusoidal variation in spatial density obtained using $f_2 = \tilde{f}_2(1 + .02\cos(3\pi x/L))$

The $f(x,v)$ values needed in integration are generally not going to directly correspond to a known nodal value, and need to be interpolated. In this example we use a first-order scheme identical to the gather used in PIC simulations to interpolate electric field onto the particle position. It is important to realize that this first-order scheme is numerically diffusive and higher order schemes are needed in production codes. We use the first order scheme due to its simplicity. This code is added to the World class,

```
double World::interp(Field &f, double x, double v) {
  double fi = (x-0)/dx;
  double fj = (v-(-v_max))/dv;

  // periodic boundaries in i
  if (periodic) {
    if (fi<0) fi+=ni-1;   // -0.5 becomes ni-1.5
    if (fi>ni-1) fi-=ni-1;
  }
  else if (fi<0 || fi>=ni-1) return 0;

  // return zero if velocity less or more than limits
  if (fj<0 || fj>=nj-1) return 0;

  int i = (int)fi;
  int j = (int)fj;
  double di = fi-i;
  double dj = fj-j;

  double val = (1-di)*(1-dj)*f[i][j];
  if (i<ni-1) val+=(di)*(1-dj)*f[i+1][j];
  if (j<nj-1) val+=(1-di)*(dj)*f[i][j+1];
  if (i<ni-1 && j<nj-1) val+=(di)*(dj)*f[i+1][j+1];
  return val;
}

// makes left and right edge values identical
```

```
void World::applyBC(Field &f) {
    if (!periodic)  return;
    for (int j=0;j<nj;j++) {
        f[0][j] = 0.5*(f[0][j]+f[ni-1][j]);
        f[ni-1][j] = f[0][j];
    }
}
```

The above code takes into account periodic boundary conditions. It is important to note that the periodicity extends only in the spatial i direction.

We also add code to evaluate the 1D variation in electron number density,

$$n_e = \int_v f \, dv \qquad (8.116)$$

This integration is performed using the trapezoidal rule,

```
for (int i=0;i<ni;i++) {
    ne[i] = 0;
    for (int j=0;j<nj-1;j++)
        ne[i]+=0.5*(fs[i][j+1]+fs[i][j])*dv;
}
```

We assume that ions form a fixed neutralizing background with $n_i = 1$. Charge density is thus given by $\rho = e(1 - n_e)$.

We next add the code to perform the integration of the VDF per algorithm in Equation 1,

```
// main loop
for (it=0;it<=1000;it++) {
    if (it%100==0)   cout<<it<<endl;
    if (it%5==0) saveVTK(it,world,scalars2D,scalars1D);

    // compute f*
    for (int i=0;i<ni;i++)
        for (int j=0;j<nj;j++) {
            double v = world.getV(j);
            double x = world.getX(i);

            fs[i][j] = world.interp(f,x-v*0.5*dt,v);
        }

    world.applyBC(fs);

    // get number density by integrating f with the trapezoidal rule
    /* ... */

    // compute the right hand side, -rho = (ne-1)
    for (int i=0;i<ni;i++)      b[i] = ne[i]-1;
    b[0] = 0.5*(b[0]+b[ni-1]);
    b[ni-1] = b[0];

    // solution of the Poisson's equation
    solvePoissonsEquationGS(world,b,phi,E);

    // compute f**
```

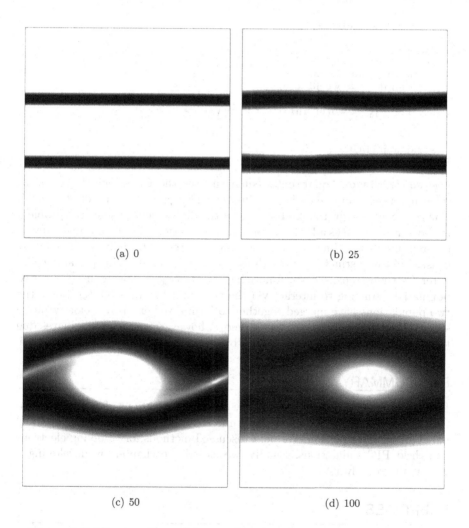

(a) 0

(b) 25

(c) 50

(d) 100

Figure 8.6: Vlasov solver solutions after the given number of time steps.

```
for (int i=0;i<ni;i++)
  for(int j=0;j<nj;j++) {
    double v = world.getV(j);
    double x = world.getX(i);
    fss[i][j] = world.interp(fs,x,v+E[i]*dt);
  }

world.applyBC(fss);

// compute f(n+1)
for (int i=0;i<ni;i++)
  for(int j=0;j<nj;j++) {
    double v = world.getV(j);
    double x = world.getX(i);
    f[i][j] = world.interp(fss,x-v*0.5*dt,v);
  }

world.applyBC(f);
}
```

The `solvePoissonsEquationGS` equation uses the Gauss-Seidel scheme to solve potential. Subsequently, the function also computes the electric field. The iterative GS solver is used instead of the direct method since the problem is ill-defined as it lacks a Dirichlet boundary condition. Results obtained after a different number of time steps are shown in Figure 8.6. Initially, the simulation consists of two distinct beams with no (apparent) variation in spatial density. After 25 time steps, we start noticing slight wiggles arising from the two populations starting to interact via the electric field. After 50 iterations, the two populations have merged together to form a vortex in the velocity phase space. This vortex then continues to move, however, it quickly dissipates due to the numerical diffusion introduced by our first order interpolation scheme.

8.6   SUMMARY

In this chapter, we first introduced the governing equations of magnetohydrodynamics. We then discussed several time marching algorithms using a model one-dimensional equation. We next discussed methods for resolving electrons in hybrid PIC simulations. Finally, we closed the chapter by developing a 1D-1V Vlasov solver.

EXERCISES

8.1 *Integration Schemes.* Modify the `ad1-ftcs.cpp` example to implement the Euler backward, Crank-Nicolson, Dufort-Frankel, Adams-Bashford, Adams-Moulton, and Runge-Kutta integration schemes. Compare results for several different values of n_i and Δt.

8.2 *Vacuum Pump.* Write a 2D X/Y code to simulate the diffusion of gas into an impermeable and initially empty rectangular vessel. The gas source is a tank large enough so that we can assume density in the tank remains constant. The time evolution of density is governed by the diffusion equation, Equation 8.84. Use the following parameters: $n_i = 41$, $n_j = 21$, $\Delta x = \Delta y = 0.1$ m, $D = 10^{-2}$ m^2/s, $\Delta t = 0.1$ s, $\rho_{source} = 10$ kg/m^3. Let the inlet span nodes $i = 0$, $j = [9 : 11]$, and place the pumps at $i = [20 : 24]$, $j = 0$ and $j = n_j - 1$. Use the FVM formulation to derive the discretized equations. Use the zero flux condition, $\nabla \rho \cdot \hat{n} = 0$ on the non-inlet and non-pump boundaries to model a solid impermeable wall. Run the simulation for 50,000 time steps. Observe the time evolution of results. You should notice that steady state is reached by time step 25,000. Next, consider a setup in which the pumps have some finite pumping efficiency f. The mass flux to the pumps on the appropriate faces becomes $f(\nabla \rho \cdot \hat{n})$. Plot results for $f = 0, 0.25, 0.5, 0.75$, and 1. For $f = 0$, you should obtain $\rho = \rho_{source}$ everywhere, while $f = 1$ should result in the solution from the first code version.

8.3 *Detailed Electron Model.* Implement potential solver based on the polytropic relationship and the detailed electron model in the example code from Chapter 3. Compare the resulting potential and ion density to cases with the Boltzmann quasi-neutral solver and the non-neutral Poisson solution.

8.4 *Landau Damping.* In [24], Chen writes that "The theoretical discovery of wave damping without energy dissipation by collisions is perhaps the most astounding result of plasma physics research." The objective of this exercise is to simulate this dissipation, known as Landau Damping, using the Vlasov code. Start by loading a spatially non-uniform electron population. Given sufficient time, the density becomes uniform despite the simulation not including collisions. Start by modifying the example code by loading $f = (1 + \cos(2\pi x / L)) f_m$ where x is the position and L is the domain length. Test your code with constant $E = 0$. You should observe the population with non-zero v to move, but the $v = 0$ region remains stationary. Next incorporate the Poisson solver to compute E. Run the code to observe that electron density becomes spatially uniform.

CHAPTER 9

Parallel Programming

9.1 INTRODUCTION

T HE CODES that we have developed so far are examples of *serial* programs. They execute only a single instruction at a time. Modern computers contain multi-core processors that can run multiple instructions concurrently. They may also include an onboard graphics card that can be used for general computing. Multiple computers can also be connected into a network, with workload distributed among individual machines. In this chapter we learn how to utilize these hardware technologies to either speed up existing codes, or to support cases that would otherwise be too large to fit onto a single machine.

9.2 ARCHITECTURES

As an example, consider my fairly standard desktop workstation. It has a 3.2 / 4.6 GHz Intel i7-8700 processor featuring 6 physical cores and supporting up to 12 concurrent *threads*. A thread is a term for a sequence of instructions executed serially. Every program starts as a single thread. It can spawn additional threads as desired. This process, known as *multithreading* is the first parallelization technique at our disposal. Its great benefit is that all threads have access to the same memory space. There is no need to distribute data to the workers.

My computer also comes with an NVIDIA Quadro P1000 graphics card, or a GPU. Consider a code to push particles. A serial version may read

```
for (size_t p=0;p<np;p++)
  part[p].x += part[p].v*dt;
```

The order of operations is not important. We just simply need to apply this increment to every item in the array. We could as well start from the end,

```
for (size_t p=np−1;p>=0;p−−)
  part[p].x += part[p].v*dt;
```

Alternatively, on a system with many cores, we could assign each array index to a different thread. GPUs are optimized for this type of *vector*, or Single Instruction Multiple Data (SIMD), operations. The Quadro P1000 contains 640 cores. This card can process, at least, 640 particles at the same time. The "at least" arises since unlike their CPU counterpart, GPU cores are optimized to run multiple threads concurrently to minimize latency associated with memory transfer. The GPU cores operate at a lower clock speed than the CPU but the fact there are so many of them suggests that moving computations to a GPU should offer a significant speedup. This is not always the case. A GPU operates, at least for the legacy architecture, on its own onboard memory. Any data needs to be first transferred to the card. This memory transfer introduces a significant overhead that can negate any speedup from the faster computation. Furthermore, GPU computations operate in groups of 32 threads known as *warps*. Every thread in a warp needs to be executing the same instruction. If the logic diverges, let's say by one thread jumping into an `if` branch, the rest of the threads are suspended until the branches merge again. Therefore, GPUs are best suited to operations that involve identical mathematical operations applied to a large set of data. We write GPU code with the help of CUDA (for NVIDIA GPUs) or OpenCL (hardware independent) language extensions [58].

We may also need additional speed up beyond what is possible with a single machine. The simulation may also require mesh dimensions or particle counts too great to fit into the available memory. My workstation comes with 32 Gb of RAM. Assuming 100 bytes per particle (about 12 double precision values), the system can handle at most 343 million particles not counting memory required for the operating system or the field data. Alternatively, again assuming 100 bytes per mesh node to store field data such as potential, the finest mesh resolution that my computer can support is $700 \times 700 \times 700$. While these limits may be more than sufficient for the majority of simulations, we can easily envision some detailed study requiring more particles or a finer mesh. We then use *distributed computing* to run the simulation on multiple computers interconnected with network cables. The individual machines are known as nodes, and the entire system is called a *cluster* or a *supercomputer*. The nodes communicate by passing messages with the help of the Message Passing Interface (or MPI) library. With distributed computing, the problem size is limited only by the available resources (in other words, budget). Historically, only major research institutions had access to large supercomputers. These days, anybody can generate a cluster on demand with the help of cloud computing providers such as Amazon Web Services (AWS). The ecosystem of available tools is shown schematically in Figure 9.1. We call the division of tasks to different MPI-interconnected machines *coarse grain* parallelization. Work division using multiple cores (or even multiple multi-core CPUs) and graphics cards located within the same machine is known as *fine grain* parallelization.

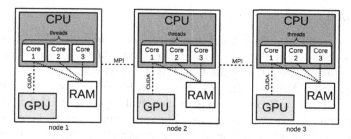

Figure 9.1: Schematic of a typical system architecture.

9.2.1 Multithreading

We now illustrate these different methods with a simple example that computes the sum $c = a + b$ using parallel architectures. We start with multithreading. Historically we had to utilize external libraries such as Boost to implement platform-independent multithreaded applications. Alternatively, *OpenMP* preprocessor directives offered a simple way to parallelize loops. These days, support for threads is included in C++11 natively. To create a thread, we simply instantiate a new object of type **thread**. The first argument to the constructor is the function, or a function-like object (a *functor* or an anonymous lambda function), to be launched in parallel. Additional arguments are passed to the function. These can be used, among other things, to provide a pointer to the memory location to hold the results. A simple example of multithreading is found in **ch9/snippets/add_thread.cpp**.

```cpp
#include <thread>
#include <iostream>
using namespace std;
void add(double a, double b, double *c) {(*c)=a+b;}
int main() {
  double c;
  thread t1(add,1,2,&c);   // run add() as a new thread
  t1.join();               // wait for the thread to finish
  cout<<"c = "<<c<<endl;
  return 0;
}
```

We compile and run it with

```
$ g++ -std=c++11 add_thread.cpp -o add_thread -lpthread
$ ./add_thread
$ c = 3
```

The **-lpthread** compiler parameter instructs the linker to include the thread library. The benefit of multithreaded applications is that they do not require any additional libraries or hardware architecture. They are executed in a manner identical to the serial counterparts.

9.2.2 MPI

MPI codes on the other hand use functions from an MPI library [32]. The two most popular implementations are MPICH and Open-MPI. These libraries need to be first installed before the code can be compiled and run. Among other things, these libraries provide the <mpi.h> header, and the mpic++ wrapper which calls g++ with the appropriate include and library directories. The program is then launched with mpirun -np N -hostfile hosts.txt ./program ..., where N is the number of processes (instances) to launch and the file hosts.txt lists the names or IP addresses of the systems to use. If the hosts file is omitted, the processes launch locally. Every MPI program must at the minimum call MPI_Init and MPI_Finalize. We also use calls to MPI_World_size and MPI_World_rank to determine the total number of processes, and the identification number (also known as *rank*) of the current instance. mpirun simply launches multiple copies of the same program. It is up to us to implement some logic to speed up the computation. This is often done by implementing *domain decomposition*, in which each process handles only a fraction of the simulation domain. By convention, the process with rank zero is known as the *root*. We let the root be in charge of distributing workload and performing output. Data is communicated using MPI_Send, MPI_Recv, MPI_Sendrecv and similar functions. The MPI library also supports data reduction, such as summing local values, with MPI_Reduce and MPI_Allreduce functions. The second variant provides the result to all nodes, while only the root gets the result with the first version. This list is far from exclusive, but it does represent the most commonly used MPI functions. The use of MPI is demonstrated in add_mpi.cpp

```cpp
#include <mpi.h>
#include <iostream>
using namespace std;

int main(int n_args, char *args[]) {
  MPI_Init(&n_args, &args);
  int mpi_size;        // number of processes
  int mpi_rank;             // my id
  MPI_Comm_size(MPI_COMM_WORLD, &mpi_size);
  MPI_Comm_rank(MPI_COMM_WORLD, &mpi_rank);
  cout<<"I am "<<mpi_rank<<" of "<<mpi_size<<endl;

  // root sends data to the worker and prints the result
  if (mpi_rank == 0) {
    double buf[2] = {1,2};
    // send an array of two doubles to rank 1 with tag set to 42
    MPI_Send(buf, 2, MPI_DOUBLE, 1, 42, MPI_COMM_WORLD);
    double c;        // receive a value from rank 1 and store in c
    MPI_Recv(&c, 1, MPI_DOUBLE, 1, 42, MPI_COMM_WORLD,
      MPI_STATUS_IGNORE);
    cout<<"c = "<<c<<endl; // show the data
  }
  // worker receives data from root, returns the result
```

```
else if (mpi_rank == 1) {
  double buf[2];
  // receive two doubles from rank 0 and store in buf
  MPI_Recv(buf, 2, MPI_DOUBLE, 0, 42, MPI_COMM_WORLD,
    MPI_STATUS_IGNORE);
  double c = buf[0] + buf[1]; // perform operation
  // send the result to rank 0
  MPI_Send(&c, 1, MPI_DOUBLE, 0, 42, MPI_COMM_WORLD);
}
MPI_Finalize();
  return 0;
}
```

The code starts by determining the total number of processes and the current rank. These operations are specific to a *communicator*. MPI_COMM_WORLD is the top level communicator that includes all processes spawned by mpi_run. The size variable is identical on each process, but rank varies from 0 to size-1. The process with rank zero then *packs* data to a buffer and sends it to rank 1. The value of 42 is an arbitrary tag. It is used to distinguish between different data packets having the same sender and receiver pair. MPI_Send and MPI_Receive are examples of *blocking* commands that suspend the sender and the receiver until the data is transferred. It is also possible to send data in a non-blocking manner. The root rank waits until the result data is received from the worker. The data is then printed to the screen. Concurrently, the thread with rank 1 starts by waiting for data from the root. The addition is then performed and the data is sent back. Any other ranks, in case the program is launched with more than two processes, simply exit after printing their initial greeting. A possible output is shown below

```
$ mpic++ add_mpi.cpp -o add_mpi
$ mpirun -np 5 ./add_mpi
I am 0 of 5
I am 1 of 5
I am 2 of 5
I am 3 of 5
c = 3
I am 4 of 5
```

While MPI codes can be run on a single machine, they introduce an overhead by needing to pack and ship data among processors. On the other hand, with domain decomposition, we can effectively parallelize every aspect of the simulation. This may not always be desired, since as discussed below, the field solver and particle push algorithms may benefit from different parallelization strategies. MPI codes also require that the MPI library is installed on the system.

9.2.3 CUDA

Finally, we demonstrate how to perform this addition on an NVIDIA graphics card using CUDA. CUDA is a language extension to C++ that introduces some new syntax for the GPU (also known as *device*) code. CUDA programs are typically given a `.cu` extension. They are compiled with `nvcc` which is provided by installing the CUDA Toolkit from the NVIDIA website. `nvcc` is a preprocessor that parses out the GPU algorithms. The rest of the CPU *host* code is compiled with `gcc`. You may want to review references [49] and [58] for a more detailed description. The CUDA version of the add code is found in `add_cuda.cu`,

```cpp
#include <cuda_runtime.h>
#include <iostream>
using namespace std;

// device "kernel" code to run on the GPU
__global__ void add(float a, float b, float *c) {
    *c = a + b;       // code executed on the GPU
}

int main(int n_args, char *args[]) {
    float *dev_c;
    cudaMalloc((void**)&dev_c, sizeof(float));
    add<<<1,1>>>(1,2,dev_c);   // run add kernel on the GPU

    float c;
    cudaMemcpy(&c, dev_c, sizeof(float), cudaMemcpyDeviceToHost);

    cout<<"c = "<<c<<endl;
    return 0;
}
```

The above code is compiled and run as

```
$ nvcc -O2 add_cuda.cu -o add_cuda
$ ./add_cuda
$ c = 3
```

The `__global__` keyword makes `add` a *kernel* function. This is the code for each single thread that runs on the GPU. Kernels are launched from the CPU with the `<<<num_blocks, threads_per_block >>>` syntax. The two parameters correspond to the number of blocks and threads per block. These values can be scalar, or can be a `dim3` object describing a three dimensional grid. The total number of threads is given by cross-multiplying these parameters. Each kernel function has access to built-in variables `threadIdx`, `blockIdx`, and `blockDim` that are used to determine the placement of the current thread in the global system. As part of the kernel call, we provide the `add` function a pointer to the memory location for the result. The memory addressed by the pointer corresponds to a memory block on the GPU - attempting to pass in a CPU pointer would cause an error. This memory is allocated using cudaMalloc. The result is transferred to the CPU host with cudaMemcpy. This memory copy is a major bottleneck in CUDA simulations.

9.3 CAVEATS

Code parallelization is not trivial. Simply throwing more computers at a problem does not necessarily lead to a faster performance. Furthermore, if we are not careful, it is possible to introduce errors that produce incorrect results. Therefore, we need to be aware of some caveats.

9.3.1 Race Condition

All CPU or GPU threads have access to the same memory space. While this is advantageous when reading data, it can lead to errors if multiple threads attempt to *write* to the same memory space. Consider the following example

```
int main() {
    for (int k=1; k<=10000; k++) {   // run 10,000 times
        int i=0;
        auto f = [&] {++i;};          // lambda function to increment i
        thread t1(f);                 // create two threads running f
        thread t2(f);
        t1.join(); t2.join();         // wait for threads to finish
        cout<<(i==2?'.':'*');         // output . if i=2, * otherwise
        if (k%100==0) cout<<endl;
    }
    return 0;
}
```

This code creates two threads that increment the value of a local integer variable i. While we may expect the final value to be 2, this is not always the case. An increment consists of three steps. First, the current value is retrieved from system memory to a processor register. The register value is then incremented. Finally, the new value is written back to RAM. The expected sequence of operations is

```
i = 0
t1 reads i into register A, A = 0
t1 increments A, sets i = A = 1
t2 reads i into register B, B = 1
t2 increments B, sets i = B = 2
```

Since the two threads run concurrently, the following sequence is also possible:

```
i = 0;
t1 reads i into register A, A = 0
t2 reads i into register B, B = 0
t1 increments A, sets i = A = 1
t2 increments B, sets i = B = 1
```

To demonstrate that this in fact happens, the code is wrapped in a loop running for 10,000 iterations. If $i = 2$, we print a dot, otherwise, we print an asterisk. While the output consists mainly of dots, we do get the occasional asterisk. This behavior is known as *race condition*. The name implies that results can change depending on which thread reaches some code point first.

9.3.2 Locks

Therefore when writing multithreaded code, be it on a CPU or a GPU, we need to carefully review all operations that modify variables to assure that each thread writes to a unique location. Alternatively, we can serialize critical code with the help of *locks*. On the CPU side, serialization is performed by locking a *mutex*,

```
#include <mutex>
std::mutex mtx;
int i = 0;   // global data to modify

void f() {
  mtx.lock();
  i++;
  mtx.unlock();
}
```

Any thread that attempts to lock an already locked mutex is suspended until the mutex unlocks. Instead of locking and unlocking mutexes manually, we can wrap them in a lock object, such as `unique_lock`. This object automatically unlocks when it goes out of scope. Locks should be used sparingly as they lead to suboptimal performance. An alternate to locks are *atomic* variables. These variables implement operations such as increment or addition in a thread safe fashion. Atomics may be provided directly by hardware, or may be implemented in software by effectively placing a lock around the read, increment, and set operations.

Even if we are not explicitly using locks, we need to be aware that they may be found in built-in library functions. A good example of this is the random number generator. The generator needs to update its sequence index after each value is sampled, and hence requires the use of locks or atomics. It is imperative that each thread has access to its own generator. This can be easily achieved if threads are encapsulated within a functor class. Otherwise, using a global generator, only a single thread can sample a random number at a time. The remaining threads are suspended. The example code in `thread_rnd.cpp` defines a Worker class that, based on a flag, uses either a global or a local generator,

```
Rnd rnd_glob(0);   // global generator

class Worker {
public:
  Worker(int id):rnd{id} {}
  void operator() (int n_samples, bool use_global) {
```

```
  double sum = 0;
  if (use_global)
    for (int i=0;i<n_samples;i++) sum += rnd_glob();
  else
    for (int i=0;i<n_samples;i++) sum += rnd();
    ave = sum/n_samples;
  }
  double ave;
  Rnd rnd;
};
```

The main function launches 12 threads, with each computing the average of 10,000 random numbers. With the single global generator, the code takes 29 ms to finish. Switching to the thread-local versions, the run time decreases to 4.2 ms, or a seven fold decrease in run time.

9.3.3 Deadlock

MPI code utilizing blocking communication commands needs to make sure that for each send command, there is a corresponding receive. Otherwise, the code will freeze. This condition is known as *deadlock*. Consider the following code from `mpi_deadlock.cpp`

```
int main(int n_args, char *args[]) {
MPI_Init(&n_args, &args);
int my_rank;
MPI_Comm_rank(MPI_COMM_WORLD,&my_rank);

int other_rank;
if (my_rank == 0) {
  MPI_Ssend(&my_rank, 1, MPI_INT, 1, 42, MPI_COMM_WORLD);
  MPI_Recv(&other_rank, 1, MPI_INT, 1, 42, MPI_COMM_WORLD,
    MPI_STATUS_IGNORE);
}
else if (my_rank == 1) {
  MPI_Ssend(&my_rank, 1, MPI_INT, 0, 42, MPI_COMM_WORLD);
  MPI_Recv(&other_rank, 1, MPI_INT, 0, 42, MPI_COMM_WORLD,
    MPI_STATUS_IGNORE);
}
cout<<"my_rank = "<<my_rank<<", other_rank = "<<other_rank<<endl;
MPI_Finalize();
return 0;
}
```

This code, when run with two processors, is supposed to send the local rank to the other process and receive that other process' rank back. However, the code hangs. The `MPI_Send` function assures that the data is not modified locally before being picked up by the receiver. In most cases, this implies that the function does not return until the data is transmitted. This is the *block*. However, some MPI implementations use a buffer to store a copy, and may, or may not, block, depending on whether the data fits into the buffer. The `MPI_Ssend` variant used above does not use buffering. It is used to demonstrate the *deadlock* regardless of the data size. In the above implementation, each rank enters the send command, and waits for the other process to receive it. This is however not possible since the receiver is suspended in its own `MPI_Send` call. Therefore, we need to reverse the order of operations on one of the ranks,

```
else if (my_rank == 1) {
    MPI_Recv(&other_rank, 1, MPI_INT, 0, 42, MPI_COMM_WORLD,
        MPI_STATUS_IGNORE);
    MPI_Ssend(&my_rank, 1, MPI_INT, 0, 42, MPI_COMM_WORLD);
}
```

This version now runs as expected. Since these types of transfers are quite common, and prone to deadlocks, MPI includes a `MPI_Sendrecv` function that sends and receives data using a single function call.

9.3.4 Overhead

Finally, use of each of the three considered technologies involves some overhead. Thread creation is computationally expensive. The code in `thread_timing.cpp` computes the time to spawn and join threads by comparing times needed to push a large array of particles directly, and using a single thread at a time. On my system, the code yields

```
Time to push one particle: 0.00192073 (us)
Time to create a thread: 11.8433 (us)
```

In other words, over 6,000 particles can be moved in the time required for the operating system to spawn a new thread. While it may be tempting to implement multihreading everywhere, it is really practical only for large problems.

Different scenario is encountered in GPU processing. Thread creation on a GPU is fast, but memory transfer from and to the host RAM is not. Consider the following code found in `cuda_timing.cu` This code attempts to speed up the particle push by performing it on the CPU.

```
struct Particle {
    double pos[3] = {0,0,0}; // note, GPU math is faster with floats
    double vel[3] = {0,0,0};
};

// kernel code to run on the GPU
__global__ void gpu_push(Particle *particles, double dt, size_t N)
{
    int p = blockIdx.x*blockDim.x + threadIdx.x;
    if (p<N) {
        Particle &part = particles[p];
        for (int i=0;i<3;i++)
            part.pos[i] += part.vel[i]*dt;
    }
}

// code to push a single particle
void push(Particle *part, double dt) {
    for (int i=0;i<3;i++) part->pos[i] += part->vel[i]*dt;
```

```
}

int main(int n_args, char *args[]) {
    cudaFree(0);   // reset CUDA context
    size_t num_particles = 1000000;
    Particle *particles = new Particle[num_particles];

    // set some initial values
    for (size_t i=0;i<num_particles;i++)
        particles[i].vel[0]=1/(double)num_particles;

    const double dt = 0.1;
    // *** CPU particle push ***
    auto start_cpu = chrono::system_clock::now();
    for (size_t i=0;i<num_particles;i++) push(&particles[i],dt);
    auto end_cpu = chrono::system_clock::now();

    // *** GPU particle push ***
    auto start_gpu = chrono::system_clock::now();
    Particle *devParticles;
    cudaMalloc((void**)&devParticles,
        sizeof(Particle)*num_particles);
    cudaMemcpy(devParticles, particles, sizeof(Particle)*num_particles,
        cudaMemcpyHostToDevice);

    const int threads_per_block = 1024;
    int num_blocks = (num_particles-1)/threads_per_block + 1;
    cout<<"Creating "<<num_blocks*threads_per_block<<"
        threads"<<endl;
    gpu_push<<<num_blocks,threads_per_block>>>(devParticles, dt,
        num_particles);
    cudaMemcpy(particles, devParticles, sizeof(Particle)*num_particles,
        cudaMemcpyDeviceToHost);

    auto end_gpu = chrono::system_clock::now();

    // output timing info
    std::chrono::duration<double,std::nano> elapsed_cpu = end_cpu -
        start_cpu;
    std::chrono::duration<double,std::nano> elapsed_gpu = end_gpu -
        start_gpu;
    cout<<"Time per particle on CPU:
        "<<elapsed_cpu.count()/num_particles<<" (ns)"<<endl;
    cout<<"Time per particle on GPU:
        "<<elapsed_gpu.count()/num_particles<<" (ns)"<<endl;

    delete[] particles;
    return 0;
}
```

CUDA threads are launched in blocks, with the maximum number of threads per block set by hardware to 1024. The

```
int num_blocks = (num_particles-1)/threads_per_block + 1;
```

line computes the minimum number of blocks needed for `num_particles` particles. For 1024 particles, this algorithm yields 1 block, as desired. For n_particles=1025, we obtain two blocks, with the second block containing only a single valid entry. Each GPU thread computes its global index from $p = i_{block} n_{blocks} + i_{thread}$, where n_{block} is the number of threads per block. The thread performs its computation only if the global thread index is within bounds, $p < N$.

Running the code on my system, I obtain

```
$ nvcc -O2 cuda_timing.cu -o cuda_timing
$ ./cuda_timing
$ Time per particle on CPU: 3.15658 (ns)
$ Time per particle on GPU: 10.5427 (ns)
```

This is quite pitiful. The GPU version took over four times longer than the CPU one! Running the code through NVIDIA's Visual Profiler, we determine that the kernel code took 1.4 ms, while the memcpy operation to and from the GPU took 4.3 and 4.4 ms, respectively. We spent six times as much time moving data between the GPU and the CPU than actually operating on it. We can obtain some improvement by using *pinned* memory. The memory on the CPU host side is divided into multiple *pages*. These pages allow the operating system to move data around as needed, and possibly to even swap it to a hard drive. To avoid corruption during data transfer, the cudaMemcpy function sends data only using pinned memory. Any data not residing in pinned memory needs to be copied there first. We can eliminate this additional copy by allocating any host data that is being transferred to or from the GPU as pinned. This is done by replacing the call to

```
Particle *particles = new Particle[num_particles];
```

with

```
Particle *particles;   // allocate pinned memory
cudaHostAlloc(&particles, sizeof(Particle)*num_particles,
   cudaHostAllocDefault);
```

We also replace delete[] particles; with cudaFreeHost(particles);. With these changes, the run time reduces to

```
Time per particle on CPU: 3.16183 (ns)
Time per particle on GPU: 9.23862 (ns)
```

While some improvement is apparent, the GPU version is still slower than the CPU one. Historically, GPUs supported only single precision values. Support for double precision is now ubiquitous, but there is a noticeable difference in performance. By modifying the Particle data structure to store single precision floats,

```
struct Particle {
   float pos[3] = {0,0,0};
   float vel[3] = {0,0,0};
};
```

and also changing the type of dt, we obtain

```
Time per particle on CPU: 2.47104 (ns)
Time per particle on GPU: 4.87881 (ns)
```

The GPU version is still slower than the serial CPU one, but the ratio has decreased to less than a factor of two. While some additional optimization could be performed, this example demonstrates a crucial downside of using GPUs for general computing: the overhead of moving data between the CPU and the GPU can often negate any benefit from the faster computation. Therefore, one possibility is to port the entire code to the GPU. While this approach eliminates the overhead associated with data transfer, it also results in the powerful CPU sitting idle.

9.4 PROFILING

Prior to diving into code parallelization, we should perform *profiling* of the serial code. A profiler is a tool that measures how much time is spent in each function, among other things. It allows us to identify algorithms that take the longest amount of time. On Linux systems, we can use command line tools such as perf or gprof. In order to use the latter, we compile the code with a -pg flag to include the appropriate *instrumentation*. We should also compile with the optimizer turned on (-O2) and debugging symbols (-g) included so the output is meaningful,

```
$ g++ -O2 -g -pg *.cpp -o sphere
$ ./sphere
$ gprof ./sphere > prof.txt
```

A program compiled with the -pg flag creates a binary file called gmon.out when running. The last line uses the gprof command to interpret the telemetry from this file and export it in a human readable form in prof.txt. Integrated development environments such as Eclipse or Netbeans contain a graphical interface that simplifies the investigation of the produced output.

To illustrate the benefit of profiling, Figure 9.2 shows measurements from the code in Chapter 3. The two plots compare the case with the "poor man's" non-linear Gauss-Seidel (GS) and with the quasineutral (QN) solver. We can see that in the first case, the vast majority of time is spent in the potential solve. Before attempting parallelization, we should review the code to determine if some serial *optimization* is warranted. This may involve implementing a different algorithm. We have in fact already done this, as we saw that switching to the Preconditioned Conjugate Gradient (PCG) solver decreased the run time almost five times. But for this example, let's assume that the code is already optimized. The profiling study indicates that we should focus our effort on parallelizing the potential solver as this is the part of the code where the processor spends the most time.

On the other hand, let's say that we do not care about resolving the

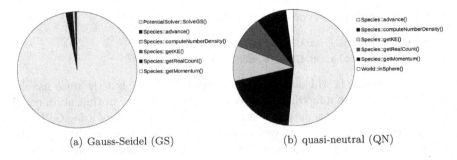

(a) Gauss-Seidel (GS)   (b) quasi-neutral (QN)

Figure 9.2: Profiling results with the Gauss-Seidel (a) and quasi-neutral (b) potential solver showing the six slowest functions.

plasma sheath, and the quasi-neutral approximation is sufficient. Then, as shown in Figure 9.2(b), there is no point in speeding up the solver, as it now takes a negligible amount of time. The `PotentialSolver::solveQN` function does not even show up in the list of the top six slowest functions. It is now the particle push that takes the most time, followed by the code to compute number density. Therefore, for a QN simulation, we should focus our effort on speeding up the particle pusher.

9.5 PARALLELIZATION STRATEGIES

The parallelization strategy depends both on the hardware technologies at our disposal, and on which parts of the code need to be accelerated. We may need to use different strategies when writing code for a massive cluster than when developing an application to be run on a single workstation. In order to write an efficient parallel code, it is crucial that each compute unit is performing approximately the same amount of work. This is known as *load balancing*. It does not make sense to utilize a twenty node cluster, if most of the work is done by a single machine. The other nineteen computers are sitting mainly idle, and despite utilizing twenty-times the total computational power, we do not obtain results any faster than if the code ran serially.

9.5.1 Domain Decomposition

Obtaining good load balancing in Particle in Cell codes is not trivial. Assume that the particle push and potential solve both take approximately the same amount of time. One common parallelization strategy is to perform *domain decomposition*, which involves splitting the computational domain into smaller chunks. We then assign each chunk to a different MPI process. In order to optimize the field solve, we need each domain to contain the same number of unknowns. Ignoring the local reduction due to internal Dirichlet boundaries, the optimal solver decomposition involves splitting the domain into equal-sized chunks. Such a decomposition may be far from optimal for the particle push. This is especially true in plume simulations, in which the vast majority of particles may be localized within a relatively narrow region. We can see this even in our flow around a sphere case. Imagine that the domain is extended in z to simulate the wake behind the sphere. The concentration of particles is the highest upstream of the sphere, since many ions are lost to surface neutralization, while others just fly out of the domain. With uniform sized domain decomposition, the processors assigned to the upstream region do more work than those downstream of it. Domain decomposition also introduces some overhead. The potential solver requires the use of *ghost cells* to capture data from the neighbor processor. Particles crossing domain boundaries need to be *packed* and transferred across processor boundaries.

9.5.2 Ensemble Averaging

The unique benefit of domain decomposition is that it allows us to run simulations that would be too large for a single machine. But if this is not required, we can obtain improved load balancing by utilizing different decomposition strategies for the particles and the solver. The particle population is divided evenly among different processors. Each node contains the full domain. Particle densities and other macroscopic properties are *ensemble averaged* across the processors. Domain decomposition, or a similar parallel matrix algorithm, is utilized for the potential solver. Since each processor solves potential only on a subset of the domain, the local solutions need to be assembled. The overhead introduced by this communication is generally comparable to the amount of data needed to be transferred when moving particles between domains.

9.5.3 GPU Concurrent Computing

We can reduce the impact of the GPU memory transfer overhead by using the GPU for concurrent computation. Typically, the CPU is sitting idle while the GPU computation proceeds. Depending on the problem setup, we may be able to assign a subset of the domain to the GPU with the CPU computing the rest. Alternatively, we can have the CPU compute macroscopic properties using the "stale" particle data from a prior time step while the GPU is computing the new particle positions and velocities. Given that these properties are used mainly for visualization, and that we average results over many time steps, the use of the old data has negligible impact on the simulation results.

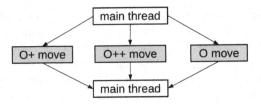

Figure 9.3: One option for implementing multithreading in multi-species simulations.

9.6 EXAMPLES

9.6.1 Multithreaded Particle Push

We now demonstrate different strategies by parallellizing various parts of the flow past the sphere code. We start by adding multithreading to the particle mover. Consider the simulation program from Chapter 3 with three particle species,

```
species.push_back(Species("O", 16*AMU, 0, 1e6, world));
species.push_back(Species("O+", 16*AMU, QE, 5e1, world));
species.push_back(Species("O++", 16*AMU, 2*QE, 5e1, world));
sources.emplace_back(species[0],world,7000,nda);    // neutrals
sources.emplace_back(species[1],world,7000,0.8*ndi); // ions
sources.emplace_back(species[2],world,7000,0.1*ndi); // doubles
```

One way to parallelize the push is to assign each species to a different thread. The push then proceeds as shown in Figure 9.3. While this approach is simple to implement, it has two disadvantages. First, since a single thread is used to move entire species, we can at most utilize only as many threads as there are species. This simulation contains three species, but my processor has six cores. Therefore, this approach utilizes at most half the available resources of my system. Load balancing is also poor. At steady state, the system contains approximately 500,000 single ions, 48,000 double ions, and 100,000 neutrals. Assuming the same computational effort for each species (not quite true since there is no force update for neutrals), the thread pushing the singly charged ions is doing majority of the work. Therefore, it is better to multi-thread the actual `Species::move()` function.

To do this, we replace the single loop over particles in `Species::advance`

```
// loop over all particles
for (Particle &part: particles) {
  double3 lc = world.XtoL(part.pos);
  /* ... */
}
```

with code to generate threads,

```
// calculate number of particles per thread
size_t np = particles.size();
int n_threads = world.getNumThreads();
```

```
size_t np_per_thread = np/n_threads;
vector<thread> threads;
for (int i=0;i<n_threads;i++) {
  size_t p_start = i*np_per_thread;
  size_t p_end = p_start + np_per_thread;
  if (i==n_threads-1) p_end = np;  // reduce block on last thread
  threads.emplace_back(advanceKernel, p_start, p_end,
    std::ref(world), std::ref(particles), charge/mass, dt);
}
// wait for threads to finish
for (thread &t: threads) t.join();
```

We start by calculating the number of particles to assign to each thread by dividing the total count by the number of threads to use. The number of particles may not be evenly divisible, and thus we reset the upper bound for the last thread. We then create the threads, with each launching `advanceKernel` function with arguments that include, among others, the index of the starting and the ending particle to process. We also pass in references to the world object and the particle list. This approach is needed since threads cannot be non-static member functions. The threads are added into a `vector`. We then call `join` on each thread to wait for it to finish. The `advanceKernel` function implements the code that was previously inside `advance`, with the exception that only particles in $[p_{start}, p_{end})$ are considered,

```
void advanceKernel(size_t p_start, size_t p_end, World &world,
  vector<Particle> &particles, double qm, double dt) {
  // loop over particles in [p_start,p_end)
  for (size_t p = p_start; p<p_end; p++) {
    Particle &part = particles[p];
    /* ... */
  }
}
```

9.6.1.1 Number of Threads

The number of threads to use is a user parameter. We default to `thread::hardware_concurrency`, which returns the maximum number of threads the system can support concurrently. Due to *hyperthreading*, this value is typically double the number of actual hardware cores. We also check for a command line argument in `main`, and if found, use it to update the count

```
int num_threads = thread::hardware_concurrency();
if (argc>1) num_threads = atoi(args[1]);  // update with command
  line arg if provided
cout<<"Running with "<<num_threads<<" threads"<<endl;
world.setNumThreads(num_threads);
```

9.6.1.2 Density Calculation

We would also like to parallelize the number density computation. This task is not as trivial as it leads to the race condition discussed in Section 9.3.1 if we let multiple threads scatter data in the same cell. To avoid it, we allocate a temporary buffer for each thread. These local densities are then added together into the global field. We add these buffers to World with the help of the setNumThreads functions called by main

```
void setNumThreads(int num_threads) {
  this->num_threads = num_threads;
  buffers = std::vector<Field>();
  for (int i=0;i<num_threads;i++)
    buffers.emplace_back(ni,nj,nk);
}
// temporary buffers for density (or other) calculation
std::vector<Field> buffers;
```

Then just as before, we modify Species::computeNumberDensity to divide the particle list into several equal-sized chunks. The density calculation is moved to function computeDensityKernel.

```
void computeDensityKernel(int thread_index, size_t p_start, size_t
  p_end, World &world, vector<Particle> &particles) {
  Field &den = world.buffers[thread_index];
  den.clear();

  for (size_t p = p_start; p<p_end; p++) {
    Particle &part = particles[p];
    double3 lc = world.XtoL(part.pos);
    den.scatter(lc, part.mpw);
  }
}
```

This code is almost identical to what we have in the serial version, except that the den variable references one of the temporary buffers. We then loop over the subset of particles. The buffers are added together once all threads finish running. Check for a serial case is included to avoid the unnecessary copy when running with only a single thread.

```
void Species::computeNumberDensity() {
  size_t np = particles.size();
  int n_threads = world.getNumThreads();
  if (n_threads == 1) {computeNumberDensitySerial();return;}

  size_t np_per_thread = np/n_threads;
  vector<thread> threads;
  for (int i=0;i<n_threads;i++) {
    size_t p_start = i*np_per_thread;
    size_t p_end = p_start + np_per_thread;
    if (i==n_threads-1) p_end = np;  // reduce block on last thread
    threads.emplace_back(computeDensityKernel, i, p_start, p_end,
      std::ref(world), std::ref(particles));
  }

  // wait for threads to finish
  for (thread &t: threads) t.join();

  // add up local fields
  den.clear();
```

```
for (int i=0;i<n_threads;i++)
    den += world.buffers[i];  // uses the overloaded + operator

// divide by node volume
    den /= world.node_vol;
}
```

9.6.1.3 Parallel Efficiency

To test this new code, I reduced the macroparticle weight by a factor of ten, and ran the simulation with the quasi-neutral (QN) solver. The code was then run with a varying number of threads, as shown below,

```
ch9/MT$ ./sphere 4
Running with 4 threads
Sphere potential: -100 V
ts: 0    0:2800    0+:44800    0++:5600
ts: 1    0:5600    0+:89600    0++:11200
...
ts: 400    0:993227    0+:10598642    0++:996731
Simulation took 102 seconds
```

The resulting times are recorded and visualized in Figure 9.4. The gray line corresponds to timing with only the particle push parallelized, while the black line also includes the multithreaded density calculation. The dashed lines plot parallel efficiency, computed as

$$\eta_{par,n} = \frac{t_1}{nt_n} \tag{9.1}$$

where t_1 is the time required for the serial run, and t_n is the time with n threads (or processors). We can see that while we are able to reduce the run time by half, we do not see much speedup with more than four cores. The parallel efficiency also is not great. While we achieve 73% efficiency with two threads, it drops to 49.5% with four. It gets even worse with 12 threads, where the efficiency drops to 17.5%. The reason for the low efficiency is two fold. First, my computer contains only six physical cores, with the support for 12 threads arising from *hyperthreading*. Secondly, multithreading has been implemented in only two functions. The rest of the code continues to run in serial. Even if the time needed to push particles and update density is reduced to zero, the remaining functions introduce a finite baseline that is independent of the number of threads.

9.6.2 MPI Domain Decomposition

The alternate paralellization strategy is known as *domain decomposition*. With this scheme, we subdivide the entire computational domain among different

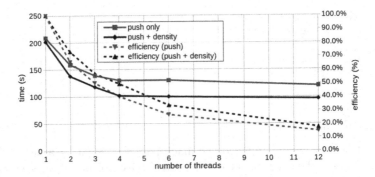

Figure 9.4: Timing study of the multithreaded simulation with the QN solver.

processors. This is the scheme most commonly used with MPI. Each processor now owns only a small subset of the entire domain. Domain decomposition introduces two major changes into our codes. First, the potential solver needs to be modified to take into account the solution from neighbor processors. We also need code to transfer particles between domains. The code described in this section can be found in ch9/MPI.

9.6.2.1 Subdomains

We start by letting the user define the desired number of subdomains in each of the three spatial dimensions, nx, ny, and nz. Since each subdomain is assigned to an individual MPI process, the total number of processes to launch must be $nx \times ny \times nz$. The decomposition is provided by the second command line argument in format ni:nj:nk

```
if (n_args >2) {
    // omitted code to split args[2] into vector<string>pieces
    doms[0] = stoi(pieces[0]);
    doms[1] = stoi(pieces[1]);
    doms[2] = stoi(pieces[2]);
    if (doms[0]*doms[1]*doms[2]!=mpi_size)
        error("Incorrect MPI size");
}
```

We then use some algorithm to convert the MPI rank to the domain index. This is really the "magic" behind MPI codes. The MPI library simply tells us the number of processes and our own rank. It is up to us to decide how to use this information. In our case, we use a consecutive numbering of the domains, with the domain index given by the same relationship used previously to flatten a 3D array to a one dimensional vector in Section 3.6.1,

$$R = k \cdot n_x \cdot n_y + j \cdot n_x + i \tag{9.2}$$

Knowing R and n_x and n_y, we obtain the i, j, and k index per

$$k = R/(n_x n_y) \tag{9.3}$$
$$j = (R - k n_x n_y)/n_x \tag{9.4}$$
$$i = (R - k n_x n_y)\% n_x \tag{9.5}$$

Note that since all values are integers, these calculations are inherently made with the integer math in which the fractional parts are discarded. The actual domain decomposition is then performed using World::initMPIDomain, given below. This function uses the above equation to determine which subdomain we correspond to. It also computes the extents of this domain for each dimension by dividing the number of global nodes by the number of subdomains. The function then resets the mesh origin $\vec{x}_0$ and node counts accordingly.

```
void World::initMPIDomain(int mpi_rank, int mpi_size, int doms[3])
{
  this->mpi_rank = mpi_rank;    // set member functions
  this->mpi_size = mpi_size;
  mpi_size_i = doms[0];
  mpi_size_j = doms[1];
  mpi_size_k = doms[2];

  // set our position in the MPI world
  mpi_k = mpi_rank/(mpi_size_i*mpi_size_j);
  int mod = mpi_rank-mpi_k*(mpi_size_i*mpi_size_j);
  mpi_j = mod/mpi_size_i;
  mpi_i = mod%mpi_size_i;

  num_nodes_global = ni*nj*nk;   // save the global node count
  int ni_global = ni;
  int nj_global = nj;
  int nk_global = nk;

  // number of nodes per process
  int proc_ni = ni/mpi_size_i;
  int proc_nj = nj/mpi_size_j;
  int proc_nk = nk/mpi_size_k;

  // set the range we are simulating
  int my_i1 = proc_ni*mpi_i;
  int my_i2 = my_i1 + proc_ni+1;
  if (my_i2>ni_global) my_i2 = ni_global;

  int my_j1 = proc_nj*mpi_j;
  int my_j2 = my_j1 + proc_nj+1;
  if (my_j2>nj_global) my_j2 = nj_global;

  int my_k1 = proc_nk*mpi_k;
  int my_k2 = my_k1 + proc_nk+1;
  if (my_k2>nk_global) my_k2 = nk_global;

  // set origin and max corner of our domain
  x0[0] = x0[0] + my_i1*dh[0];
```

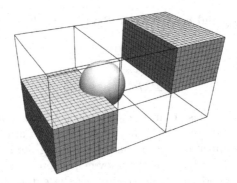

Figure 9.5: Example of a $2 \times 1 \times 2$ domain decomposition.

```
x0[1] = x0[1] + my_j1*dh[1];
x0[2] = x0[2] + my_k1*dh[2];
xd[0] = x0[0] + (my_i2−my_i1−1)*dh[0];
xd[1] = x0[1] + (my_j2−my_j1−1)*dh[1];
xd[2] = x0[2] + (my_k2−my_k1−1)*dh[2];

// save number of nodes*/
ni = my_i2−my_i1;
nj = my_j2−my_j1;
nk = my_k2−my_k1;
}
```

We next modify the particle source to sample particles only if mpi_k=0, in other words, if our domain is one of the domains along the global inlet face,

```
void ColdBeamSource::sample() {
  // only sample on global k=0 face
  if (world.mpi_k!=0) return;
}
```

Finally, the output subroutines need to be changed. Historically, only the root process with rank of zero would write out to the file system. This was especially useful when running on a cluster without a shared file system. These days, it is more common to have each rank write out its data, and then use VTK's *parallel file format* or Paraview's *collection* to group them into a single data set. We modify Output::fields to save to a file following the results_ts_rank.vti naming convention. Code is also added to generate a .pvd Paraview collection file. For brevity, this code is not included here. Figure 9.5 then visualizes an example $2 \times 1 \times 2$ domain decomposition.

9.6.2.2 Particle Transfer

Previously, any particle leaving the computational domain was removed from the simulation. But now, many domains have another domain as their neighbor. We thus need to check for particles crossing these *processor boundaries*.

These particles need to be transferred to the neighbor processor. This code is added to `Species::advanceKernel`.

```
if (part.mpw>=0 && !world.inBounds(part.pos)) {   // out of bound
  // check if particle crossed processor boundary
  if (world.mpi_i>0 && lc[0]<0)
    transfer[0].emplace_back(&part);
  else if (world.mpi_i<world.mpi_size_i-1 && lc[0]>=world.ni-1)
    transfer[1].emplace_back(&part);
  else if (world.mpi_j>0 && lc[1]<0)
    transfer[2].emplace_back(&part);
  else if (world.mpi_j<world.mpi_size_j-1 && lc[1]>=world.nj-1)
    transfer[3].emplace_back(&part);
  else if (world.mpi_k>0 && lc[2]<0)
    transfer[4].emplace_back(&part);
  else if (world.mpi_k<world.mpi_size_k-1 && lc[2]>=world.nk-1)
    transfer[5].emplace_back(&part);

  // kill particles leaving the domain (includes MPI boundary)
  part.mpw = 0;
}
```

Here `transfer` is a six-component array of `vector<Particle>` vectors. The six indexes correspond to the faces of a cube, with index 0 corresponding to x^-. Index 1 is x^+, index 2 is y^-, and so on. The above logic is placed after the check for particles hitting the sphere. If the particle left the computational domain, we check if the exit face is *processor boundary*. If so, we copy the particle into the appropriate `transfer` array. The particle is then killed as it no longer belongs in our domain. The actual shipment of data occurs in `Species::transferParticles`. This function iterates over the six faces, and if that face has a neighbor MPI domain, we communicate with it the number of particles to send. We similarly obtain the number of particles to receive from that face. We then allocate temporary *buffers*. Advanced MPI functions allow us to define custom data types, but for simplicity, here we limit ourselves to arrays of basic types. The information stored in the `Particle` struct is serialized into two buffers, with one holding floating point values and the other used for integers. This step is known as *packing*. Let's say the particle data structure consists of

```
struct Particle {
  double3 x;
  double3 v;
  double mpw;
  int id;
}
```

Our buffers then look like:

```
dbuffer = [x0,y0,z0,u0,v0,w0,mpw0,x1,y1,z1,...]
ibuffer = [id0,id1,id2,...]
```

The packed data is transferred with the help of `MPISendrecv`. The received data is then *unpacked*. The entire algorithm is included below

```
void Species :: transferParticles () {
  const int TAG_COUNT = 20;
  const int TAG_DATA = 22;

  for (int face=0;face<6;face++) {
    int target = world.getNeighborRank(face);

    // transfer if we have a neighbor
    if (target>=0) {
      // first send particle counts
      int send_size = transfer[face].size();
      int recv_size = 0;

      MPI_Sendrecv(&send_size, 1, MPI_INT, target, TAG_COUNT,
        &recv_size, 1, MPI_INT, target, TAG_COUNT, MPI_COMM_WORLD,
        MPI_STATUS_IGNORE);

      // allocate memory buffers for sending and receiving data
      double *send_dbuffer=nullptr, *recv_dbuffer=nullptr;
      int *send_ibuffer=nullptr, *recv_ibuffer=nullptr;

      // since we can't allocate zero memory
      if (send_size>0) {
        send_dbuffer = new double[send_size*7];  // 7 doubles each
        send_ibuffer = new int[send_size*1];  // 1 int each
      }
      if (recv_size>0) {
        recv_dbuffer = new double[recv_size*7];
        recv_ibuffer = new int[recv_size*1];
      }

      // pack particles to a buffer
      auto it = transfer[face].begin();
      for (int i = 0;i<send_size;i++,it++) {
        Particle &part = **it;  // *it is a pointer
        send_dbuffer[6*i+0] = part.pos[0];
        send_dbuffer[6*i+1] = part.pos[1];
        send_dbuffer[6*i+2] = part.pos[2];
        send_dbuffer[6*i+3] = part.vel[0];
        send_dbuffer[6*i+4] = part.vel[1];
        send_dbuffer[6*i+5] = part.vel[2];
        send_dbuffer[6*i+6] = part.mpw;
        send_ibuffer[i] = part.id;
      }
      // transfer particles
      MPI_Sendrecv(send_dbuffer, send_size*7, MPI_DOUBLE, target,
        TAG_DATA, recv_dbuffer, recv_size*7, MPI_DOUBLE, target,
        TAG_DATA, MPI_COMM_WORLD, MPI_STATUS_IGNORE);
      MPI_Sendrecv(send_ibuffer, send_size*1, MPI_INT, target,
        TAG_DATA, recv_ibuffer, recv_size*1, MPI_INT, target,
        TAG_DATA, MPI_COMM_WORLD, MPI_STATUS_IGNORE);

      // append particles from the buffer to our list
      for (int i = 0;i<recv_size;i++) {
        double3 pos, vel;

        pos[0] = recv_dbuffer[6*i+0];
```

```
pos [1] = recv_dbuffer [6*i+1];
pos [2] = recv_dbuffer [6*i+2];
vel [0] = recv_dbuffer [6*i+3];
vel [1] = recv_dbuffer [6*i+4];
vel [2] = recv_dbuffer [6*i+5];
double mpw = recv_dbuffer [6*i+6];
int id = recv_ibuffer [i];

// only add the particle if it belongs to us
if (world.inDomain(pos))
    addParticle(pos, vel, mpw, id);
else {
    // do nothing, should add to a new transfer array
}
}

// memory cleanup
if (send_size >0) {
    delete [] send_dbuffer;
    delete [] send_ibuffer;
}
if (recv_size >0) {
    delete [] recv_dbuffer;
    delete [] recv_ibuffer;
}
}

transfer [face].clear(); // remove data
} // for face
}
```

As can be seen, most of the code involves packing and unpacking data. For each received particle, we first need to make sure it truly belongs to us. It is possible for a particle to jump across multiple domains when leaving near corners or edges. The proper algorithm is to determine the new exit face, and add it to that face's **transfer** array, if that face corresponds to a processor boundary. The above loop over faces should then repeat until all particles find a new home. The code as implemented above simply discards received particles that are out of domain, and hence introduces a non-physical mass sink. The **getNeighborRank** function uses **World::rankFromOffset** to evaluate Equation 9.2 per

```
int rankFromOffset(int di, int dj, int dk) {
    return (mpi_k+dk)*(mpi_size_j*mpi_size_i) +
        (mpi_j+dj)*mpi_size_i+(mpi_i+di));
}
```

9.6.2.3 Boundary Update

We also need to update domain boundaries. Consider the computation of number density on boundary nodes. Each processor adds in only its own contribution without knowing there are particles in the adjacent cell belonging to a different processor. We therefore need to sum up mesh fields along domain boundaries. This is done using the `Field::updateBoundaries` function,

```
void updateBoundaries() {
  const int TAG_DATA = 20;
  int ni=world.ni;
  int nj=world.nj;
  int nk=world.nk;

  for (int face=0;face<6;face++) {
    int target = world.getNeighborRank(face);
    if (target>=0) {
      int size;          // number of nodes on this face
      if (face/2==0) size=nj*nk;
      else if (face/2==1) size=ni*nk;
      else if (face/2==2) size=ni*nj;

      // allocate buffers
      double *send_buffer = new double[size];
      double *recv_buffer = new double[size];

      // pack data
      double *s=send_buffer;   // pointer to the first entry
      switch (face) {
        case 0: for(int j=0;j<nj;j++) for(int k=0;k<nk;k++)
          (*s++)=data[0][j][k];break;
        case 1: for(int j=0;j<nj;j++) for(int k=0;k<nk;k++)
          (*s++)=data[ni-1][j][k];break;
        case 2: for(int i=0;i<ni;i++) for(int k=0;k<nk;k++)
          (*s++)=data[i][0][k];break;
        case 3: for(int i=0;i<ni;i++) for(int k=0;k<nk;k++)
          (*s++)=data[i][nj-1][k];break;
        case 4: for(int i=0;i<ni;i++) for(int j=0;j<nj;j++)
          (*s++)=data[i][j][0];break;
        case 5: for(int i=0;i<ni;i++) for(int j=0;j<nj;j++)
          (*s++)=data[i][j][nk-1];break;
      }

      // transfer values
      MPI_Sendrecv(send_buffer, size, MPI_DOUBLE, target,
        TAG_DATA, recv_buffer, size, MPI_DOUBLE, target, TAG_DATA,
        MPI_COMM_WORLD, MPI_STATUS_IGNORE);

      // add to our values
      double *r=recv_buffer;  // pointer to start
      switch (face) {
        case 0:  for (int j=0;j<nj;j++) for (int k=0;k<nk;k++)
          data[0][j][k]+=(*r++);break;
        case 1:  for (int j=0;j<nj;j++) for (int k=0;k<nk;k++)
          data[ni-1][j][k]+=(*r++);break;
        case 2:  for (int i=0;i<ni;i++) for (int k=0;k<nk;k++)
          data[i][0][k]+=(*r++);break;
        case 3:  for (int i=0;i<ni;i++) for (int k=0;k<nk;k++)
          data[i][nj-1][k]+=(*r++);break;
        case 4:  for (int i=0;i<ni;i++) for (int j=0;j<nj;j++)
          data[i][j][0]+=(*r++);break;
```

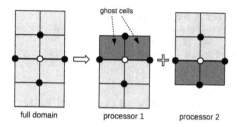

ghost cells

full domain   processor 1   processor 2

Figure 9.6: Illustration of ghost cells.

```
case 5:    for (int i=0;i<ni;i++) for (int j=0;j<nj;j++)
    data[i][j][nk-1]+=(*r++);break;
}

delete[] send_buffer;
delete[] recv_buffer;
}
} // for face }
```

9.6.2.4 Potential Solve

The finite difference discretization of Poisson's equation leads to a 7-point stencil along each regular node. Along processor boundaries, we have nodes with stencils pointing to nodes outside the local domain. The typical approach is to define boundary *ghost cells* (or *ghost nodes*), as shown in Figure 9.6. The ghost cells contain potential solution along the remote node from the previous solver iteration. Using ghost cells, we end up with the following algorithm for the MPI Poisson solver:

1. Perform one iteration of Gauss-Seidel solver on the local domain to obtain ϕ^{k+1}

2. Transfer new values of ϕ^{k+1} along the first internal face ($i = 1$ or $i = n_i - 2$ in the x direction) to the neighbor

3. Receive neighbor values and use them to update own ghost cells

4. At some frequency, compute global residue across all processors

5. Repeat, until convergence

We need to decide how to store the ghost cells data. In some of my older codes, I defined the finite difference stencil for each node through pointers to the potential values. This allowed me to use a separate buffer to store the ghost data. However, in this example, we use the more conventional approach of increasing the matrix size by 2 nodes in each direction, and then mapping global i, j, k to $i + 1, j + 1, k + 1$ entry in the resized field. Most effort to parallelize the solver then involves simply shifting the indexes when building the coefficient matrix,

```
void PotentialSolver :: buildMatrix () {
    for (int i=1;i<ni−1;i++)      // process internal, non ghost nodes
        for (int j=1;j<nj−1;j++)
            for (int k=1;k<nk−1;k++) {
                int u = ijkn(i,j,k);  // index in our shifted system

                // object using global indexing
                object[u] = world.object[i−1][j−1][k−1];
                if (object[u]>0) {
                A.d[u]=1;
                continue;
                }
                /* ... */
```

We also add a call to **updateGhosts** at the beginning of each Gauss-Seidel iteration. The following snippet highlights the changes involving non-local communication.

```
bool PotentialSolver :: solveGS () {
    // solve potential
    for (int solver_it=0;solver_it <max_it; solver_it++) {
        updateGhosts(phi);      // update ghost nodes
        /* perform local update ... */

        // check for convergencefadd a call
        if (solver_it%25==0) {
            double my_sum = 0;
            /* compute my local residue */

            // compute global sum
            double glob_sum;
            MPI_Allreduce(&my_sum,&glob_sum,1,MPI_DOUBLE,MPI_SUM,
                MPI_COMM_WORLD);

            L2 = sqrt(glob_sum/world.num_nodes_global);
            if (L2<tol) {converged=true; break;}
        }
    }
}
```

The code to compute residue is also updated. Each domain computes the local contribution to the global residue, $R_r = \sum(A\vec{x} - \vec{b})^2$. These residues are then summed up across all MPI processes using **MPI_Allreduce** and the global norm is computed using the total number of nodes. This calculation is performed on all ranks, and hence all processes terminate the solver in unison.

The code to update ghost nodes in **PotentialSolver::updateGhosts** is similar to **Species::updateBoundaries**, except that the data is replaced instead of accumulated. We also need to pay attention to the difference in indexing due to the expanded grid. The first internal grid node in the x direction is $i = 2$. It gets stored in the recipients $i = ni - 1$ position. Then neighbor then sends us their data from $i = ni - 3$ which we store in $i = 0$,

```
void PotentialSolver :: updateGhosts(double *phi) {
    int size;
    if (face/2==0) size=(nj−2)*(nk−2);
```

```
else if  (face/2==1)  size=(ni-2)*(nk-2);
else if  (face/2==2)  size=(ni-2)*(nj-2);

double *s=send_buffer;  // pointer to start
switch (face) {
  case 0:   for (int j=1;j<nj-1;j++) for (int k=1;k<nk-1;k++)
    (*s++)=phi[ijkn(2,j,k)];break;
  case 1:   for (int j=1;j<nj-1;j++) for (int k=1;k<nk-1;k++)
    (*s++)=phi[ijkn(ni-3,j,k)];break;
  case 2:   for (int i=1;i<ni-1;i++) for (int k=1;k<nk-1;k++)
    (*s++)=phi[ijkn(i,2,k)];break;
  /* ... */
}
}
```

9.6.3   CUDA Potential Solve

Finally, we briefly discuss the use of CUDA to solve the potential. When it comes to GPU programming, the phrase of the day is "profile, profile, profile". While porting CPU code to a GPU is not too difficult, writing code that fully utilizes the GPU capabilities is. Often new data access patterns result in a significant speed up as they let the processor take advantage of data locality. GPUs contain a special type of *texture* memory that is particularly suited for stencil-like data operations encountered in the potential solver. For simplicity, we do not use textures in this example. As discussed in Section 9.2.3, GPU programming involves writing a special *kernel* that is executed in parallel by hundreds, if not thousands, of GPU threads. A Jacobi matrix solver loops over all mesh nodes, and on each, evaluates Equation 2.14. At some interval, we also need to compute residue, which also involves looping over all nodes and on each computing the local error. The internals of these two loops form the body of GPU kernels. Then instead of using a loop, we launch $n_i \times n_j \times n_k$ threads. Example of a simple "Gauss-Seidel" GPU solver is shown below.

```
__global__ void cudaGSupdate (devSeptaD *A, double *phi, double
*b, char *object) {
int i = blockIdx.x*blockDim.x+threadIdx.x;
int j = blockIdx.y*blockDim.y+threadIdx.y;
int k = blockIdx.z*blockDim.z+threadIdx.z;
int ni = A->ni;
int nj = A->nj;
int nk = A->nk;
int u = k*ni*nj+j*ni+i;       // unknown index

if (i>0 && i<ni-1 && j>0 && j<nj-1 && k>0 && k<nk-1) {
  double rhoe = 0;
  if (object[u]==0) // open node
    rhoe = (QE*dev_n0*exp((phi[u] - dev_phi0)/dev_kTe0))/EPS_0;

  double g = ((b[u] + rhoe) - A->a[u]*phi[u-ni*nj] -
              A->b[u]*phi[u-ni] - A->c[u]*phi[u-1] -
              A->e[u]*phi[u+1] - A->f[u]*phi[u+ni] -
              A->g[u]*phi[u+ni*nj])/A->d[u];
```

```
    phi[u] = g;
  }
}
```

Here **septaD** is an alternate implementation of a sparse matrix that stores the seven diagonal vectors for the Poisson equation coefficient matrix. Note that, despite the name, this kernel does not implement the correct form of the Gauss-Seidel algorithm. Due to the random order in which the threads execute, there is no longer any guarantee that nodes with smaller i, j, and k indexes have already been updated. For some problems, this random access pattern may lead to divergence. In that case, a Jacobi solver may be needed, which utilizes a secondary buffer to store the "new" values. To avoid the need to copy down the new buffer, $\phi = \phi_{new}$, we can have our solver alternate between the two buffers,

```
__global__ void cudaGSupdate (devSeptaD *A, double *phi_in , double
    *phi_out double *b, char *object , int it) {
  if (it%2==0) {double *r=phi_in; phi_in=phi_out; phi_out=r;} //
    swap
  /* ... */
  double g = ((b[u] + rhoe) − A−>a[u]*phi_in [u−ni*nj] − ...
  phi_out [u] = g;
}
```

The second kernel **cudaGSresidue** computes the residue,

```
__global__ void cudaGSresidue(double *res , devSeptaD *A, double
    *phi, double *b, char *object){
  __shared__ float my_res[1024];      // 1024 is max threads per
    block
  double R = 0;

  int tx = threadIdx.x;
  int ty = threadIdx.y;
  int tz = threadIdx.z;
  int i = blockIdx.x*blockDim.x+tx;
  int j = blockIdx.y*blockDim.y+ty;
  int k = blockIdx.z*blockDim.z+tz;
  int ni = A−>ni;
  int nj = A−>nj;
  int nk = A−>nk;
  int u = k*ni*nj+j*ni+i; // global unknown index

  if (i>0 && i<ni−1 && j>0 && j<nj−1 && k>0 && k<nk−1) {
    double rhoe = 0;

    if (object [u]==0)
      rhoe = (QE*dev_n0*exp((phi[u] − dev_phi0)/dev_kTe0))/EPS_0;

    R = (b[u] + rhoe) − A−>a[u]*phi[u−ni*nj] − A−>b[u]*phi[u−ni] −
      A−>c[u]*phi[u−1] − A−>d[u]*phi[u] − A−>e[u]*phi[u+1] −
      A−>f[u]*phi[u+ni] − A−>g[u]*phi[u+ni*nj];
  }

  // store my local residue
```

```
my_res[tz*blockDim.x*blockDim.y+ty*blockDim.x+tx] = R*R;

// wait for all threads from block to finish
__syncthreads();

// if this is "root", sum up, slow way
if (tx==0 && ty==0 && tz==0) {
  double sum = 0;
  for (int i=0;i<blockDim.x*blockDim.y*blockDim.z;i++)
    sum+=my_res[i];
  // save in global memory
  res[blockIdx.z*gridDim.x*gridDim.y+blockIdx.y*gridDim.x+
    blockIdx.x] = sum;
}
}
```

This kernel takes advantage of a special GPU *shared* memory to sum up residues on each node. CUDA threads are launched in *blocks*, with each block containing up to 1024 blocks. The actual problem decomposition into blocks and threads per block is specified in the

```
kernel_fun <<<num_blocks, threads_per_block >>>(/* ... */);
```

kernel launch command. All threads within a single block have access to the same shared memory array. We use the predefined **threadIdx** and **blockDim** parameters to compute the linear index of our thread within its block. We then store the local R^2 term into that element. We then wait for all threads within the block to finish using the **__syncthreads**— barrier. Subsequently we add the per-node residues to obtain the total residue for the entire block. Sanders [49] illustrates an efficient mechanism that uses multiple threads to perform this summation. Here for simplicity, we use an inefficient algorithm in which only the "root" (thread with block index of 0) processes the data. The R^2_{block} value is then stored in a global memory array. CUDA does not support synchronization across blocks, and hence the final summation is performed on the CPU.

The solver is called from the CPU as shown below. This code is found in `SolverCUDA.cu`

```
for (solver_it=0;solver_it<max_it;solver_it++){
  // launch threads
  cudaGSupdate<<<num_blocks3,threads_per_block3>>>
    (dev_devA, dev_phi, dev_b);

  if (solver_it%25==0) {
    cudaGSresidue<<<num_blocks3,threads_per_block3>>>
      (dev_res, dev_devA, dev_phi, dev_b, dev_object);
    cudaMemcpy(res_pinned, dev_res, num_blocks*sizeof(double),
      cudaMemcpyDeviceToHost);
    double sum=0;
    for (int i=0;i<num_blocks;i++) sum+=res_pinned[i];
    L2 = sqrt(sum/(A.nu));
    if (L2<tol) {converged=true;break;}
  }
}
```

At each solver iteration, we call `cudaGSupdate` to obtain new estimates for ϕ. After every 25 iterations, we also call `cudaGSresidue` to compute the residue. The per-block values of the residue are copied to a host (CPU) *pinned* memory. This special page-locked array improves performance of the `cudaMemcpy` operation. The total residue is then computed, and the process continues until convergence. At this point, we have a choice to make. Do we use the CPU or the GPU to compute the electric field? There is no clear front runner and your choice may very well be platform dependent. GPUs are optimized for performing exactly the types of operations found in the `computeEF` function. This function loops over all the nodes, and on each, performs computation using near-by neighbor data. But this function also effectively converts a scalar field into a three-component vector field. The performance gained by the faster application of the finite difference equations may be lost by the need to transfer three times more data.

9.7 SUMMARY

In this final chapter we introduced concepts needed to develop parallel codes. We started by learning how to create a simple parallel program to add two numbers using multithreading, MPI, and CUDA. We then discussed parallelization strategies related to Particle in Cell plasma simulations. We saw how to parallelize the particle push using multithreading, and also developed an MPI version that uses domain decomposition to speed up the potential solver. We also discussed particle transfer between domains. We closed by introducing a simple GPU-based potential solver.

EXERCISES

9.1 *Julia Set.* The Julia set is a fractal and also a good example of a trivially parallelizable program. The image is created by letting the two coordinates map to the complex plane. Let the horizontal axis correspond to real numbers, while the vertical axis consists of imaginary numbers. Then for each $Z = R + \rangle I$ combination, we determine if the sum $Z^{k+1} = (Z^k)^2 + C$ converges. C is a constant with $C = -0.8 + 0.156i$ producing a nice image. This convergence check involves determining, for every point (pixel) on the 2D plane, whether the corresponding $Z = x + yi$ results in the sum converging or not. In order to check for convergence, we simply loop for some fixed number of iterations (such as 200) while checking if the magnitude exceeds some threshold (for instance, 1000). We terminate the search when this magnitude or the maximum number of iterations is reached. The final iteration number is used to set the corresponding pixel value. This data is then output to an VTK image file for visualization. As you can see, computing the Julia set involves a large number of calculations, but the calculation for each pixel is independent of the other pixels. Write a multithreaded, MPI, and CUDA version of the Julia set. Perform scaling studies. You should observe approximately linear scaling.

9.2 *CUDA Taylor Series.* From Taylor (or MacLaurin) series, we can derive an approximate expression for $\cos x$ as

$$\cos(x) = \sum_{n=0}^{\infty} \frac{(-1)^n}{(2n)!} x^{2n} = 1 - \frac{x^2}{2!} + \frac{x^4}{4!} - \cdots \qquad (P9.1)$$

Write a CPU and a GPU code to compute the value of $\cos(x)$ for $x \in [0, 2\pi]$ discretized into 1000 data points. At each point, compute 20 terms of the series ($n \in [0, 19]$) Double precision should be used since 38! is a very big number. Plot results for the CPU and the GPU calculation and also compare the run times. For the GPU case, make sure to include the memory transfer.

9.3 *GPU Electric Field Solve.* Write the GPU code to compute electric field. With this approach, you do not need to copy potential to the CPU except when the solution is written to a file for visualization. Which version (CPU or GPU) is faster?

9.4 *MPI Sheath Code.* Write an MPI one-dimensional code to model the plasma sheath. Start by loading equal ion and electron densities in a region $x = [0, L]$ with $kT_e > kT_i$. Set $\phi(0) = \phi(L) = 0$. While you may want to explore domain decomposition, for this problem it may be preferred to utilize ensemble averaging for the particle densities. With this scheme, each processor runs an effectively serial domain, but the scattered densities are averaged across the ranks. Potential is then computed from $\epsilon_0 \partial^2 \phi / \partial x^2 = \bar{\rho}$. Remember to seed the random number generator on each processor to obtain unique numbers. Compare results on a different number of processors to a serial run. You should observe the formation of a neutral bulk region with $E = 0$, neutral presheath with $E \neq 0$, and a non-neutral sheath with $E \neq 0$.

9.5 *CUDA Sheath Code.* Now write a CUDA version of the 1D sheath code. Compare performance to the MPI version and serial version. As a bonus, combine the CUDA and serial version so that simulation runs concurrently on the CPU and the GPU.

Bibliography

[1] LAPACK -Linear Algebra PACKage. www.netlib.org/lapack.

[2] Multigrid solver. https://www.particleincell.com/2018/multigrid-solver/.

[3] SDRL, file format storehouse. http://www.sdrl.uc.edu/sdrl/sdrl/sdrl/sdrl/referenceinfo/universalfileformats/file-format-storehouse.

[4] Sphere point picking. http://mathworld.wolfram.com/SpherePointPicking.html.

[5] Tridiagonal matrix algorithm. https://en.wikipedia.org/wiki/Tridiagonal_matrix_algorithm.

[6] JC Adam, A Héron, and G Laval. Study of stationary plasma thrusters using two-dimensional fully kinetic simulations. *Physics of Plasmas*, 11(1):295–305, 2004.

[7] Patrick R Amestoy, Iain S Duff, Jean-Yves L'Excellent, and Jacko Koster. MUMPS: a general purpose distributed memory sparse solver. In *International Workshop on Applied Parallel Computing*, pages 121–130. Springer, 2000.

[8] Dale Anderson, John C. Tannehill, and Richard H. Pletcher. *Computational Fluid Mechanics and Heat Transfer*. Taylor & Francis, 2 edition, 1997.

[9] Samuel J. Araki and Richard E. Wirz. Cell-centered particle weighting algorithm for PIC simulations in a non-uniform 2D axisymmetric mesh. *Journal of Computational Physics*, 272:218–226, 2014.

[10] Rutherford Aris. *Vectors, Tensors, and the Basic Equations of Fluid Mechanics*. Dover Publications, 1962.

[11] Satish Balay, Shrirang Abhyankar, Mark Adams, Jed Brown, Peter Brune, Kris Buschelman, Lisandro Dalcin, Alp Dener, Victor Eijkhout, W Gropp, et al. *PETSc Users Manual*. Argonne National Laboratory, 2019.

[12] Graeme A. Bird. *Molecular Gas Dynamics and the Direct Simulation of Gas Flows*. Clarendon Press, 1994.

[13] Graeme A. Bird. *The DSMC Method*. CreateSpace, 2013.

[14] Charles K Birdsall. Particle-in-Cell charged-particle simulations, plus Monte Carlo Collisions with neutral atoms, PIC-MCC. *IEEE Transactions on Plasma Science*, 19(2):65–85, 1991.

[15] Charles K. Birdsall and A Bruce Langdon. *Plasma Physics via Computer Simulation*. CRC Press, 2004.

[16] Jay P. Boris. Relativistic plasma simulation-optimization of a hybrid code. In *Proceedings of 4th Conference on Numerical Simulations of Plasmas*, pages 3–67, 1970.

[17] William E. Boyce and Richard C. DiPrima. *Elemenatary Differential Equations and Boundary Value Problems*. John Wiley & Sons, 6 edition, 1997.

[18] Iain D Boyd. Conservative species weighting scheme for the direct simulation monte carlo method. *Journal of Thermophysics and Heat Transfer*, 10(4):579–585, 1996.

[19] Iain D. Boyd and Thomas E. Schwartzentruber. *Nonequilibrium Gas Dynamics and Molecular Simulation*. Cambridge University Press, 2017.

[20] Iain D Boyd and John T Yim. Modeling of the near field plume of a Hall thruster. *Journal of Applied Physics*, 95(9):4575–4584, 2004.

[21] Lubos Brieda. *Development of the DRACO ES-PIC code and fully-kinetic simulation of ion beam neutralization*. PhD thesis, Virginia Tech, 2005.

[22] Lubos Brieda. Numerical model for molecular and particulate contamination transport. *Journal of Spacecraft and Rockets*, 56(2):485–497, 2018.

[23] Richard L. Burden and Faires J. Douglas. *Numerical Analysis*. Brooks/-Cole, 7 edition, 2001.

[24] Francis F Chen. *Introduction to Plasma Physics and Controlled Fusion*, volume 1. Springer, 1984.

[25] Chio-Zong Cheng and Georg Knorr. The integration of the Vlasov equation in configuration space. *Journal of Computational Physics*, 22(3):330–351, 1976.

[26] Andrew J. Christlieb. Grid-free plasma simulation techniques. *IEEE Transactions on Plasma Science*, 34(2), 2006.

[27] Said Doss and Keith Miller. Dynamic ADI methods for elliptic equations. *SIAM Journal on Numerical Analysis*, 16(5):837–856, 1979.

[28] J.H. Ferziger and M. Perić. *Computational Methods for Fluid Dynamics*. Springer, 3 edition, 2002.

[29] John Michael Fife. *Hybrid-PIC modeling and electrostatic probe survey of Hall thrusters*. PhD thesis, Massachusetts Institute of Technology, 1998.

[30] Justin Fox. *Advances in Fully-Kinetic PIC Simulations of a Near-Vacuum Hall Thruster and other Plasma Systems*. PhD thesis, Massachusetts Institute of Technology, 2007.

[31] Matthew R Gibbons and Dennis W Hewett. The Darwin direct implicit particle-in-cell (dadipic) method for simulation of low frequency plasma phenomena. *Journal of Computational Physics*, 120(2):231–247, 1995.

[32] William Gropp, William D Gropp, Argonne Distinguished Fellow Emeritus Ewing Lusk, Ewing Lusk, and Anthony Skjellum. *Using MPI: portable parallel programming with the message-passing interface*, volume 1. MIT press, 1999.

[33] Daniel Hastings and Henry Garrett. *Spacecraft-Environment Interactions*. Cambridge University Press, 2004.

[34] R. W. Hockney and J.W. Eastwood. *Computer Simulation Using Particles*. IOP Publishing, 1988.

[35] Thomas J.R. Hughes. *The Finite Element Method: Linear Static and Dynamic Finite Element Analysis*. Courier Corporation, 2012.

[36] John David Jackson. *Classical Electrodynamics*. John Wiley & Sons, 3 edition, 1999.

[37] Robert G. Jahn. *Physics of Electric Propulsion*. McGraw-Hill, 1968.

[38] Revathi Jambunathan and Deborah A Levin. CHAOS: An octree-based PIC-DSMC code for modeling of electron kinetic properties in a plasma plume using MPI-CUDA parallelization. *Journal of Computational Physics*, 373:571–604, 2018.

[39] Stephen Jardin. *Computational Methods in Plasma Physics*. CRC Press, 2010.

[40] Raed Kafafy and Joseph Wang. A hybrid grid immersed finite element particle-in-cell algorithm for modeling spacecraft–plasma interactions. *IEEE transactions on plasma science*, 34(5):2114–2124, 2006.

[41] A Bruce Langdon, Bruce I Cohen, and Alex Friedman. Direct implicit large time-step particle simulation of plasmas. *Journal of Computational Physics*, 51(1):107–138, 1983.

[42] Michael A Lieberman and Allan J Lichtenberg. *Principles of Plasma Discharges and Materials Processing*. Wiley-Interscience, 2 edition, 2005.

[43] Robert Martin. Conservative bin-to-bin fractional collisions. In *AIP Conference Proceedings*, volume 1786, page 090003. AIP Publishing, 2016.

[44] Makoto Matsumoto and Takuji Nishimura. Mersenne twister: a 623-dimensionally equidistributed uniform pseudo-random number generator. *ACM Transactions on Modeling and Computer Simulation (TOMACS)*, 8(1):3–30, 1998.

[45] A.I. Morozov, Yu. V. Esinchuk, G.N. Tilinin, A. V. Trofimov, Yu. A. Sharov, and Shchepkin G. Ya. Plasma accelerator with closed electron drift and extended acceleration zone. *Soviet Physics - Technical Physics*, 17(1), 1972.

[46] John F O'Hanlon. *A User's Guide to Vacuum Technology*. John Wiley & Sons, 2005.

[47] Robert Osada, Thomas Funkhouser, Bernard Chazelle, and David Dobkin. Shape distributions. *ACM Transactions on Graphics (TOG)*, 21(4):807–832, 2002.

[48] William H Press, Saul A Teukolsky, William T Vetterling, and Brian P Flannery. *Numerical Recipes: The art of Scientific Computing*. Cambridge university press, 3 edition, 2007.

[49] Jason Sanders and Edward Kandrot. *CUDA by Example*. Addison-Wesley, 2011.

[50] Mark Michael Santi. *Hall thruster plume simulation using a hybrid-PIC algorithm*. PhD thesis, Massachusetts Institute of Technology, 2003.

[51] Will Schroeder, Martin Ken, and Lorensen Bill. *The Visualization Toolkit*. Kitware, 4 edition, 2006.

[52] Lyman Jr. Spitzer. *Physics of Fully Ionized Gases*. Dover Publications, 2 edition, 1962.

[53] Bjarne Stroustrup. *A Tour of C++*. Addison-Wesley, 2 edition, 2018.

[54] W Sutton, George and Arthur Sherman. *Engineering Magnetohydrodynamics*. Dover Publications, 1965.

[55] Toshiki Tajima. *Computational Plasma Physics*. Westview Press, 2004.

[56] Jiyuan Tu, Yeoh Guan-Heng, and Liu Chaoqun. *Computational Fluid Dynamics*. Elsevier, 2 edition, 2013.

[57] Joseph Wang, Raed Kafafy, and Lubos Brieda. An ife-pic simulation model for plume-spacecraft interactions. In *39th AIAA/ASME/SAE/ASEE Joint Propulsion Conference and Exhibit*, page 4874, 2003.

[58] Nicholas Wilt. *The CUDA handbook.* Pearson Education, 2013.

[59] Zhuomin M. Zhang. *Nano/Microscale Heat Transfer.* McGraw-Hill, 2007.

Index

Printed in the United States
by Baker & Taylor Publisher Services